Methods in Molecular Biology

For further volumes:
http://www.springer.com/series/7651

For over 35 years, biological scientists have come to rely on the research protocols and methodologies in the critically acclaimed *Methods in Molecular Biology* series. The series was the first to introduce the step-by-step protocols approach that has become the standard in all biomedical protocol publishing. Each protocol is provided in readily-reproducible step-by-step fashion, opening with an introductory overview, a list of the materials and reagents needed to complete the experiment, and followed by a detailed procedure that is supported with a helpful notes section offering tips and tricks of the trade as well as troubleshooting advice. These hallmark features were introduced by series editor Dr. John Walker and constitute the key ingredient in each and every volume of the *Methods in Molecular Biology* series. Tested and trusted, comprehensive and reliable, all protocols from the series are indexed in PubMed.

Cell Reprogramming for Immunotherapy

Methods and Protocols

Edited by

Samuel G. Katz and Peter M. Rabinovich

Department of Pathology, Yale School of Medicine, New Haven, CT, USA

Humana Press

Editors
Samuel G. Katz
Department of Pathology
Yale School of Medicine
New Haven, CT, USA

Peter M. Rabinovich
Department of Pathology
Yale School of Medicine
New Haven, CT, USA

ISSN 1064-3745 ISSN 1940-6029 (electronic)
Methods in Molecular Biology
ISBN 978-1-0716-0202-7 ISBN 978-1-0716-0203-4 (eBook)
https://doi.org/10.1007/978-1-0716-0203-4

This Humana imprint is published by the registered company Springer Science+Business Media, LLC, part of Springer Nature.
The registered company address is: 233 Spring Street, New York, NY 10013, U.S.A.

Preface

After nearly a half a century of experiments on gene therapy, the first application of a cell reprogrammed with a foreign gene was approved in the Fall of 2017. More remarkably, who would have predicted that the first successful procedure of gene therapy would come in the form of immunotherapy for cancer, a field with its own long-awaited promises. The reprogramming of T cells with chimeric antigen receptors (CARs) is the leading edge of numerous other methods of manipulating the immune system to counteract cancer. The coming decades will see numerous modalities of immune cell reprogramming, manipulation of multiple arms of the immune system, various methods of in vitro and in vivo monitoring, and the translation of the accumulated knowledge into clinical trials.

This edition of Methods in Molecular Biology, *Cell reprogramming for Immunotherapy*, will provide some key protocols towards developing strategies in immunotherapy. The first section of the book (Chapters 1–9) discusses issues related to reprogramming T cells and NK cells. After a review of T cell receptor and CAR signaling, we have protocols on T cells dedicated towards finding new surface targets, improving viral transduction, analyzing cytokines released by single cells, and analyzing T cell and metabolic states. For NK cells, we include protocols on reprogramming with mRNA as well as by lentivirus, and then assessment of activity in vitro and in vivo. The second section (Chapters 10–16) will discuss reprogramming of various immune cells. After a brief review, we present protocols for working with dendritic cells, exosomes, and macrophages. We also include protocols for CRISPR screening strategies for cytotoxicity and working with oncolytic viruses. The third section (Chapters 17 and 18) is concerned with monitoring the engineered cells in vivo. Finally, the fourth section (Chapters 19–21) will discuss some clinical considerations insofar as analyzing patients treated with CAR-T cells, factors in cell therapy with tumor-infiltrating lymphocytes (TILs), and functions of an academic GMP facility.

We are grateful to the leaders of this burgeoning field, who have taken the time to write a chapter in this book and share their expertise with the broader community. We also thank each of the readers, who will benefit from these protocols and improve upon them for the next generation.

New Haven, CT, USA

Samuel G. Katz
Peter M. Rabinovich

Contents

Contributors

ALEX M. ABEL • *Laboratory of Molecular Immunology and Immunotherapy, Blood Research Institute, Blood Center of Wisconsin, Milwaukee, WI, USA; Department of Pediatrics, Medical College of Wisconsin, Milwaukee, WI, USA; Department of Microbiology and Immunology, Medical College of Wisconsin, Milwaukee, WI, USA*

RAMI ALZHRANI • *Use-Inspired Biomaterials and Integrated Nano Delivery (U-BiND) Systems Laboratory, Department of Pharmaceutical Sciences, Wayne State University, Detroit, MI, USA*

WILL BAILIS • *Division for Protective Immunity, Department of Pathology and Laboratory Medicine, The Children's Hospital of Philadelphia, Philadelphia, PA, USA*

REUBEN BENJAMIN • *King's College London, London, UK; King's Hospital London, London, UK*

ALEXEY BERSENEV • *Clinical Laboratory Medicine, Yale University, New Haven, CT, USA; Advanced Cell Therapy Laboratory, Yale New Haven Hospital, New Haven, CT, USA*

KETKI BHISE • *Use-Inspired Biomaterials and Integrated Nano Delivery (U-BiND) Systems Laboratory, Department of Pharmaceutical Sciences, Wayne State University, Detroit, MI, USA*

DARIO CAMPANA • *Departments of Pediatrics, Yong Loo Lin School of Medicine, National University of Singapore, Singapore, Singapore*

QINQIN CHENG • *Department of Pharmacology and Pharmaceutical Sciences, School of Pharmacy, University of Southern California, Los Angeles, CA, USA*

BRYDEN CONSIDINE • *Yale Comprehensive Cancer Center, New Haven, CT, USA*

DIANA C. DELUCIA • *Division of Human Biology, Fred Hutchinson Cancer Research Center, Seattle, WA, USA; Division of Clinical Research, Fred Hutchinson Cancer Research Center, Seattle, WA, USA; Division of Medical Oncology, Department of Medicine, University of Washington School of Medicine, Seattle, WA, USA*

CAROLINA I. DOMAICA • *Laboratorio de Fisiopatología de la Inmunidad Innata, Instituto de Biología y Medicina Experimental (IBYME), Consejo Nacional de Investigaciones Científicas y Técnicas (CONICET), Buenos Aires, Argentina*

ALAN DUNLOP • *King's College London, London, UK; Viapath, London, UK*

RICHARD L. EDELSON • *Department of Dermatology, Yale School of Medicine, New Haven, CT, USA*

RONG FAN • *Department of Biomedical Engineering, Yale University, New Haven, CT, USA*

ANDREW FESNAK • *Clinical Pathology and Laboratory Medicine, University of Pennsylvania, Philadelphia, PA, USA*

RENATA FILLER • *Department of Dermatology, Yale School of Medicine, New Haven, CT, USA*

AMANDA FINCK • *Department of Biomedical Engineering, Yale University, New Haven, CT, USA*

XIAOYING FU • *Section of Hematology, Department of Internal Medicine and Yale Comprehensive Cancer Center, Yale University School of Medicine, New Haven, CT, USA; Department of Laboratory Medicine, Shenzhen Children's Hospital, Shenzhen, People's Republic of China*

MERCEDES B. FUERTES • *Laboratorio de Fisiopatología de la Inmunidad Innata, Instituto de Biología y Medicina Experimental (IBYME), Consejo Nacional de Investigaciones Científicas y Técnicas (CONICET), Buenos Aires, Argentina*

RANA GBYLI • *Section of Hematology, Department of Internal Medicine and Yale Comprehensive Cancer Center, Yale University School of Medicine, New Haven, CT, USA*

STEPHANIE HALENE • *Section of Hematology, Department of Internal Medicine and Yale Comprehensive Cancer Center, Yale University School of Medicine, New Haven, CT, USA*

PATRICK HAN • *Department of Dermatology, Yale School of Medicine, New Haven, CT, USA; Department of Chemical & Environmental Engineering, Yale University, New Haven, CT, USA*

DOUGLAS HANLON • *Department of Dermatology, Yale School of Medicine, New Haven, CT, USA*

MICHAEL E. HURWITZ • *Yale Comprehensive Cancer Center, New Haven, CT, USA*

ARUN K. IYER • *Use-Inspired Biomaterials and Integrated Nano Delivery (U-BiND) Systems Laboratory, Department of Pharmaceutical Sciences, Wayne State University, Detroit, MI, USA; Molecular Imaging Program, Karmanos Cancer Institute, Detroit, MI, USA*

AGNIESZKA JOZWIK • *King's College London, London, UK; King's Hospital London, London, UK*

SAMUEL G. KATZ • *Department of Pathology, Yale School of Medicine, New Haven, CT, USA*

NOUR KIBBI • *Department of Dermatology, Yale School of Medicine, New Haven, CT, USA*

WILBUR A. LAM • *Division of Pediatric Hematology/Oncology, Department of Pediatrics, Aflac Cancer Center and Blood Disorders Service of Children's Healthcare of Atlanta, Emory University School of Medicine, Atlanta, GA, USA; The Wallace H. Coulter Department of Biomedical Engineering, Georgia Institute of Technology & Emory University, Atlanta, GA, USA*

JOHN K. LEE • *Division of Human Biology, Fred Hutchinson Cancer Research Center, Seattle, WA, USA; Division of Clinical Research, Fred Hutchinson Cancer Research Center, Seattle, WA, USA; Division of Medical Oncology, Department of Medicine, University of Washington School of Medicine, Seattle, WA, USA*

SUBRAMANIAM MALARKANNAN • *Laboratory of Molecular Immunology and Immunotherapy, Blood Research Institute, The Blood Center of Wisconsin, Milwaukee, WI, USA; Department of Pediatrics, The Medical College of Wisconsin, Milwaukee, WI, USA; Department of Microbiology and Immunology, The Medical College of Wisconsin, Milwaukee, WI, USA; Department of Medicine, The Medical College of Wisconsin, Milwaukee, WI, USA*

ROBERT E. MEANS • *Department of Pathology, Yale School of Medicine, New Haven, CT, USA*

MICHAEL C. MILONE • *Center for Cellular Immunotherapies, Perelman School of Medicine at the University of Pennsylvania, Philadelphia, PA, USA; Department of Pathology and Laboratory Medicine, Perelman School of Medicine at the University of Pennsylvania, Philadelphia, PA, USA*

Amer M. Najjar • *Division of Pediatrics, The University of Texas MD Anderson Cancer Center, Houston, TX, USA*
Arash Nanbakhsh • *Laboratory of Molecular Immunology and Immunotherapy, Blood Research Institute, The Blood Center of Wisconsin, Milwaukee, WI, USA*
Roddy S. O'Connor • *Center for Cellular Immunotherapies, Perelman School of Medicine at the University of Pennsylvania, Philadelphia, PA, USA; Department of Pathology and Laboratory Medicine, Perelman School of Medicine at the University of Pennsylvania, Philadelphia, PA, USA*
Peter M. Rabinovich • *Department of Pathology, Yale School of Medicine, New Haven, CT, USA*
Eve Robinson • *Department of Dermatology, Yale School of Medicine, New Haven, CT, USA*
Sounak Ghosh Roy • *Department of Pathology, Yale School of Medicine, New Haven, CT, USA*
Katy Sanchez • *King's College London, London, UK; Viapath, London, UK*
Samaresh Sau • *Use-Inspired Biomaterials and Integrated Nano Delivery (U-BiND) Systems Laboratory, Department of Pharmaceutical Sciences, Wayne State University, Detroit, MI, USA*
Nathan J. Schloemer • *Laboratory of Molecular Immunology and Immunotherapy, Blood Research Institute, Blood Center of Wisconsin, Milwaukee, WI, USA; Department of Pediatrics, Medical College of Wisconsin, Milwaukee, WI, USA*
Xiaojing Shi • *Department of Pharmacology and Pharmaceutical Sciences, School of Pharmacy, University of Southern California, Los Angeles, CA, USA*
Noriko Shimasaki • *Departments of Pediatrics, Yong Loo Lin School of Medicine, National University of Singapore, Singapore, Singapore*
Jessica M. Sierra • *Laboratorio de Fisiopatología de la Inmunidad Innata, Instituto de Biología y Medicina Experimental (IBYME), Consejo Nacional de Investigaciones Científicas y Técnicas (CONICET), Buenos Aires, Argentina*
Olga Sobolev • *Department of Dermatology, Yale School of Medicine, New Haven, CT, USA*
Yuanbin Song • *Section of Hematology, Department of Internal Medicine and Yale Comprehensive Cancer Center, Yale University School of Medicine, New Haven, CT, USA*
Stephan S. Späth • *University of Tübingen, Tübingen, Germany*
Kazuki Tatsuno • *Department of Dermatology, Yale School of Medicine, New Haven, CT, USA*
Monica S. Thakar • *Laboratory of Molecular Immunology and Immunotherapy, Blood Research Institute, Blood Center of Wisconsin, Milwaukee, WI, USA; Department of Pediatrics, Medical College of Wisconsin, Milwaukee, WI, USA*
Reginald Tran • *Division of Pediatric Hematology/Oncology, Department of Pediatrics, Aflac Cancer Center and Blood Disorders Service of Children's Healthcare of Atlanta, Emory University School of Medicine, Atlanta, GA, USA; The Wallace H. Coulter Department of Biomedical Engineering, Georgia Institute of Technology & Emory University, Atlanta, GA, USA*
Aaron Vassall • *Department of Dermatology, Yale School of Medicine, New Haven, CT, USA*

ALESSANDRA VENTURA • *Department of Dermatology, Yale School of Medicine, New Haven, CT, USA; Dermatology Department, University of Rome Tor Vergata, Rome, Italy*
SHERMAN M. WEISSMAN • *Department of Genetics, Yale School of Medicine, New Haven, CT, USA*
ALP YURTER • *Department of Dermatology, Yale School of Medicine, New Haven, CT, USA*
JIALING ZHANG • *Department of Pathology, Yale School of Medicine, New Haven, CT, USA; Department of Genetics, Yale School of Medicine, New Haven, CT, USA*
YONG ZHANG • *Department of Pharmacology and Pharmaceutical Sciences, School of Pharmacy, University of Southern California, Los Angeles, CA, USA; Department of Chemistry, Dornsife College of Letters, Arts and Sciences, University of Southern California, Los Angeles, CA, USA; Norris Comprehensive Cancer Center, University of Southern California, Los Angeles, CA, USA; Research Center for Liver Diseases, University of Southern California, Los Angeles, CA, USA*
NORBERTO W. ZWIRNER • *Laboratorio de Fisiopatología de la Inmunidad Innata, Instituto de Biología y Medicina Experimental (IBYME), Consejo Nacional de Investigaciones Científicas y Técnicas (CONICET), Buenos Aires, Argentina*

Part I

Reprogramming T and NK Cells

Chapter 1

T Cell Reprogramming Against Cancer

Samuel G. Katz and Peter M. Rabinovich

Abstract

Advances in academic and clinical studies during the last several years have resulted in practical outcomes in adoptive immune therapy of cancer. Immune cells can be programmed with molecular modules that increase their therapeutic potency and specificity. It has become obvious that successful immunotherapy must take into account the full complexity of the immune system and, when possible, include the use of multifactor cell reprogramming that allows fast adjustment during the treatment. Today, practically all immune cells can be stably or transiently reprogrammed against cancer. Here, we review works related to T cell reprogramming, as the most developed field in immunotherapy. We discuss factors that determine the specific roles of αβ and γδ T cells in the immune system and the structure and function of T cell receptors in relation to other structures involved in T cell target recognition and immune response. We also discuss the aspects of T cell engineering, specifically the construction of synthetic T cell receptors (synTCRs) and chimeric antigen receptors (CARs) and the use of engineered T cells in integrative multifactor therapy of cancer.

Key words T cell, T cell receptor (TCR), Chimeric antigen receptor (CAR), alpha beta T cells, gamma delta T cells, Memory T cells, Immune synapse, Reprogramming, Adoptive cell therapy, Signal transduction, TCR clustering

1 Introduction

Progress in immunotherapy has reached a critical point where available funding and efforts can provide practical improved clinical outcomes for patients. These advances are based on findings in academic and clinical studies in immunology, adoptive immunotherapy, gene editing, and stem cell modulation, among other fields. Despite our rapidly increasing understanding of tumor–immune system interactions, there are profound limits to our knowledge. Nonetheless, the urgent need for therapeutic improvements facilitates the development of new drugs and modified cells in parallel with new methods of their clinical evaluation. Particularly important is the opportunity to exploit combinatorial multifactor treatment protocols based on protein and cell engineering.

Samuel G. Katz and Peter M. Rabinovich (eds.), *Cell Reprogramming for Immunotherapy: Methods and Protocols*, Methods in Molecular Biology, vol. 2097, https://doi.org/10.1007/978-1-0716-0203-4_1,

Immune cells can be programmed with molecular modules that increase their therapeutic potency and specificity. Although in its infancy, modern immunotherapy strives to provide personalized therapy that is modifiable during the course of treatment based on the patient's baseline characteristics and ongoing accurate evaluation of the course of the disease. The development of immune modulation by cell reprogramming already has been translated into patient cures. This highlights both a fascinating discovery and our relative ignorance about how to prevent high morbidity, off-target effects, and other complications.

Fortunately, some of the gaps in our knowledge are being addressed rapidly. It has become obvious that successful immunotherapy must take into account the full complexity of the immune system and when possible include the use of different types of immunocytes and multifactor cell reprogramming, and apply flexible methods that allow fast adjustment during the treatment depending on the patient's conditions and needs. Today, practically all immune cells can be stably or transiently reprogrammed against cancer. The most developed field is T cell reprogramming, although quite promising results have been achieved with natural killer (NK) cells, macrophages, and others [1–5]. Several chapters in this volume exemplify these different sources of cells that can be reprogrammed (e.g., Chapters 6–9 for NK cells, Chapter 11 for dendritic cells (DCs), Chapter 14 for macrophages).

Challenges in cell engineering appear at many levels. At the subcellular level, the design of novel proteins or other molecules that can be expressed and function in accord with endogenous cell systems is nontrivial. At the cellular level, the complexity of cell-to-cell interaction dictates accurate construct adjustment and modification. In addition, there may be the need to introduce additional molecules of different classes that optimize the cognate cell function. At the organismal level, there is a need to evaluate multiple reactions by an integral combinatorial approach where cell engineering synergistically couples with other therapies.

During last 20 years, adoptive cell therapy (ACT) was developed based on two main premises: (1) Cytotoxic T cells eliminate diseased cells, and (2) artificial modular protein constructs can be designed to recognize specific antigens on the surface of target cells and trigger T cell target killing. Since 2011, the number of patents related to chimeric antigen receptor (CAR)-mediated immunotherapy has grown exponentially [6]. Today, ACT is best demonstrated in the treatment of blood B cell tumors with chimeric antigen receptor T cell (CAR-T) therapy products: Kymriah (Novartis) and Yescarta (Kite Pharma/Gilead), which are approved by the US Food and Drug Administration [7] and the European Medicines Agency [8]. Other hematopoietic cancers and solid cancers have been more challenging to target, because T cell function is impeded by the absence of specific tumor antigens, multiple barriers of tumor accessibility, and immunosuppressive conditions. Increased knowledge of the processes that take place in the tumor microenvironment, metastasis development, and immune tuning on both systemic and local levels will be necessary to improve cell engineering and resolve both the fundamental and technical problems. In this review, we briefly address the main areas in T cell reprogramming relevant to ACT of cancer and describe some obvious underdeveloped areas important for building better integrative personalized therapies.

2 T Cells

2.1 T Cell Diversity

Functional diversity of T cell populations and their T cell receptor (TCR) repertoire are important factors that determine immune health. In cancer research, attention is often focused on the rather narrow task of finding a T cell population, among a weakened immune system, that is robust enough to yield a sufficient amount and be reprogrammed to kill cancer targets and then maintained in patients long enough to achieve efficacy. However, cross-communication among the subsets of T cells, dendritic cells (DCs), and other immunocytes is also an important part of immune response. That explains growing attention to various subsets of immune cells.

In humans, it is estimated that ~7 billion T cells are present in the peripheral blood, 25 billion in the bone marrow, 30 billion in the spleen, and 150 billion in the lymph nodes. Taken together, these three organs contain >200 × 10^9 T cells, which constitutes the majority of total T cells [9].

T cells consist of many subtypes, the largest of which are the "conventional" αβ T cells with "classic" major histocompatibility complex (MHC) restriction. These T cells are part of the sophisticated adoptive immune system with a relatively slow response. Some other T cells are part of the innate immune system. They are characterized by a limited TCR diversity, are either "non-classic MHC" restricted or MHC independent, and exhibit a fast immune response. They include γδ T cells, natural killer T (NKT) cells, CD1- and MHC class Ib-restricted T cells, MR-1-restricted mucosal-associated invariant T cells (MAIT), and intraepithelial lymphocytes (IELs) [10–12]. Although the subtypes of T cells are functionally different, this difference is not inflexible. For example, human peripheral γδ T cells can be transdifferentiated ex vivo into αβ T cells [13]. γδ T cells are a minor subset of peripheral lymphocytes in humans (<5%) [14] but are enriched in epithelial and mucosal tissues [15].

2.2 T Cell Development

T cells originate in the bone marrow, and most of them develop in the thymus. These cells arise from immature CD4 and CD8 double-negative thymocytes and express either the αβ or γδ TCR

[14, 16]. αβ T cells undergo positive and negative selection through recognition of self-peptide–MHC (p–MHC). Most αβ T cells with high affinity to self-peptides undergo negative selection and die. However, some of them survive and after positive selection form certain subtypes, such as IELs, NKTs, and Foxp3+ regulatory T cells (Tregs) [17]. αβ T cells with low affinity to self-p–MHC develop into "conventional" CD8+ cytotoxic lymphocytes (CTL) or CD4+ helper T cells that recognize foreign or "diseased" peptide antigens presented by MHC class I and class II molecules, respectively. This difference is not rigid, as a substantial part of CD4+ T cells can be cytotoxic similar to CD8+ T cells [18–22].

After selection in the thymus, αβ T cells proceed to the periphery where they circulate as naïve CD4+ and CD8+ T cells between blood and secondary lymphoid organs. Positive selection of naïve T cells continues as cyclical "tonic" activation as they circulate between the blood and secondary lymphoid organs where they encounter self-p–MHC [23–25].

γδ T cells recognize antigens in a "classic MHC"-independent manner. They recognize a range of structurally different moieties, such as non-classic MHC molecules, proteins, peptides, and phospholipids [26]. Studies in mice have demonstrated that γδ T cells undergo negative selection in thymus, but the role of positive selection in their development may not be obligatory [27].

2.3 Naïve, Effector, and Memory $\alpha\beta$ T Cells

Naïve T cells are maintained in a state of quiescence, which is characterized by low cell volume, low metabolic rate, and low homeostatic proliferation. Gradual naïve T cell propagation without differentiation is upheld by IL7 signaling and tonic, low-level activity of the TCR that monitors "normal" self-p–MHC presented by surrounding cells [28, 29]. Conventional activation of naïve T cells by antigen-presenting cells (APCs) requires the simultaneous engagement of three receptors on the T cells: TCR ligation with agonist p–MHC, CD28 or another costimulatory receptor ligation with cognate ligand, and cytokine receptor ligation with IL2 or another homeostatic cytokine [30–32].

Sometimes, naïve T cells can be activated without TCR ligation. For example, in vivo T cells become activated when their population is severely depleted [33], whereas in vitro T cells can be activated in the presence of high concentrations of interleukins IL2 or IL7 [34].

Upon APC-mediated activation, naïve T cells undergo clonal expansion and differentiation. They activate mTORC1 [35], shift from catabolic to anabolic metabolism, increase glucose and amino acid (AA) uptake, remodel their mitochondrial function, and increase their cell volume and their rate of proliferation [29]. In the initial expansion phase, naïve T cell doubling time decreases from 500–1000 days [36, 37] to a few hours [38, 39]. This process may occur in different tissues and is highly dependent on receptor–ligand affinity and density, as well as environmental factors, such as pH, redox potential, and the availability of nutrients. The activation leads to clonal expansion by sequential asymmetric cell division. In this process, cell divisions convert activated naïve T cells into terminally differentiated effector T cells (T_{EFF}) and self-renewing memory lineages (T_{MEM}) [40]. Some of the progeny cells become more activated and directed toward T_{EFF} , while the others become less metabolically active T_{MEM} [41]. Thus, T_{MEM} cells are metabolically more active than naïve T cells, with higher mitochondrial mass and respiratory capacity [42], but less active than T_{EFF} cells and maintain predominately catabolic metabolism [43]. This contributes to the prolonged persistence of T_{MEM} cells, such that their longevity is between naïve and effector T cells.

Activated T_{EFF} cells include both CD8+ CTLs and CD4+ helper T cells that provide acute protection from immune challenges [44]. Activated CTLs recognize an appropriate target, bind to it, and secrete killing molecules toward the target, before detaching from the dying target [45]. In humans, the main mechanism of T cell cytotoxicity is the granzyme–perforin pathway [46–49]. Other killing systems, such as Fas/FasL, TNF-α/TNF receptor 1, and TRAIL/DR4/DR5, are usually employed in T cell activation-induced cell death (AICD), which is an essential part of T cell homeostasis regulation [50]. However, these systems can also cause target killing [45, 51, 52].

After the immediate immune response and foreign antigen is cleared, the T cell population enters a contraction phase, and most of the T_{EFF} subsequently die by apoptosis [53].

Long-term immune protection is provided by subsets of T_{MEM} cells that include central memory (T_{CM}) and effector memory (T_{EM}) cells. T_{CM} are circulating cells that are prevalent in lymph nodes and have enhanced longevity and proliferative potential. Although T_{CM} lack effector functions themselves, they generate T_{EFF} and T_{EM} cells. T_{EM} cells are circulating cells that are more prevalent in nonlymphoid tissues [54–57]. They possess immediate effector functions; they rapidly migrate toward targets and provide antigen elimination [58]. Differentiation from naïve T cells, to T_{CM}, to T_{EM} is associated with decreasing expression of the Wnt/β catenin transcription factors, LEF-1, and TCF-1 [59–61]. Clonal development is a flexible process, and to some degree, it can be reversed. For example, in mice, long-lived memory CD8+ T_{CM} cells can develop from effector T_{EFF} cells through a process of dedifferentiation [62].

T_{MEM} population is diversified, and their discovery and classification are especially difficult if they are not circulating. Recently, such tissue-resident memory (T_{RM}) cells were located in the lungs, salivary glands, female reproductive tract, skin, and liver, where they orchestrate the response to different pathogens [63, 64].

An important subset within the T_{CM} population generated during the primary immune response has stem cell-like characteristics, defined as T_{SCM}. These cells exhibit a high capacity for homeostatic proliferation and can give rise to other memory subsets. In normal homeostasis, the pool of T_{SCM} cells in humans is believed to comprise 2–3% of all circulating T lymphocytes [61, 65].

Some CD8+ T_{MEM} cells are virtual memory cells that originate from naïve T cells under strong IL15 signaling but without agonist antigen stimulation [66–68]. The role of these cells is debatable, but one theory is that they provide innate-like protection during the earliest stages of bacterial or viral infection [68, 69].

Once created, conventional T_{MEM} cells also become independent of further antigenic stimulation, which allows their maintenance after resolving an acute infection. T_{MEM} homeostasis depends on paracrine and autocrine IL15 signaling, and T_{MEM} cells can proliferate in response to IL15 in a TCR-independent fashion [70, 71]. Because the overall number of T_{MEM} cells remains constant over long periods of time, the observed continual proliferation of T_{MEM} cells must be accompanied by a nearly equal death rate, probably as result of asymmetric division, where half of the offspring becomes apoptotic [72].

2.4 Persistence of Circulating αβ T Cells

The accurate evaluation of T cell subpopulations is difficult and varies in different studies. Vrisekoop et al. estimated that human naïve CD4 and CD8 T cells have half-lives of 4.2 and 6.5 years, respectively, whereas memory CD4 and CD8 T cells have half-lives of 0.4 and 0.7 years [73]. However, these measurements were made on circulating cells, whereas T_{MEM} in nonlymphoid tissues and bone marrow can have much longer life spans that allow long clonal survival. In fact, some memory T cells persist for more than 10 years [74, 75].

The age-related decline in T cell population is caused by thymic involution, impaired peripheral T cell maintenance, repeated antigen exposure, and persistent inflammation. Although healthy aging individuals can maintain a sufficient T cell content over time, naïve CD4+ and CD8+ T cell numbers and repertoire gradually decline [76], and the peripheral T cell pool becomes dominated by memory T cells [77, 78].

3 The T Cell Receptor (TCR)

3.1 Formation of the αβ TCR

Present on more than 90% of the T cells, the αβ TCR comprises the predominant TCR complex on the surface of human T cells [16]. Interestingly, it is also present on some neutrophils [79, 80] and macrophages [81, 82].

To date, no intrinsic enzymatic activity has been found for any of the TCR proteins, but rather, it is provided by the set of cytosolic, transmembrane (TM), and membrane-bound (e.g., myristoylated) enzymes that associate with the TCR. These proteins are organized in a highly cooperative system where the inter-protein signaling is quite often regulated by phosphorylation–dephosphorylation. Other interactions include a variety of enzymatic reactions and cytoskeleton and membrane rearrangements, which result in the attraction of scaffold proteins and adaptors, and signal amplification through multiple downstream metabolic pathways.

The conventional αβ TCR is located in membrane lipid raft subdomains. It contains two antigen recognition proteins, α chain and β chain, and four "signal transduction" CD3 proteins, δ, γ, ε, and ζ. The α and β subunits, bound together by a disulfide bridge, are sequence-variable proteins, with very short cytoplasmic tails, that can recognize agonist peptide in context of the MHC complex. The αβ heterodimer is flanked by the CD3 proteins as non-covalently associated heterodimers of εγ, εδ, and a disulfide-linked homodimer of ζ- [83–85]. The CD3 proteins are invariant, with relatively longer cytosolic domains that contain binding sites for cytosolic adaptors and enzymes, and maintain an invariant path for information of TCR/p–MHC recognition through the cell membrane down to the cytoplasm, to achieve an adequate cellular response [86].

Subunit cooperativity in the TCR is substantially determined by coordination of ionizable residues in their transmembrane (TM) domains, which form a specific hydrophobic/ionic interface between αβ, δε, γε, and ζζ [87]. Each of the ζ, ε, and δ molecules possesses an ionizable aspartic acid in their TM regions, while γ possesses a glutamic acid residue. Together, this gives six acidic residues in the TM region of the TCR complex. The α TM domain possesses two basic residues (arginine and lysine), while the β TM domain possesses one basic residue (lysine) [88]. Therefore, an αβγεδεζζ complex has three extra negative charges in the TM region leading to a "charge imbalance." This led to a search for an alternative bivalent TCR structure, γεαβζζαβδε, with a neutral TM region [89, 90] or even structures with higher valency [91–93]. Purification of the TCR with different detergents or without detergents has led to various deduced TCR structures [16, 90, 92]. In fact, it is possible that the actual TCR structure may "resonate" between mono-, di-, and higher variants of valency.

TCR proteins emerge on the plasma membrane as a fully built complex. ζ is the last subunit to be associated with TCR complex in the Golgi [16, 88]. In resting human T cells, a portion of ζ associates with the actin cytoskeleton. This interaction, mediated by a sequence in the C-terminus of ζ, might be involved in the localization of the TCR into lipid raft structures as well as play a role in TCR recycling [94]. In the cytosol, ζ is present in excess, compared to the other TCR proteins, and can participate in reactions not related to TCR activity. For example, ζ interacts with the transferrin receptor (TfR), and a TfR-ζ complex is expressed on the cell surface independently from TCR. ζ is also expressed in cells other than T cells, like NK cells and neurons. In NK cells, ζ is associated with NK FcgRIII (CD16) and may participate in cell surface expression of this receptor complex [95]. ζ is also associated with NKp46 and NKp30 receptors on NK cells, and its phosphorylation is required for transmission of activating signals upon antigen binding to these receptors. ζ expression in retinal ganglion cells and brain neurons regulates neuronal development by reducing the size of the dendritic arbor [96].

3.2 TCR/p–MHC Ligation

Upon TCR/p–MHC ligation, the shift in αβ conformation determines transmission of antigen-binding energy on the cell surface down into the CD3 intracellular tails [97]. A plausible interpretation of the available data is that the signaling domains of CD3 in the inactive TCR are submersed in the inner leaflet of the cell membrane which screens them from cytosolic adaptors [16, 83, 98, 99]. The mechanical force applied during TCR/p–MHC ligation weakens intra-subunit associations, changes membrane composition, and leads to subunit rearrangement, moving the CD3 intracellular domains out of the membrane and exposing them to cytosolic signaling adaptors [83, 98, 100–102].

CD3 proteins bind cytosolic adaptors by specific docking motifs. One such motif, tyrosine-based activation motifs (ITAM), is present as a single copy in γ, δ, and ε and as three copies in ζ, for a total of ten ITAMs in the CD3 complex [16]. This distinguishes the TCR from other cell receptors with ITAMs, which contain only one or two. A plausible view on ITAMs' composition is that ITAM tyrosines in CD3 are not redundant and might have specific roles [103]. However, the exact role of each ITAM in the CD3 subunits, including the detailed dynamics of ITAM phosphorylation, remains to be elucidated.

ITAM multiplicity determines the correct signaling for T cells developing in the thymus [104]. Decreasing the number of CD3 ITAMs to less than seven in mice impedes T cell development. These animals developed autoimmune disease probably due to inefficient signaling under negative selection in the thymus [104]. In these animals, studied prior to the onset of autoimmunity, CD3s with slightly decreased numbers of ITAMs still activated signaling cascades including cytokine production and T cell proliferation. Further decrease of CD3 ITAM multiplicity to two to four ITAMs per CD3 resulted in only a limited response with cytokine production, but not T cell proliferation [105].

For developed T cells, ITAM multiplicity appears to have minimal influence on signal amplitude [106]. Instead, it might help with the coordinated activation of all the subunits in the CD3 complex, such that there is a switch-like "all or nothing" response upon TCR binding to p–MHC [107]. Therefore, ITAM multiplicity may determine potency for synchronous activation of the T cell population [106, 107].

The ITAM motif (YXXL/IX_{6-8}YXXL/I) contains two tyrosines, which are usually phosphorylated by the key TCR activator, lymphocyte-specific protein tyrosine kinase (Lck). Lck associates with CD3 proteins as well as with the coreceptors CD4 and CD8 [108]. Lck exists in open (active) and closed (inactive) forms, which is determined by the state of phosphorylation. The transmembrane phosphatase CD45 activates or inhibits Lck, depending on location of dephosphorylation. Lck can also be inhibited by the cytosolic kinase Csk or by the cytosolic phosphatase SHP-1. Also, Lck can be activated through serine/threonine phosphorylation by extracellular signal-regulated kinase (ERK), and this may interfere with SHP-1 recruitment to the TCR complex [109].

Phosphorylation of both tyrosines in an ITAM by Lck attracts the cytosolic adaptor ZAP-70. After TCR/p–MHC ligation, ζ phosphorylation proceeds from the most membrane-distal ITAM toward the membrane [110], and their affinity to Zap-70 increases in the same direction [111]. When bound to the CD3 ITAM, Zap-70 is not activated, but released from its autoinhibited conformation [112]. This opens Zap-70 to be activated through phosphorylation by Lck or by autophosphorylation [106]. Zap-70 phosphorylation can occur by Lck or Zap-70 molecules that are bound to the same protein or by those associated with neighboring CD3 subunits [106], thereby enhancing CD3 cooperativity.

Two CD3 proteins, ε and ζ, deserve special attention. Both are non-glycosylated proteins and present in the TCR as two copies: ε as heterodimers with δ and γ and ζ as a homodimer with the possibility to undergo covalent S–S binding. ε and ζ subunits contain other docking sites besides ITAMs for cytosolic adaptors and can be involved in regulation beyond the ITAM–Zap-70 reaction [113–115]. For example, ε contains a proline-rich motif that can bind the adaptor protein Nck [116]. Nck participates in actin reorganization, cell adhesion, and movement. Nck binds to a partially phosphorylated ε ITAM, which contains a non-phosphorylated Y39 and a phosphorylated Y50. The Nck–ε interaction peaks in the beginning of TCR/p–MHC ligation and then decreases after 10 min as the phosphorylation of ε increases [103]. The ζ subunit has the longest cytosolic tail protein among the CD3 proteins and the highest involvement in multiplex signaling. It can bind adaptor proteins Shc and Grb2, the p85 subunit of PI3K [86], SLP-76, Vav, and negative regulators such as SLAP [117], SLAP-2 [118, 119], TRIM, CTLA4, and Unc119, as well as actin [113]. These proteins as well as GADS, phospholipase PLCγ1, Nck, p38 MAPK, and ADAP are recruited into a complex with linker for the activation of T cells (LAT) upon its phosphorylation by ZAP-70 [16, 120, 121].

The ensuing reactions may lead to signal amplification or abruption, depending on multiple factors involved in the regulation of T cell activity. When the TCR senses self-p–MHC, it enters a state of "tonic activity" with only partial phosphorylation of CD3 ITAMs and the associated cytosolic adaptors. The weak reaction on p–MHC may trigger a negative feedback, through recruitment of Csk, CD45, and SHP-1, which inactivate Lck and lead to receptor desensitization. In contrast, a stronger reaction may turn on a positive feedback, involving further activation and phosphorylation by Lck and prevention of SHP-1 and CD45 recruitment [109]. The activation triggers further signaling in local and distant protein networks and recruitment of the TCR coreceptors, resulting in a full cellular response [122–124].

3.3 TCR Coreceptors

Upon binding agonist p–MHC on target cell, the αβ TCR attracts transmembrane TCR coreceptors CD8 or CD4 to the p–MHC and checkpoint receptors that recognize additional non-MHC proteins on the target cell membrane. These additional interactions can either augment or undermine TCR signaling outcome.

Coreceptors CD4 or CD8 bind to the p–MHC cooperatively with the TCR, which may increase the overall binding force. However, the most important role of CD4 and CD8 proteins is to deliver activated Lck, associated with coreceptors, into the area of the TCR/p–MHC interaction so that CD3 cytoplasmic tails move out of membrane, become exposed to cytosol, and can be phosphorylated, to attract Zap-70 [100, 125]. Whereas the CD4 receptor is a single protein that spans the cell membrane and binds Lck [126], the CD8 receptor contains two proteins: CD8α and CD8β. While both CD8α and CD8β span the cell membrane, only CD8α binds Lck. Thus, it is likely that CD8α is involved in both ligand recognition and signaling, and CD8β participates only in the recognition. The CD8αβ heterodimer is expressed on CD8 T cells [127]; the CD8αα homodimer is expressed on some γδ T cells, NK cells, and IELs; and the CD8ββ homodimer has not been found on lymphocytes [100]. The binding affinity of CD8 to MHC is independent of TCR specificity or affinity; therefore, the impact of coreceptors on p–MHC binding and signaling decreases with increasing TCR affinity [128].

Coreceptor involvement in TCR/p-MHC ligation is followed by Zap-70-mediated phosphorylation of the scaffold protein LAT, which in turn nucleates multiple downstream pathways. Phosphorylated LAT binds additional adaptors GADS, GRB2, and phospholipase PLCγ1(NF-kB). This LAT complex subsequently recruits other adapters and enzymes, including SLP76, VAV1, Nck, p38 MAPK, and ADAP [16, 120, 121]. Phosphorylated ZAP-70 also has a noncatalytic function as a scaffold phosphoprotein that facilitates the high-affinity state of the integrin LFA-1, which in turn increases T cell adhesion by binding ICAM-1 on antigen-presenting cells [129].

Further T cell activation turns on a cascade of reactions, including structural changes with actin rearrangement [130], attraction of the centrosome, increasing of endocytosis, and further accumulation of TCR receptors in the synapse by lateral diffusion or exocytosis (*see* also Subheading 3.6). In addition, there is inositol phospholipid hydrolysis and mobilization of Ca^{2+} through activation of phospholipase C-gamma 1 and serine/threonine kinases [120]. Finally, distant signaling pathways are induced including PI3K/Akt/mTOR, Myc [44, 105, 131–133], NFAT [134], NF-κB, and AP-1 [135]. Overall, the signal cooperativity of CD3 proteins with the coreceptors may include cross-phosphorylation among ITAMs, synergism in adaptors' binding, and cross-activation among CD3 complexes in TCR clusters.

3.4 The Immune Synapse

The structure and specific activity of immune synapses are determined by the type of T cells (cytotoxic, helper, Treg, NKT), TCR (αβ TCR and γδ TCR), coreceptors (CD4 or CD8), and the set of checkpoint receptors that bind to various ligands outside the p–MHC and add either positive or negative cooperativity. For example, the synapse between a helper CD4+ T cell and B cell exists longer and leads to different outcomes than the synapse between a cytotoxic CD8+ T cell and B cell [136]. As a second example, the synapse with DCs primes naïve CD8+ T cells to proliferate and differentiate into CTLs over the course of several days, whereas it primes CTLs to kill diseased cells by secretion of cytolytic granules at the point of TCR signaling [137].

Target cells also determine synapse structure and function. Potential target cells include "professional" APCs, such as a dendritic cell (DC), macrophage, or B cell [138]; "atypical" APCs, such as a granulocyte [139], lymphatic epithelial cell [140, 141], basophil, mast cell, or eosinophil [138]; or "true target" diseased cells that should be eliminated. Synapses between T cells and different APCs have different organizations [142, 143]. CTLs attached to dendritic cells are less toxic toward their target than CTLs attached to B cells [144, 145].

Wild-type TCRs usually have low affinity for their p–MHC targets with a dissociation equilibrium constant (K_D) of 1–100 μM [136, 146, 147]. The precise number of p–MHC target antigens per cell required for optimal αβ T cell activation can vary, but in principle, T cells can be activated in response to only a very few p–MHC antigens [148–151]. In fact, the αβ TCR accurately recognizes relatively infrequent agonist peptides among "self"-peptides in the MHC class I, with a ratio as low as $\sim 1 \times 10^4$ self-peptides: one agonist peptide [152]. Technically, it occurs by constant p–MHC monitoring that provides low-level T cell tonic stimulation. It has been proposed that naïve T cells have the ability to adjust their activation threshold by a mechanism dependent not only on specific TCR affinity to p–MHC [153], but includes dynamic tuning with participation of coreceptors CD8 and CD4 and proteins with TCR inhibitory activity, like CD5, CD6, and CD45 [153, 154]. Intrinsic TCR affinity for self-p–MHC ligands in mice correlates with expression of CD5 [154, 155] and inversely correlates with expression of CD8 [156].

Serial activation of TCRs by p–MHC ligation may elevate reactivity to low-affinity antigens and also help to discriminate between different high-affinity ligands. For example, it is important to discriminate between an acute infection that usually results in expression of high-affinity antigens at high density and a normal high-affinity antigen that escaped thymus presentation, which is often expressed at low density. Comparison of TCRs with different affinities showed that a TCR with an affinity greater than the physiologic range mediated stronger and faster responses than wild-type TCR. Paradoxically, this leads to an inability to recognize such an antigen presented in low density on target cells [157]. Ligation of high-affinity p–MHC with TCR can be relatively stable and impede ligation of this p–MHC with other TCRs. In this case, p–MHC in high density still may provide strong serial triggering. However, serial triggering should be blocked if p–MHC is presented in low density. In this fashion, TCR affinity and clustering can be considered part of peripheral immune tolerance.

The initial TCR/p–MHC binding is sterically limited because the exodomains of the TCR and p–MHC are short (~7 nm) requiring a relatively small intracellular cleft of ~15 nm to make contact [151]. This interaction is difficult because in the cell membrane the TCR neighbors highly abundant large surface proteins like CD45, whose extracellular segment ranges from 20 to 50 nm depending on the isoform [158]. Individual TCR/p–MHC interactions are short-lived (seconds), but for T cell activation, binding should continue for minutes to hours [151]. Therefore, a close contact between the T cell and target cell within a cleft of ~15 nm must be built and persist by exclusion of large surface moieties.

T cells use microvilli to create a close contact with the interrogated antigen-presenting cells (APCs) and target cells [24, 159]. These ~150 nm-diameter cell membrane protrusions contain adhesive receptors and form contact with target membranes with a short ~15 nm cleft, which sterically excludes CD45 and other bulky membrane proteins [151]. Microvilli usually last for seconds, but if a TCR complex presented in it binds agonist p–MHC, this can lead to cytoskeleton rearrangement that stabilizes the microvilli for a longer period of time [160]. The effective binding involves redistribution of adhesion receptors CD2 and LFA-1 that bind CD58 and ICAM-1 target proteins, respectively. The relatively long LFA-1/ICAM-1 interaction (~40 nm) moves to the periphery of the contact zone, while the shorter CD2/CD58 interaction (~15 nm) is placed in close proximity to the TCR/p–MHC binding [161].

The synapse is a dynamic structure that may contain various numbers of TCRs. Often it undergoes a transition from nanoclusters (~20) of TCR to microclusters (~300). These can proceed further to increased concentric membrane aggregates that in some conditions comprise up to 20% of the cell surface. Such complexes contain three concentric supramolecular activation clusters (SMACs): a central TCR/p–MHC cluster (cSMAC) with a narrow ~15 nm cleft [137], a peripheral ring of LFA-1/ICAM-1 (pSMAC) with a ~40 nm cleft, and a distal ring that includes CD45 and F-actin (dSMAC) in a bigger cleft [162]. This structure has been observed in helper, cytotoxic, and regulatory T cells [163].

The SMAC complex is a dynamic structure that can persist over a period of hours and provides signaling that can change in a timely fashion [16]. While cSMAC was initially considered only an activation domain, the recent identification of nearby late endosomal compartments suggests it can also function as a domain of TCR downregulation [161].

After completion of the immune response, T cell activation caused by TCR ligation should be downregulated. This can occur simply by exhaustion of available antigens, by negative feedback with synapse-associated cytosolic enzymes, or by specific checkpoint inhibitory receptors, such as PD-1, CTLA-4, B7-H3, DGK-α, LAG-3, and Tim-3, that are activated in parallel with the TCR and can work in the synapse by binding cognate target cell ligands [164]. PD-1 has two ligands, PD-L1 and PD-L2, which are members of the B7 family [165]. PD-1 is located in the immune synapse interface and recruits cytosol phosphatase SHP-2 to dephosphorylate CD28 [166]. PD-1 may also dephosphorylate phosphotyrosines in other TCR-associated proteins, such as Zap-70 and CD3 [165]. CTLA-4 shares two ligands, CD80 and CD86, with the stimulatory receptor, CD28, and can downregulate CD28 activity by binding its ligands [167, 168].

3.5 Signal Transduction Downstream of the p–MHC

Not only T cells but also target cells ("true targets" and APCs) can sense the immune synapse and react accordingly. For professional APCs, TCR/self-p–MHC contact during routine immune monitoring may cause signaling on the APC side to prevent its killing. After T cell attachment, DCs activate signaling mechanisms that facilitate cell–cell communication and actin and membrane remodeling [24, 169].

T cells can activate APCs by depositing their membrane fragments that contain TCRs, costimulatory and adhesion molecules, and cytokines on cognate antigen-bearing APCs. CD4+ T cells transfer to DCs membrane proteins by trogocytosis and the budding of T cell microvilli particles (TMP). The TMPs contain CD2, CD28, CD4, CD25, and activating cytokines and, upon uptake, initiate DC activation, including a calcium response and expression of costimulatory proteins such as CD40, CD80, and CD86 [24]. Memory

CD8+ T cells release the DC-activating factor TNF-α, which induces the expression of an endogenous granzyme B inhibitor, PI-9, that protects DCs from killing by CD8+ effector T cells [170].

The reactions developed by cancer cells in response to T cell binding may include various pathways, such as induction of inhibitory ligands and cytokines [171, 172], abnormal tumor angiogenesis [173], downregulation of MHC expression [174], secretion of inhibitory exosomes [175], and complex modulation of the tumor microenvironment [176].

3.6 αβ TCR Clustering

Clustering and spatial cooperation of proteins in membranes are observed in many signaling pathways. Although TCR clustering may exist without antigen activation [100], such activation leads to a spatial reorganization of TCRs into signaling-competent clusters. In turn, initial TCR clustering may cooperatively facilitate further cluster development attracting additional resources from both membrane and cytosolic compartments [177].

The efficiency of the T cell reaction can be achieved either if the TCR has a relatively high affinity to p–MHC or if the affinity is relatively low, by cooperative involvement of neighboring TCRs. TCRs on the T cell membrane are usually localized in lipid rafts as 2D nanocluster aggregates, and the reaction on a single TCR may be amplified by lateral (horizontal) activation extension in the cluster. Therefore, structural rearrangement of activated TCR adaptors in the underlying cytosol might facilitate similar processes in its vicinity and promote cooperative lateral activation without additional p–MHC ligation. A strong positive cooperativity between individual TCRs has been detected in nanoclusters containing up to 20 TCRs [90, 178, 179]. In these experiments, binding of only two p–MHC (a p–MHC dimer) could stabilize 20 TCRs in the signaling-competent state [180]. Such allosteric reactions between TCRs can be an important factor of TCR selectivity that allows the detection of a relatively rare and weak "signal" (foreign p–MHC l) in the presence of abundant "noise" (self-p–MHC) [181]. Clustering of TCR complexes can be mediated by extracellular domain oligomerization, intracellular domain interactions, and attached cytoplasmic scaffold proteins [100].

In contrast to naïve T cells, T_{EFF} and T_{MEM} cells have a lower threshold of activation and response and also an elevated level of TCR clustering [178]. Unlike B cells, T cells lack the capacity to undergo "affinity maturation" after antigen engagement. However, "functional avidity maturation" can be achieved by TCR clustering [178, 182].

3.7 Structure and Signaling of the γδ TCR

The structure and function of the TCR in γδ T cells appear to be fairly different from αβ T cells. While the γδ TCR sometimes contains the same CD3 complex as the αβ TCR, for the recognition of homodimer, it uses TCR-γ and TCR-δ chains instead of TCR-α and TCR-β chains. In addition, Hayes et al. report that some γδ TCRs lack the CD3 δ chain and instead have the stoichiometry: γδ, γε, γε, and ζζ [183–186]. Similar to αβ TCRs, γδ TCRs are contained in lipid rafts and can form clusters on the cell surface [187]. Distinct from αβ T cells, γδ T cells do not require "classic MHC" molecules to recognize antigens, and they do not require CD4 or CD8 coreceptors.

Given the underlying differences, it is not surprising that signaling mechanisms in γδ T cells are distinct from αβ T cells. The γδ TCR has a stronger signaling capacity, which may be due to the fact that they constitutively express approximately twofold more of the TCR/CD3 complex than αβ T cells [188–190]. Also, the γδ TCR may provide signal transduction without a conformation shift of the CD3 complex. This quite unexpected divergence from the αβ TCR may be caused by differences in the TCR-γ and TCR-δ amino acid content, their glycosylation and orientation in the membrane, the pattern of TCR clustering, or the complement of associated kinases [191, 192]. For example, B lymphoid kinase (Blk), an Src family kinase expressed primarily in B cells, is expressed in γδ T cells but not in αβ T cells [193]. In addition, a subpopulation of γδ T cells, but not αβ T cells, has been detected in Lck-deficient and Zap-70-deficient mice, suggesting that Lck and Zap-70 are necessary for αβ T cell viability but not for γδ T cells [192]. In primary murine γδ T cells, TCRs contain ζζ homodimers. However, following ex vivo activation and expansion, one or both ζ subunits are replaced with FcεR1γ proteins [183].

4 Synthetic Receptors

4.1 Engineered TCRs

Synthetic TCRs (synTCRs) can potentially recognize all peptides processed and presented in the context of MHC molecules, thus allowing TCRs to target both surface and intracellular antigens. On the other hand, TCRs only recognize peptides in the context of the MHC complex. Therefore, this approach is hindered by some factors, including the need for MHC matching, MHC downregulation by cancer cells, suppressive tumor environment, and off-target/off-tumor killing [194]. In addition, the activity of TCR-transduced cells may be affected by the formation of mixed dimers between exogenous and endogenous α and β proteins which may decrease activity or lead to nonspecific reactivity [195, 196]. To prevent such dimerization, the selective binding between α and β proteins in synTCRs can be achieved by rearranging specifically interacting amino acid sequences in α and β constant domain interface [197] or by adding a second disulfide bond [198]. Another approach is to construct synTCRs with a murine constant region in place of the human constant region. This resulted in preferential pairing of the murine constant domains in α and β subunits and higher expression of the engineered TCR on the surface of the human lymphocytes [199]. One theoretical drawback in the use of mouse TCR domains might be the development of human anti-mouse TCR immune responses; however, this was not observed in clinical trials [200]. Expression of cancer-reactive γδ TCRs in αβ T cells prevents formation of mixed dimers [147]. This reprogramming accompanied by CRISPR-mediated elimination of the endogenous αβ TCR led to increased γδ TCR expression and efficiency irrespective of patient MHC type [201].

Besides optimization of amino acid content, several other factors should be considered in constructing a synthetic TCR. Adequate expression of exogenous αβ heterodimer depends on the configuration of the vector and expression cassettes. It also depends on availability of CD3 proteins (ε, ζ, δ, γ) for correct assembly in the Golgi [195]. Co-transfer of CD3 and αβ genes into primary murine T cells enhanced TCR expression and antigen-specific T cell function in vitro and in vivo [202]. TCRs with high affinity, not available in normal T cells because of negative selection in the thymus, may enhance the ability to kill target cells in the cancer environment [203]. Engineered T cells containing "high-affinity" TCRs showed efficiency in treating myeloma and synovial cell sarcomas [204, 205]. Also, generation of synthetic TCRs able to recognize specific cancer neo-antigens or known cancer-specific antigen/MHC combinations can be advantageous for developing individualized anticancer therapy [206, 207].

Another approach is to create synTCRs with antibody recognition domains. In these receptors, exodomains of α and β subunits of the TCR are modified by replacing their variable domains with antibody domains that can recognize cancer-associated antigens. It can be just variable domains, V_H and V_L [208], or Fab fragments with V_H–C_H domains fused over the TCR-α constant domain and V_L–C_L domains fused over the TCR-β constant domain [209]. The resulting chimeric TCR is expressed on the surface of cytotoxic T lymphocytes, recognizes antigen in a non-MHC-restricted manner, and transmits the signal through the CD3 complex for T cell activation [208]. The absence of p–MHC in the synapse excludes CD8 co-signaling; however, the affinity of Fab can be increased to a level that sufficiently compensates for the absence of CD8 [128].

4.2 Chimeric Antigen Receptors (CARs)

Basic chimeric antigen receptors (CARs) contains three elements: (1) a recognition domain that is a surface ligand-binding domain; (2) a transmembrane (TM) domain that is a structure ~20 amino acid long, enriched with hydrophobic amino acids (AAs) and forming an alpha helix in the cell membrane; and (3) an intracellular effector part that can contain various signaling domains needed for sustained effector cell ability to kill and propagate [208, 210].

Upon ligation to cognate antigens on the target cell, it is thought that the CAR dimerizes at the site of recognition and undergoes a conformational shift in its cytoplasmic domains, which leads to their phosphorylation, binding, and activation of Zap-70 with sequential activation of multiple signaling cascades [211]. In support of dimerization occurring, a mutation in a CAR's ectodomain that facilitates spontaneous dimerization (even without antigen recognition) increased its functionality [212], whereas mutations in a CAR transmembrane domain that resulted in disruption of dimerization led to decreased CAR-T cell activation and cytolytic activity [213].

First-generation CARs contain a single-strand antibody (ScFv), TM, and CD3-ζ signaling domain [210, 214]. Second-generation CARs include a coactivator cytoplasmic domain in *cis* to provide additional T cell co-stimulation. The most widespread is a CD28 or 4-1BB signaling domain inserted between the TM and ζ domains. CD28 signals through activation of LCK, PI3K-Akt [215], Grb2, and Gads [216] and induces Bcl-X_L [217] and IL2 [216]. 4-1BB signaling upon aggregation (trimerization) of 4-1BB ligand attracts TNF receptor-associated factors and forms a "signalosome" that activates T cell proliferation and survival [218]. This leads to phosphorylation of CD3 proteins ε and ζ, Lck, and LAT [219]. Other costimulatory domains, like ICOS, OX40, and CD27, can also function in CARs between the TM and ζ domains [220–222]. Third-generation CARs include two costimulatory domains, like CD28 and 4-1BB inserted between TM and ζ [223]. This additional co-stimulation apparently increases the basal activity of CARs and can be counterproductive due to baseline activation and auto-toxicity [224, 225].

CARs can apparently function in many different cytotoxic immunocytes [1, 226]. For human CD8+ T cells, the granzyme–perforin pathway seems to be the most common activated by the CAR, as this is the predominant cytotoxic mechanism in human T cells [46, 47, 227]. However, other pathways are also used as Hong et al. demonstrated Fas-mediated killing by CD30 CAR-T cells [228]. Because some CD4+ T cells possess cytotoxic activity, they also can be reprogrammed for CAR-mediated killing [227, 229]. Beyond conventional αβ T cells, CAR-mediated killing has also been shown in NK cells [230, 231], γδ T cells [232, 233], NKT cells [234, 235], and neutrophils [236]. While the mechanisms of killing by other effector cells reprogrammed with CARs might be more diverse, it is assumed that upon target recognition, CARs can activate the natural cytotoxic signaling pathways present in a host cell. Interestingly, for macrophages, a CAR that contains the cytosolic domains of Fc receptor instead of the ζ-signaling domain leads to phagocytosis upon target recognition instead of cytotoxicity [5].

In T cells, analysis of CAR-mediated targeting showed that affinity to cognate antigen in the interval of 10 μM to 1 mM allows for both effective recognition and dissociation when the T cell action is completed [237, 238]. However, lower affinity might be preferable to prevent off-tumor killing [239]. Steric hindrance both inside and outside the cell should be taken into consideration when designing a CAR. The length of the extracellular segment should be comparable with the optimal TCR/p–MHC distance at ~15 nm (*see* Subheading 3.4) [212, 240]. Likewise, steric limitations should be applicable to the cytosolic part of the CAR, because the signaling adaptor proteins should act in a certain distance from the cell membrane. Of course, the sophisticated TCR architecture makes it difficult to easily deduce the ideal CAR sequence. That is why during CAR construction, combinations of ecto-, TM, and endodomain amino acid contents have to be tested in parallel to determine variations in target affinity and signaling. It is important to note that even small differences in amino acid content can dramatically change the tertiary structure of the CAR, with obvious consequences for protein stability and function [241].

4.3 Comparing CARs with αβ TCRs

In engineering T cells, the simplicity of a CAR compared to a synTCR leads to both advantages and disadvantages. One advantage is that CARs recognize and bind targets independently of MHC and coreceptors like CD4 or CD8. In addition, CARs can recognize a wide spectrum of ligands on the cell surface, including proteins, carbohydrates [242], glycolipids, and other moieties [243], that are usually not recognized by TCRs. The CAR recognition motif structure appears flexible as scFvs, ligands (e.g., CD70 that binds CD27 receptor) [244], and single-strand avidin (that binds biotinilated targets) [245, 246] are all functional. Compared to TCRs, CARs provide a faster killing dynamic [149] and can be used in the presence of other CARs and TCRs in the host, both independently of them and in cooperation [247].

The main disadvantages of CARs are that they cannot target intracellular antigens and they do not communicate through the balanced system of CD3 proteins and coreceptors leading to less regulation in cytosolic signaling. That can undermine some important functions such as antigen recognition proofreading and adequate dynamics of the cell response. For example, first-generation CARs with a sole ζ chain as a signaling domain and without a coactivator signal were unable to maintain robust T cell viability in the presence of cognate antigen. In contrast to the TCR, the conventional CAR is not involved in cell monitoring and combinatorial antigen evaluation. Rather, it works as a binary operator that turns on the response as soon as it recognizes the target. Compared to conventional CTLs, CAR-T cells are less sensitive to p–MHC density. Whereas a CTL's response may need only a few agonist p–MHCs per target cell [148–151], a CAR-T cell response may need about ~200 antigens per cell [248, 249].

Another major flaw of CARs compared to TCRs is that they usually produce a high basal signal, which can be deleterious for T cell viability [250]. Whereas the TCR emerges from the cell membrane as a tightly cooperative complex with accurate regulation of its subunits' conformation, the CAR emerges as a single protein prone to specific and nonspecific reactions with surrounding molecules. Thus, while TCR tonic signal is caused by ligation with self-p–MHC and is a part of T cell homeostasis, the basal signal activity of the CAR is independent of antigen presentation and may disrupt T cell homeostasis. This basal activity is correlated with the density of CAR proteins on the cell membrane [250, 251]. In fact, CAR proteins can spontaneously aggregate in the cell membrane independent of external ligands, potentially because of thermodynamic driving factors and variation in physicochemical properties of CARs and surrounding proteins [252]. Multiple factors may determine CAR aggregation and toxicity including the spacer connecting the CAR's recognition and TM domains [253], the configuration of the CAR's active ITAMs [254], and the activity of T cell death signaling pathways, Fas and DR5 [255].

The comparison of CD19 CARs containing 4-1BB-ζ and CD28-ζ cytoplasmic tails showed that CD28-ζ CAR had higher spatial aggregation and more "basal" ζ phosphorylation and was more toxic for T cells [250]. However, basal CAR toxicity also has been shown for 4-1BB-ζ CAR [251].

4.4 Complications in the Clinical Use of CAR-T Therapy

Although CAR-T cells have been very successful in some clinical trials, this therapy is associated with serious complications. Among other factors, problems associated with CAR-T cells include uncontrollable activation, expansion, and persistence, as well as on-target/off-tumor and off-target/off-tumor killing. Upon introduction in patients, CART-19 cells can achieve rapid proliferation (up to 10^4-fold expansion), which may result in tumor lysis syndrome (TLS), cytokine release syndrome (CRS), and neurotoxicity [256–259]. In addition, CART-19 therapy kills all CD19+ cells leading to B cell aplasia [256]. On the other hand, B cell tumors with a mutated CD19 can escape CART-19 killing [260–262].

5 Next-Generation Strategies for T Cell Engineering

This quickly developing field has extended in many directions in an attempt to improve upon the original CAR and TCR approaches. Designs to improve both safety and efficacy include modifications of the CAR itself, combining multiple CARs, and adding multiple factors in addition to the CAR.

5.1 Construction of CARs with New Domains

Since the first description of a CAR, an impressive number of structural modifications with a wide number of variations have been introduced. Considerable attention has been focused on the discovery of novel scFv recognition domains to target different antigens. However, any structure able to bind a cancer cell is a theoretically viable alternative.

Several "universal" CAR systems have been constructed. In one strategy, the antigen-recognizing domain is replaced with a monomeric avidin moiety that binds biotin. Biotinylated tumor-specific molecules, such as a monoclonal antibody, can then direct the CAR-T cell to different target cells. Simply changing the biotinylated antibody redirects the CAR-T cell to recognize and trigger killing of cancer cells that are "stained" with the biotinylated monoclonal antibody [245, 246]. In a second strategy, Cho et al. replaced the scFv with a leucine zipper domain, such that the CAR could bind to a second chimeric protein composed of the cognate leucine interaction domain fused to an scFv. Introduction of such second chimeric proteins can continuously redirect the CAR-T cell activity [263]. A third strategy fuses a tumor antigen-specific Fab with a peptide that binds to a CAR. Peptide-associated Fabs that can have different specificities can be systemically delivered in vivo to connect the CAR-T with cognate antigen on the target cells. In this system, the antigenic diversity of Fabs and their dose determine CAR-T antigen specificity and the level of activation [212].

In addition to universal systems, another approach attempts to develop a CAR that can emulate the recognition ability of a TCR for agonist p–MHC. Here, a two-gene system is used. The first gene encodes a CAR where the scFv is replaced with a TCR-α chain that is truncated at the TM region and contains a cysteine [264]. The second gene encodes a TCR-β chain that is also truncated at the TM region and contains a cysteine to mediate disulfide bridging. When expressed, the TCR-β chain binds to the TCR-α chain of CAR, creating an αβ TCR/CAR hybrid. This approach increases the spectrum of antigen recognition, by inclusion of intracellular antigens, although in the context of MHC. It can be especially important in the treatment of cancer cells with neo-antigen markers as an alternative to TCR-mediated therapy.

Whereas the previous approach attempts to bring together TCR binding with CAR signaling, others have tried to create CARs with scFv binding that directly engage with TCR signaling. One approach to directly utilize TCR signaling is a "T cell antigen coupler" (TAC) [265]. In this design, the CAR contains an anticancer scFv recognition domain attached to the CD4 TM and signaling domains. In addition, a CD3-ε-binding domain is inserted in the CAR between the scFv and the TM, which results in CAR attachment to the CD3-ε subunit of the TCR. When the CAR's scFv binds cognate antigen, the signal transduction goes through the CD4–TCR complex. In mouse models, this approach yielded increased antitumor efficacy with reduced toxicity. A second approach, called T cell receptor fusion constructs (TRuCs), was developed by sequentially attaching scFv domains to each subunit of the TCR [266]. In TRuC-T cells, the scFv is incorporated into the TCR and binding an antigen in an MHC-independent manner engages the signaling capacity of the entire TCR. Among the different TRuCs tested, fusing the scFv to the ε subunit showed the highest level of functionality. This is potentially due to its stoichiometric advantage as well as its specific cytoplasmic docking sites for Nck [116], GRK2 [267], CAST [268], and phospholipid-binding motif [269].

A separate tactic to increase CAR-T cell functionality is to insert in the CAR's cytoplasmic tail domains that can emulate the signaling by homeostatic interleukins. Kagoya et al. introduced binding motifs for STAT5 and STAT3 in CD19 CAR and showed that the new CAR strengthened the activation of JAK kinase and STAT3 and STAT5 signals, elevated in vivo persistence of CAR-T cells, and increased their antitumor activity [270].

Finally, significant improvements can be made even without modifying any of the functional domains but only taking into account the tertiary structure of CAR protein and the distances between recognition domains and signaling domains that could be crucial for signaling. Using CD19 CAR, Ying et al. modified non-enzymatic "scaffold segments" of the CAR by increasing the length of the hinge and the distance between TM and signaling domains. By computer modeling, they showed that such variations can dramatically change tertiary structure of the protein. Then, by adding additional amino acids to the hinge and two amino acids between the TM and CD28-ζ, they were able to decrease the level of CAR signaling and the rate of CAR-T propagation in the presence of targets; and by that means virtually eliminate the development of chemokine shock in a mouse model and a clinical trial [241].

5.2 Combining More Than One CAR in a Cell

Expression of multiple constructs in T cells potentially can provide recognition of multiple antigens on the targets to discriminate between cancer cells and normal cells, and obtain adequate T cell signaling by switching both activation and inhibition pathways [221, 271–273].

This approach may increase CAR-T cell efficiency in situations where tumors downregulate the expression of cognate antigens. For example, failure of the CD19 CAR against B cell tumors is sometimes caused by internal deletions in CD19, which remove the recognized epitope [274]. In such a situation, the expression of an additional CAR that recognizes other B cell markers can prevent tumor escape [275]. This has been accomplished both by supplying two separate CARs and by supplying one CAR that contains two tandem scFvs, each for a different antigen [276]. In the latter case, because different antigens may have different length, the

CAR should have appropriate sterical configuration of its recognition domains that not only allows simultaneous ligation with antigens but also maintains an appropriate distance to the target membrane for synapse formation.

A two-CAR system also may increase CAR-T cell selectivity. For example, consider the scenario where a normal cell has only one of two surface antigens present on a cancer cell. A T cell expressing two CARs, one that recognizes each antigen with sufficiently low affinity, may proceed with killing only when both CARs are simultaneously ligated with the target [273, 277]. In another scenario, a cancer cell may have only one of two surface antigens present on a normal cell. In this case, one CAR can recognize the shared antigen promoting killing, while a second CAR, known as an inhibitory CAR (iCAR), can recognize only normal cells and contain a signal inhibitory domain instead of the activation ζ domain. Fedorov et al. have shown that an iCAR with a segment of the inhibitory PD-1 cytoplasmic tail can prevent the stimulation of the other CARs when the T cell interacts with a cell containing both antigens [278].

Playing off the theme of converting positive to negative signals, switch receptors were designed to convert a negative signal to a positive one. Here, the extracellular domain of PD-1 is combined with the cytoplasmic domain of CD28. When the switch CAR binds the inhibitory ligand PD-L1, it activates T cells through the CD28-mediated pathway. In solid tumor models, T cells reprogrammed with both CAR and the switch receptor showed augmented efficacy compared to CAR-T cells [279–282].

5.3 Using Additional Genes to Reprogram the T Cells

The fourth generation of CARs, armored CARs, has been made by combining two expression cassettes: one coding a second-generation CAR and another coding an additional metabolically active protein, such as a cytokine, antibody, or another ligand [283]. Since CAR-T cells accumulate in the tumor, the active proteins are delivered locally to the site of disease, minimizing the toxicities often associated with active proteins delivered systemically. T cells loaded with a CAR construct armored with the proinflammatory cytokine IL12 showed elevated antitumor efficacy [284, 285]. CAR-T cells armored with IL18 increased CAR-T cell survival and enhanced immune response by modulating tumor microenvironment [286]. A CD20 CAR-T cell armored with IL7 and the chemokine ligand CCL19 improved immune cell infiltration and CAR-T cell survival in the tumor [287]. CAR-T cell persistence and efficiency can also be enhanced when cells are reprogrammed with constitutively active homeostatic receptors, such as IL2, IL7, or IL15 receptors, or with chimeric cytokine receptors that switch a negative signal produced by inhibitory cytokines, such as IL4, to a positive signal [288]. CD19 CAR-T cells, armored with the immune activator protein CD40L, exhibited increased cytotoxicity against CD40+ tumors and extended the survival of tumor-bearing mice in a xenotransplant model of CD19+ systemic lymphoma [289]. CAR-T cells loaded with a PD-1-blocking scFv enhanced the survival of PD-L1+ tumor-bearing mice in syngeneic and xenogeneic mouse models through both autocrine and paracrine mechanisms [290].

Suicide switch constructs can eliminate CAR-T cells upon systemic delivery of a signaling molecule. For example, this construct may contain caspase-9 protein fused with a protein that can be dimerized by a drug. In the presence of the drug, the caspase-9 dimerizes and activates the intrinsic apoptotic pathway [291]. A second strategy introduces a truncated epidermal growth factor receptor (EGFRt) that is recognized by the antibody cetuximab, which mediates antibody-dependent cellular cytotoxicity (ADCC) against the CAR-T cells [82]. However, kill switches do not allow control of the rate of T cell activation and expansion. Rather, they are turned on after the recognition of a problem and once activated cannot be reversed. In addition, leaky expression of an inducible suicide gene can undermine efficacy, while incomplete activation or targeting can undermine the purpose of the suicide switch.

Lim et al. improved the functionality of CAR-T cells by adding a synthetic Notch (synNotch) receptor that contains an antigen recognition domain (scFv), fused to the Notch regulatory core domain and a transcription activator domain. Binding of the cognate antigen stimulates cleavage of the receptor and releases the transcriptional activator, which can enter the nucleus and drive ectopic expression of genes inserted in the T cell DNA under an activator-specific promoter. One synNotch receptor, upon antigen binding, can activate multiple genes regulated by the same promoter [271, 292].

6 Challenges and Potential Future Solutions

A combinatorial approach for cancer treatment is important to emulate the complexity of the immune system in fighting tumors. Here, we will consider ways to (1) better emulate the TCR, (2) use other cells instead of αβ T cells, and (3) use other cells in combination with αβ T cells.

6.1 Better Approximation of the TCR

Conceptually, CAR-mediated activity may emulate TCR function with similar intercellular synapses and signaling pathways. Further improvement of CAR functionality might lay in better understanding the relationship between CARs and TCRs. CAR-mediated synapses are structurally and functionally different from TCR synapses, being smaller, less structured, and shorter-lived [149]. The CAR-T synapse, created without a TCR, is obviously deficient in coreceptors (CD8 or CD4) and in underlying processes related to coreceptor binding to MHC such as delivering additional Lck activator for Zap-70 (Subheading 3.3). In addition, they do not bind to MHC on target cells, so the target cell membrane state also can be different.

An important goal is to attain CARs that build a cooperative complex with CD3 (*see* Subheading 5.1). If the cytoplasmic tail of the CAR is not included in a tight TCR complex, it may constantly be in a "loose" conformation in the cytosol, prone to some degree of nonspecific aggregation or phosphorylation by Lck and basal signaling.

Optimizing the spatial and functional relationship of the CAR to the TCR is predicted to boost activity. Current experimental data suggests that the CAR–TCR relation can be quite flexible. T cells with defective TCR expression are still able to provide CAR-mediated killing [263, 293]. On the other hand, cells with an active TCR may contain TCRs as part of the CAR synapse [266]. A more integrated relationship between CARs and TCRs may be advantageous for CAR-T cell efficiency.

6.2 Expanding Beyond $\alpha\beta$ T Cells

Another direction comes out of the fact that TCRs and CARs can work not only in αβ T cells but also in other immune cells involved in the innate response. In such cells, ectopic TCR and CAR activity can be combined with innate modalities to create a more complex immune response. As already mentioned, expression of endogenous TCRs has been shown in various myeloid cells, including eosinophils [294], neutrophils [79], monocytes, and macrophages [81, 294–296]. Ectopic TCRs and CARs that redirect immune cells against cognate targets can be expressed in NK cells [297, 298], NKT cells [299], γδ T cells, and cytokine-induced killer (CIK) cells [2]. In contrast to αβ T cells, certain subsets of innate immune cells have natural attraction to tumors through the recognition of stress-related tumor antigens.

γδ T cells serve as a particularly illustrative example. One reason they are attractive for ACT is that they have natural tropism for several types of cancer by monitoring stress-induced and inflammatory markers, such as lipopeptides, pyrophosphates, microorganism-derived proteins, and self-proteins through their γδ TCR, Toll-like, NK, and CD16 receptors [191, 192, 300]. In addition, they detect alterations in cell surface molecules, such as the MHC class I chain-related ligands A and B (MICA and MICB), and cell-associated antibodies [301, 302]. Their antitumor activity involves cytokine and chemokine secretion [303] and cytotoxicity. In contrast to αβ T cells, γδ T cells do not have strong lineage separation between helper and cytotoxic subsets, but do include T_{EFF}, T_{CM}, and T_{EM} cells [303–307]. Cytotoxicity is accomplished through perforin–granzyme, TRAIL, FasL, and ADCC [308]. In addition to cytotoxicity, γδ T cells are able to phagocytose and present tumor antigens to CD8+ αβ T cells, as well as induce DC maturation by TNF-α secretion [309]. Finally, as an added advantage, γδ T cells are mainly not alloreactive and do not induce GVHD [310].

Both ectopic TCRs and CARs can be used to reprogram γδ T cells for adoptive therapy [301]. Transfer of ectopic αβ TCRs into γδ T cells can be provided without TCR mispairing and formation of mixed TCR heterodimers. γδ T cells engineered to express human αβ TCRs exhibited high levels of antitumor cytotoxic activity and cytokine release [311, 312]. Both GD2 and CD19 CARs have been used in peripheral blood-derived γδ T cells and shown to exhibit target-specific IFN-γ secretion and cytotoxicity [313, 314]. In the case of a GD2–CD28–CD3-ζ CAR, the reprogrammed γδ T cells showed enhanced GD2-specific killing beyond the γδ T cells without the CAR, which also recognized the tumor cells with endogenous "stress receptors." Expanded CAR-T cells retained the ability to take up tumor antigens and cross present the processed peptide to responder αβ T cells [233]. To refine the specificity of γδ T cell cytotoxicity, Fisher et al. created a GD2 CAR in which the ζ domain had been replaced with innate NKG2D signaling molecule DAP10 [315]. These γδ CAR-T cells killed GD2+ glioblastoma cells, but not GD2+ control cells, by working with the endogenous γδ TCR targeting glioblastoma "stress receptors." One way to augment the expression of tumor stress-related antigens can be by less specific treatments, like chemo- and radiotherapy [316].

Two obstacles to overcome in using γδ T cells are relatively low cell numbers, and they can also promote cancer progression by inhibiting antitumor responses, enhancing cancer angiogenesis, and increasing the population of myeloid-derived suppressor cells (MDSCs) [14, 317]. Thus, accurate amplification, evaluation, and selection of γδ T cell subsets will be an important part of their optimization for therapy. To some extent, the number and distinct activity of γδ T cells can be modulated with cytokine stimulation [14].

6.3 Total Immune System Engagement

The human immune system is very complex with various types of cells, proteins, and subcellular particles working synergistically against various diseases, including infection, cancer, and aging. The ability of pathological agents for fast propagation, modification, and population plasticity dictates the general immune system works as a multifactor combinatorial defense. Although today immune cell reprogramming is still limited to relatively simple combinations tested in clinical studies, it is clear that in the future, multi-cell, multi-ligand treatment will need to be developed for an efficient and flexible therapeutic approach. This can be especially important for the treatment of solid tumors that are often resistant to conventional CAR-T therapy [2] because of poor tumor recognition, penetration, and inhibition by the tumor microenvironment.

In addition to using T cells expressing multiple CARs and other chimeric receptors, B cells, NK cells, and other combinations of reprogrammed immune cells could be used. Most of the anticancer therapeutic designs are not mutually exclusive and can be applied in combinations to maximize the outcome. The improvement of therapy employing CARs and ectopic TCRs probably will depend on their cross talk with tumor-resident immunocytes. For example, the treatment may include co-introduction of αβ and γδ T cells reprogrammed with CARs and TCRs, DCs loaded ex vivo with tumor-specific antigens, cytokines, and checkpoint inhibitor ligands delivered locally to the tumors by "armored" T or NK cells. The versatile treatment will integrate the most efficacious combination of the helper, effector, and memory cells with their modulators that allow to maximize the therapeutic specificity and safety.

As a second example, tumor-infiltrating lymphocytes (TILs) are cells that can recognize and penetrate tumors. These cells can be extracted from the tumor, activated and propagated ex vivo, and then used for ACT. However, some solid tumors contain very low numbers of infiltrating T cells. So, combining therapies that increase T cell recruitment with subsequent TIL treatment might be beneficial. Although T cell recruitment can depend on chemokine gradients created by tumor-resident dendritic cells, the recruitment of DC in tumors is a process that is not fully understood. It may depend on multiple factors, including both activation and inhibition influenced by other tumor-resident cells such as macrophages and NK cells [318]. Here again, understanding better the mechanisms of intratumoral immune cell signaling might lead to approaches to more fully engage and reprogram multiple arms of the immune system.

7 Conclusions

T cell reprogramming has demonstrated that methods of genetic and cell engineering can be broadly used to great clinical benefit by augmenting the body's own immune defense. Simple designs combining a few domains are starting to give way to more complicated approaches that either more closely emulate actual TCR signaling or integrate secondary signaling pathways. Similar approaches have been successfully used in reprogramming other immunocytes and have rather universal applicability in cell biology for other cell types and other diseases. This is especially encouraging for the development of combination therapies that employ multiple reprogramming factors and reprogrammed immune cells, with accurate monitoring of clinical outcome and fast adjustments of the treatment. The accumulation of knowledge from numerous academic, biopharmaceutical, and clinical results is rapidly translating our thought processes and ability to conquer cancer.

Acknowledgment

Sources of Support: NIH R21CA198561, NIH R21AI121993, and Alliance for Cancer Gene Therapy.

References

1. Harrer DC, Dorrie J, Schaft N (2018) Chimeric antigen receptors in different cell types: new vehicles join the race. Hum Gene Ther 29(5):547–558. https://doi.org/10.1089/hum.2017.236
2. Rotolo R, Leuci V, Donini C, Cykowska A, Gammaitoni L, Medico G, Valabrega G, Aglietta M, Sangiolo D (2019) CAR-based strategies beyond T lymphocytes: integrative opportunities for cancer adoptive immunotherapy. Int J Mol Sci 20(11):2839. https://doi.org/10.3390/ijms20112839
3. Mehta RS, Rezvani K (2018) Chimeric antigen receptor expressing natural killer cells for the immunotherapy of cancer. Front Immunol 9:283. https://doi.org/10.3389/fimmu.2018.00283
4. Cassetta L, Kitamura T (2018) Macrophage targeting: opening new possibilities for cancer immunotherapy. Immunology 155(3):285–293. https://doi.org/10.1111/imm.12976
5. Morrissey MA, Williamson AP, Steinbach AM, Roberts EW, Kern N, Headley MB, Vale RD (2018) Chimeric antigen receptors that trigger phagocytosis. elife 7:e36688. https://doi.org/10.7554/eLife.36688
6. Jurgens B, Clarke NS (2019) Evolution of CAR T-cell immunotherapy in terms of patenting activity. Nat Biotechnol 37(4):370–375. https://doi.org/10.1038/s41587-019-0083-5
7. Boyiadzis MM, Dhodapkar MV, Brentjens RJ, Kochenderfer JN, Neelapu SS, Maus MV, Porter DL, Maloney DG, Grupp SA, Mackall CL, June CH, Bishop MR (2018) Chimeric antigen receptor (CAR) T therapies for the treatment of hematologic malignancies: clinical perspective and significance. J Immunother Cancer 6(1):137. https://doi.org/10.1186/s40425-018-0460-5
8. EMA (2018) First two CAR-T cell medicines recommended for approval in the European Union. European Medicines Agency, Amsterdam
9. Di Rosa F, Pabst R (2005) The bone marrow: a nest for migratory memory T cells. Trends Immunol 26(7):360–366. https://doi.org/10.1016/j.it.2005.04.011
10. Dhodapkar MV, Kumar V (2017) Type II NKT cells and their emerging role in health and disease. J Immunol 198(3):1015–1021
11. Fan X, Rudensky AY (2016) Hallmarks of tissue-resident lymphocytes. Cell 164(6):1198–1211. https://doi.org/10.1016/j.cell.2016.02.048
12. Godfrey DI, Uldrich AP, McCluskey J, Rossjohn J, Moody DB (2015) The burgeoning family of unconventional T cells. Nat Immunol 16(11):1114–1123
13. Ziegler H, Welker C, Sterk M, Haarer J, Rammensee HG, Handgretinger R, Schilbach K (2014) Human peripheral CD4(+) Vdelta1(+) gammadeltaT cells can develop into alphabetaT cells. Front Immunol 5:645. https://doi.org/10.3389/fimmu.2014.00645
14. Zou C, Zhao P, Xiao Z, Han X, Fu F, Fu L (2017) gammadelta T cells in cancer immunotherapy. Oncotarget 8(5):8900–8909. https://doi.org/10.18632/oncotarget.13051
15. Khairallah C, Chu TH, Sheridan BS (2018) Tissue adaptations of memory and tissue-resident gamma delta T cells. Front Immunol 9:2636. https://doi.org/10.3389/fimmu.2018.02636
16. Alcover A, Alarcon B, Bartolo VD (2018) Cell biology of T cell receptor expression and regulation. Annu Rev Immunol 36:103–125. https://doi.org/10.1146/annurev-immunol-042617-053429
17. Stritesky GL, Jameson SC, Hogquist KA (2012) Selection of self-reactive T cells in the thymus. Annu Rev Immunol 30:95–114
18. Serroukh Y, Gu-Trantien C, Hooshiar Kashani B, Defrance M, Vu Manh TP, Azouz A, Detavernier A, Hoyois A, Das J, Bizet M, Pollet E, Tabbuso T, Calonne E, van Gisbergen K, Dalod M, Fuks F, Goriely S, Marchant A (2018) The transcription factors Runx3 and ThPOK cross-regulate acquisition of cytotoxic function by human Th1 lymphocytes. Elife 7:e30496. https://doi.org/10.7554/eLife.30496
19. Mucida D, Husain MM, Muroi S, van Wijk F, Shinnakasu R, Naoe Y, Reis BS, Huang Y, Lambolez F, Docherty M, Attinger A, Shui JW, Kim G, Lena CJ, Sakaguchi S, Miyamoto C, Wang P, Atarashi K, Park Y, Nakayama T, Honda K, Ellmeier W, Kronenberg M, Taniuchi I, Cheroutre H (2013) Transcriptional reprogramming of

mature CD4(+) helper T cells generates distinct MHC class II-restricted cytotoxic T lymphocytes. Nat Immunol 14(3):281–289. https://doi.org/10.1038/ni.2523
20. Takeuchi A, Saito T (2017) CD4 CTL, a cytotoxic subset of CD4+ T cells, their differentiation and function. Front Immunol 8:194
21. Oja AE, Vieira Braga FA, Remmerswaal EBM, Kragten NAM, Hertoghs KML, Zuo J, Moss PA, van Lier RAW, van Gisbergen KPJM, Hombrink P (2017) The transcription factor Hobit identifies human cytotoxic CD4+ T cells. Front Immunol 8:325
22. Hombach A, Kohler H, Rappl G, Abken H (2006) Human CD4+ T cells lyse target cells via granzyme/perforin upon circumvention of MHC class II restriction by an antibody-like immunoreceptor. J Immunol 177(8):5668–5675
23. Raeber ME, Zurbuchen Y, Impellizzieri D, Boyman O (2018) The role of cytokines in T-cell memory in health and disease. Immunol Rev 283(1):176–193. https://doi.org/10.1111/imr.12644
24. Kim HR, Mun Y, Lee KS, Park YJ, Park JS, Park JH, Jeon BN, Kim CH, Jun Y, Hyun YM, Kim M, Lee SM, Park CS, Im SH, Jun CD (2018) T cell microvilli constitute immunological synaptosomes that carry messages to antigen-presenting cells. Nat Commun 9(1):3630. https://doi.org/10.1038/s41467-018-06090-8
25. Myers DR, Zikherman J, Roose JP (2017) Tonic signals: why do lymphocytes bother? Trends Immunol 38(11):844–857. https://doi.org/10.1016/j.it.2017.06.010
26. Born WK, Kemal Aydintug M, O'Brien RL (2013) Diversity of gammadelta T-cell antigens. Cell Mol Immunol 10(1):13–20. https://doi.org/10.1038/cmi.2012.45
27. Qi Q, Xia M, Hu J, Hicks E, Iyer A, Xiong N, August A (2009) Enhanced development of CD4+ gammadelta T cells in the absence of Itk results in elevated IgE production. Blood 114(3):564–571. https://doi.org/10.1182/blood-2008-12-196345
28. Brownlie RJ, Zamoyska R (2013) T cell receptor signalling networks: branched, diversified and bounded. Nat Rev Immunol 13(4):257–269
29. Geltink RIK, Kyle RL, Pearce EL (2018) Unraveling the complex interplay between T cell metabolism and function. AnnuRev Immunol 36:461–488. https://doi.org/10.1146/annurev-immunol-042617-053019
30. Kershaw MH, Westwood JA, Darcy PK (2013) Gene-engineered T cells for cancer therapy. Nat Rev Cancer 13(8):525–541. https://doi.org/10.1038/nrc3565
31. Mescher MF, Curtsinger JM, Agarwal P, Casey KA, Gerner M, Hammerbeck CD, Popescu F, Xiao Z (2006) Signals required for programming effector and memory development by CD8+ T cells. Immunol Rev 211:81–92. https://doi.org/10.1111/j.0105-2896.2006.00382.x
32. Kolumam GA, Thomas S, Thompson LJ, Sprent J, Murali-Krishna K (2005) Type I interferons act directly on CD8 T cells to allow clonal expansion and memory formation in response to viral infection. J Exp Med 202(5):637–650. https://doi.org/10.1084/jem.20050821
33. Wrzesinski C, Restifo NP (2005) Less is more: lymphodepletion followed by hematopoietic stem cell transplant augments adoptive T-cell-based anti-tumor immunotherapy. Curr Opin Immunol 17(2):195–201
34. Brenchley JM, Douek DC, Ambrozak DR, Chatterji M, Betts MR, Davis LS, Koup RA (2002) Expansion of activated human naive T-cells precedes effector function. Clin Exp Immunol 130(3):432–440. https://doi.org/10.1046/j.1365-2249.2002.02015.x
35. Jones RG, Pearce EJ (2017) MenTORing immunity: mTOR signaling in the development and function of tissue-resident immune cells. Immunity 46(5):730–742. https://doi.org/10.1016/j.immuni.2017.04.028
36. McLean AR, Michie CA (1995) In vivo estimates of division and death rates of human T lymphocytes. Proc Natl Acad Sci U S A 92(9):3707–3711. https://doi.org/10.1073/pnas.92.9.3707
37. Bains I, Antia R, Callard R, Yates AJ (2009) Quantifying the development of the peripheral naive CD4+ T-cell pool in humans. Blood 113(22):5480–5487. https://doi.org/10.1182/blood-2008-10-184184
38. De Boer RJ, Homann D, Perelson AS (2003) Different dynamics of CD4+ and CD8+ T cell responses during and after acute lymphocytic choriomeningitis virus infection. J Immunol 171(8):3928–3935
39. Yoon H, Kim TS, Braciale TJ (2010) The cell cycle time of CD8+ T cells responding in vivo is controlled by the type of antigenic stimulus. PLoS One 5(11):e15423
40. Arsenio J, Metz PJ, Chang JT (2015) Asymmetric cell division in T lymphocyte fate diversification. Trends Immunol 36(11):670–683. https://doi.org/10.1016/j.it.2015.09.004
41. Pollizzi KN, Sun I-H, Patel CH, Lo Y-C, Oh M-H, Waickman AT, Tam AJ, Blosser RL, Wen J, Delgoffe GM, Powell JD (2016) Asymmetric inheritance of mTORC1 kinase activity during division dictates CD8(+) T cell differentiation. Nat Immunol 17(6):704–711

42. van der Windt GJW, Everts B, Chang C-H, Curtis JD, Freitas TC, Amiel E, Pearce EJ, Pearce EL (2012) Mitochondrial respiratory capacity is a critical regulator of CD8+ T cell memory development. Immunity 36(1):68–78
43. Sukumar M, Liu J, Ji Y, Subramanian M, Crompton JG, Yu Z, Roychoudhuri R, Palmer DC, Muranski P, Karoly ED, Mohney RP, Klebanoff CA, Lal A, Finkel T, Restifo NP, Gattinoni L (2013) Inhibiting glycolytic metabolism enhances CD8+ T cell memory and antitumor function. J Clin Invest 123(10):4479–4488
44. Chapman NM, Chi H (2018) Hallmarks of T-cell exit from quiescence. Cancer Immunol Res 6(5):502–508. https://doi.org/10.1158/2326-6066.CIR-17-0605
45. Martinez-Lostao L, Anel A, Pardo J (2015) How do cytotoxic lymphocytes kill cancer cells? Clin Cancer Res 21(22):5047–5056. https://doi.org/10.1158/1078-0432.CCR-15-0685
46. Masaki Y, Ohminami H, Arai J, Kasahara Y, Ishida Y, Fujita S (2002) Granule exocytosis, and not the Fas/Fas ligand system, is the main pathway of cytotoxicity mediated by alloantigen-specific CD41 as well as CD81 cytotoxic T lymphocytes in humans. Blood 95(7):2352–2356
47. Benmebarek MR, Karches CH, Cadilha BL, Lesch S, Endres S, Kobold S (2019) Killing mechanisms of chimeric antigen receptor (CAR) T cells. Int J Mol Sci 20(6):E1283. https://doi.org/10.3390/ijms20061283
48. Cullen SP, Martin SJ (2008) Mechanisms of granule-dependent killing. Cell Death Differ 15(2):251–262. https://doi.org/10.1038/sj.cdd.4402244
49. de Saint Basile G, Menasche G, Fischer A (2010) Molecular mechanisms of biogenesis and exocytosis of cytotoxic granules. Nat Rev Immunol 10(8):568–579. https://doi.org/10.1038/nri2803
50. Zhan Y, Carrington EM, Zhang Y, Heinzel S, Lew AM (2017) Life and death of activated T cells: how are they different from naive T cells? Front Immunol 8:1809. https://doi.org/10.3389/fimmu.2017.01809
51. Malyshkina A, Littwitz-Salomon E, Sutter K, Zelinskyy G, Windmann S, Schimmer S, Paschen A, Streeck H, Hasenkrug KJ, Dittmer U (2017) Fas Ligand-mediated cytotoxicity of CD4+ T cells during chronic retrovirus infection. Sci Rep 7(1):7785. https://doi.org/10.1038/s41598-017-08578-7
52. Mirandola P, Ponti C, Gobbi G, Sponzilli I, Vaccarezza M, Cocco L, Zauli G, Secchiero P, Manzoli FA, Vitale M (2004) Activated human NK and CD8+ T cells express both TNF-related apoptosis-inducing ligand (TRAIL) and TRAIL receptors but are resistant to TRAIL-mediated cytotoxicity. Blood 104(8):2418–2424
53. Marrack P, Scott-Browne J, MacLeod MK (2010) Terminating the immune response. Immunol Rev 236:5–10. https://doi.org/10.1111/j.1600-065X.2010.00928.x
54. Sallusto F, Lenig D, Forster R, Lipp M, Lanzavecchia A (1999) Two subsets of memory T lymphocytes with distinct homing potentials and effector functions. Nature 401(6754):708–712
55. Masopust D, Vezys V, Marzo AL, Lefrancois L (2001) Preferential localization of effector memory cells in nonlymphoid tissue. Science 291(5512):2413–2417
56. Macallan DC, Wallace D, Zhang Y, De Lara C, Worth AT, Ghattas H, Griffin GE, Beverley PCL, Tough DF (2004) Rapid turnover of effector-memory CD4(+) T cells in healthy humans. J Exp Med 200(2):255–260
57. Amsen D, van Gisbergen KPJM, Hombrink P, van Lier RAW (2018) Tissue-resident memory T cells at the center of immunity to solid tumors. Nat Immunol 19(6):538–546
58. Farber DL, Yudanin NA, Restifo NP (2014) Human memory T cells: generation, compartmentalization and homeostasis. Nat Rev Immunol 14(1):24–35
59. Willinger T, Freeman T, Herbert M, Hasegawa H, McMichael AJ, Callan MF (2006) Human naive CD8 T cells down-regulate expression of the WNT pathway transcription factors lymphoid enhancer binding factor 1 and transcription factor 7 (T cell factor-1) following antigen encounter in vitro and in vivo. J Immunol 176(3):1439–1446. https://doi.org/10.4049/jimmunol.176.3.1439
60. Gattinoni L, Zhong XS, Palmer DC, Ji Y, Hinrichs CS, Yu Z, Wrzesinski C, Boni A, Cassard L, Garvin LM, Paulos CM, Muranski P, Restifo NP (2009) Wnt signaling arrests effector T cell differentiation and generates CD8+ memory stem cells. Nat Med 15(7):808–813. https://doi.org/10.1038/nm.1982
61. Gattinoni L, Lugli E, Ji Y, Pos Z, Paulos CM, Quigley MF, Almeida JR, Gostick E, Yu Z, Carpenito C, Wang E, Douek DC, Price DA, June CH, Marincola FM, Roederer M, Restifo NP (2011) A human memory T cell subset with stem cell-like properties. Nat Med 17(10):1290–1297. https://doi.org/10.1038/nm.2446
62. Youngblood B, Hale JS, Kissick HT, Ahn E, Xu X, Wieland A, Araki K, West EE, Ghoneim HE, Fan Y, Dogra P, Davis CW, Konieczny BT, Antia R, Cheng X, Ahmed R (2017) Effector

CD8 T cells dedifferentiate into long-lived memory cells. Nature 552(7685):404–409
63. Schenkel JM, Masopust D (2014) Tissue-resident memory T cells. Immunity 41(6):886–897
64. Martin MD, Badovinac VP (2018) Defining memory CD8 T cell. Front Immunol 9:2692. https://doi.org/10.3389/fimmu.2018.02692
65. Lugli E, Gattinoni L, Roberto A, Mavilio D, Price DA, Restifo NP, Roederer M (2013) Identification, isolation and in vitro expansion of human and nonhuman primate T stem cell memory cells. Nat Protoc 8(1):33–42
66. Kawabe T, Jankovic D, Kawabe S, Huang Y, Lee PH, Yamane H, Zhu J, Sher A, Germain RN, Paul WE (2017) Memory-phenotype CD4(+) T cells spontaneously generated under steady-state conditions exert innate TH1-like effector function. Sci Immunol 2(12). https://doi.org/10.1126/sciimmunol.aam9304
67. Demissie E, Mahajan VS, Alsufyani F, Kumari S, Yuen GJ, Viswanadham V, Tran JQ, Moon JJ, Irvine DJ, Pillai S (2019) DOCK2 sets the threshold for entry into the virtual memory CD8+ T cell compartment by negatively regulating tonic TCR triggering. bioRXiv. https://doi.org/10.1101/582486
68. White JT, Cross EW, Burchill MA, Danhorn T, McCarter MD, Rosen HR, O'Connor B, Kedl RM (2016) Virtual memory T cells develop and mediate bystander protective immunity in an IL-15-dependent manner. Nat Commun 7:11291
69. Anthony SM, Howard ME, Hailemichael Y, Overwijk WW, Schluns KS (2015) Soluble interleukin-15 complexes are generated in vivo by type I interferon dependent and independent pathways. PLoS One 10(3):e0120274
70. Geginat J, Sallusto F, Lanzavecchia A (2001) Cytokine-driven proliferation and differentiation of human naive, central memory, and effector memory CD4(+) T cells. J Exp Med 194(12):1711–1719
71. Westermann J, Bode U, Sahle A, Speck U, Karin N, Bell EB, Kalies K, Gebert A (2005) Naive, effector, and memory T lymphocytes efficiently scan dendritic cells in vivo: contact frequency in T cell zones of secondary lymphoid organs does not depend on LFA-1 expression and facilitates survival of effector T cells. J Immunol 174(5):2517–2524
72. Nolz JC, Rai D, Badovinac VP, Harty JT (2012) Division-linked generation of death-intermediates regulates the numerical stability of memory CD8 T cells. Proc Natl Acad Sci U S A 109(16):6199–6204
73. Vrisekoop N, den Braber I, de Boer AB, Ruiter AFC, Ackermans MT, van der Crabben SN, Schrijver EHR, Spierenburg G, Sauerwein HP, Hazenberg MD, de Boer RJ, Miedema F, Borghans JAM, Tesselaar K (2008) Sparse production but preferential incorporation of recently produced naive T cells in the human peripheral pool. Proc Natl Acad Sci U S A 105(16):6115–6120
74. Biasco L, Scala S, Basso Ricci L, Dionisio F, Baricordi C, Calabria A, Giannelli S, Cieri N, Barzaghi F, Pajno R, Al-Mousa H, Scarselli A, Cancrini C, Bordignon C, Roncarolo MG, Montini E, Bonini C, Aiuti A (2015) In vivo tracking of T cells in humans unveils decade-long survival and activity of genetically modified T memory stem cells. Sci Transl Med 7(273):273ra213
75. Oliveira G, Ruggiero E, Stanghellini MTL, Cieri N, D'Agostino M, D'Agostino M, Fronza R, Lulay C, Dionisio F, Mastaglio S, Greco R, Peccatori J, Aiuti A, Ambrosi A, Biasco L, Bondanza A, Lambiase A, Traversari C, Vago L, von Kalle C, Schmidt M, Bordignon C, Ciceri F, Bonini C (2015) Tracking genetically engineered lymphocytes long-term reveals the dynamics of T cell immunological memory. Sci Transl Med 7(317):317ra198
76. Fagnoni FF, Vescovini R, Passeri G, Bologna G, Pedrazzoni M, Lavagetto G, Casti A, Franceschi C, Passeri M, Sansoni P (2000) Shortage of circulating naive CD8(+) T cells provides new insights on immunodeficiency in aging. Blood 95(9):2860–2868
77. Naylor K, Li G, Vallejo AN, Lee W-W, Koetz K, Bryl E, Witkowski J, Fulbright J, Weyand CM, Goronzy JJ (2005) The influence of age on T cell generation and TCR diversity. J Immunol 174(11):7446–7452
78. Yager EJ, Ahmed M, Lanzer K, Randall TD, Woodland DL, Blackman MA (2008) Age-associated decline in T cell repertoire diversity leads to holes in the repertoire and impaired immunity to influenza virus. J Exp Med 205(3):711–723
79. Puellmann K, Kaminski WE, Vogel M, Nebe CT, Schroeder J, Wolf H, Beham AW (2006) A variable immunoreceptor in a subpopulation of human neutrophils. Proc Natl Acad Sci U S A 103(39):14441–14446. https://doi.org/10.1073/pnas.0603406103
80. Fuchs T, Puellmann K, Scharfenstein O, Eichner R, Stobe E, Becker A, Pechlivanidou I, Kzhyshkowska J, Gratchev A, Ganser A, Neumaier M, Beham AW, Kaminski WE (2012) The neutrophil recombinatorial TCR-like immune receptor is expressed across the entire human life span but repertoire diversity declines in old age. Biochem Biophys

Res Commun 419(2):309–315. https://doi.org/10.1016/j.bbrc.2012.02.017
81. Fuchs T, Puellmann K, Emmert A, Fleig J, Oniga S, Laird R, Heida NM, Schafer K, Neumaier M, Beham AW, Kaminski WE (2015) The macrophage-TCRalphabeta is a cholesterol-responsive combinatorial immune receptor and implicated in atherosclerosis. Biochem Biophys Res Commun 456(1):59–65. https://doi.org/10.1016/j.bbrc.2014.11.034
82. Kao RL, Truscott LC, Chiou TT, Tsai W, AM W, De Oliveira SN (2019) Cetuximab-mediated suicide system in chimeric antigen receptor-modified hematopoietic stem cells for cancer therapy. Hum Gene Ther 30(4):413–428
83. Brazin KN, Mallis RJ, Boeszoermenyi A, Feng Y, Yoshizawa A, Reche PA, Kaur P, Bi K, Hussey RE, Duke-Cohan JS, Song L, Wagner G, Arthanari H, Lang MJ, Reinherz EL (2018) The T cell antigen receptor alpha transmembrane domain coordinates triggering through regulation of Bilayer immersion and CD3 subunit associations. Immunity 49(5):829–841.e826. https://doi.org/10.1016/j.immuni.2018.09.007
84. Wucherpfennig KW, Gagnon E, Call MJ, Huseby ES, Call ME (2010) Structural biology of the T-cell receptor: insights into receptor assembly, ligand recognition, and initiation of signaling. Cold Spring Harb Perspect Biol 2(4):a005140. https://doi.org/10.1101/cshperspect.a005140
85. Call ME, Wucherpfennig KW (2004) Molecular mechanisms for the assembly of the T cell receptor-CD3 complex. Mol Immunol 40(18):1295–1305. https://doi.org/10.1016/j.molimm.2003.11.017
86. Love PE, Hayes SM (2010) ITAM-mediated signaling by the T-cell antigen receptor. Cold Spring Harb Perspect Biol 2(6):a002485. https://doi.org/10.1101/cshperspect.a002485
87. Park S, Krshnan L, Call MJ, Call ME, Im W (2018) Structural conservation and effects of alterations in T cell receptor transmembrane interfaces. Biophys J 114(5):1030–1035. https://doi.org/10.1016/j.bpj.2018.01.004
88. Deswal S, Schamel WWA (2012) CD3ζ. In: Choi S (ed) Encyclopedia of signaling molecules. Springer, New York, pp 294–314. https://doi.org/10.1007/978-1-4419-0461-4
89. Rojo JM, Portoles P (1991) A symmetrical view of the T-cell receptor-CD3 complex. Immunol Today 12(10):377–378
90. Schamel WW, Alarcon B (2013) Organization of the resting TCR in nanoscale oligomers. Immunol Rev 251(1):13–20. https://doi.org/10.1111/imr.12019
91. Exley M, Wileman T, Mueller B, Terhorst C (1995) Evidence for multivalent structure of T-cell antigen receptor complex. Mol Immunol 32(11):829–839
92. Fernandez-Miguel G, Alarcon B, Iglesias A, Bluethmann H, Alvarez-Mon M, Sanz E, de la Hera A (1999) Multivalent structure of an alphabetaT cell receptor. Proc Natl Acad Sci U S A 96(4):1547–1552. https://doi.org/10.1073/pnas.96.4.1547
93. Balagopalan L, Sherman E, Barr VA, Samelson LE (2011) Imaging techniques for assaying lymphocyte activation in action. Nat Rev Immunol 11(1):21–33. https://doi.org/10.1038/nri2903
94. Lee K-H, Dinner AR, Tu C, Campi G, Raychaudhuri S, Varma R, Sims TN, Burack WR, Wu H, Wang J, Kanagawa O, Markiewicz M, Allen PM, Dustin ML, Chakraborty AK, Shaw AS (2003) The immunological synapse balances T cell receptor signaling and degradation. Science 302(5648):1218–1222
95. Eleftheriadis T, Kartsios C, Yiannaki E, Kazila P, Antoniadi G, Liakopoulos V, Markala D (2008) Chronic inflammation and CD16+ natural killer cell zeta-chain downregulation in hemodialysis patients. Blood Purif 26(4):317–321. https://doi.org/10.1159/000130068
96. Baudouin SJ, Angibaud J, Loussouarn G, Bonnamain V (2008) The signaling adaptor protein CD3ζ is a negative regulator dendrite development young neurons. Mol Biol Cell 19:2444–2456. https://doi.org/10.1091/mbc.E07-09-0947
97. Krshnan L, Park S, Im W, Call MJ, Call ME (2016) A conserved alphabeta transmembrane interface forms the core of a compact T-cell receptor-CD3 structure within the membrane. Proc Natl Acad Sci U S A 113(43):E6649–E6658. https://doi.org/10.1073/pnas.1611445113
98. Xu C, Gagnon E, Call ME, Schnell JR, Schwieters CD, Carman CV, Chou JJ, Wucherpfennig KW (2008) Regulation of T cell receptor activation by dynamic membrane binding of the CD3epsilon cytoplasmic tyrosine-based motif. Cell 135(4):702–713. https://doi.org/10.1016/j.cell.2008.09.044
99. Gagnon E, Schubert DA, Gordo S, Chu HH, Wucherpfennig KW (2012) Local changes in lipid environment of TCR microclusters regulate membrane binding by the CD3epsilon cytoplasmic domain. J Exp Med 209(13):2423–2439. https://doi.org/10.1084/jem.20120790
100. Wang JH, Reinherz EL (2012) The structural basis of alphabeta T-lineage immune recognition: TCR docking topologies, mechanotransduction, and co-receptor function.

Immunol Rev 250(1):102–119. https://doi.org/10.1111/j.1600-065X.2012.01161.x
101. Minguet S, Schamel WW (2008) A permissive geometry model for TCR-CD3 activation. Trends Biochem Sci 33(2):51–57. https://doi.org/10.1016/j.tibs.2007.10.008
102. Zhang H, Cordoba SP, Dushek O, van der Merwe PA (2011) Basic residues in the T-cell receptor zeta cytoplasmic domain mediate membrane association and modulate signaling. Proc Natl Acad Sci U S A 108(48):19323–19328. https://doi.org/10.1073/pnas.1108052108
103. Paensuwan P, Hartl FA, Yousefi OS, Ngoenkam J, Wipa P, Beck-Garcia E, Dopfer EP, Khamsri B, Sanguansermsri D, Minguet S, Schamel WW, Pongcharoen S (2016) Nck binds to the T cell antigen receptor using its SH3.1 and SH2 domains in a cooperative manner, promoting TCR functioning. J Immunol 196(1):448–458
104. Holst J, Wang H, Eder KD, Workman CJ, Boyd KL, Baquet Z, Singh H, Forbes K, Chruscinski A, Smeyne R, van Oers NSC, Utz PJ, Vignali DAA (2008) Scalable signaling mediated by T cell antigen receptor-CD3 ITAMs ensures effective negative selection and prevents autoimmunity. Nat Immunol 9(6):658–666
105. Guy CS, Vignali KM, Temirov J, Bettini ML, Overacre AE, Smeltzer M, Zhang H, Huppa JB, Tsai Y-H, Lobry C, Xie J, Dempsey PJ, Crawford HC, Aifantis I, Davis MM, Vignali DAA (2013) Distinct TCR signaling pathways drive proliferation and cytokine production in T cells. Nat Immunol 14(3):262–270
106. James JR (2018) Tuning ITAM multiplicity on T cell receptors can control potency and selectivity to ligand density. Sci Signal 11(531):eaan1088. https://doi.org/10.1126/scisignal.aan1088
107. Mukhopadhyay H, Cordoba SP, Maini PK, van der Merwe PA, Dushek O (2013) Systems model of T cell receptor proximal signaling reveals emergent ultrasensitivity. PLoS Comput Biol 9(3):e1003004. https://doi.org/10.1371/journal.pcbi.1003004
108. Salmond RJ, Filby A, Qureshi I, Caserta S, Zamoyska R (2009) T-cell receptor proximal signaling via the Src-family kinases, Lck and Fyn, influences T-cell activation, differentiation, and tolerance. Immunol Rev 228(1):9–22
109. Stefanova I, Hemmer B, Vergelli M, Martin R, Biddison WE, Germain RN (2003) TCR ligand discrimination is enforced by competing ERK positive and SHP-1 negative feedback pathways. Nat Immunol 4(3):248–254
110. van Oers NS, Tohlen B, Malissen B, Moomaw CR, Afendis S, Slaughter CA (2000) The 21- and 23-kD forms of TCR zeta are generated by specific ITAM phosphorylations. Nat Immunol 1(4):322–328. https://doi.org/10.1038/79774
111. Vely F, Nunes JA, Malissen B, Hedgecock CJ (1997) Analysis of immunoreceptor tyrosine-based activation motif (ITAM) binding to ZAP-70 by surface plasmon resonance. Eur J Immunol 27(11):3010–3014. https://doi.org/10.1002/eji.1830271138
112. Wang H, Kadlecek TA, Au-Yeung BB, Goodfellow HE, Hsu LY, Freedman TS, Weiss A (2010) ZAP-70: an essential kinase in T-cell signaling. Cold Spring Harb Perspect Biol 2:a002279. https://doi.org/10.1101/cshperspect.a002279
113. Pitcher LA, van Oers NSC (2003) T-cell receptor signal transmission: who gives an ITAM? Trends Immunol 24(10):554–560. https://doi.org/10.1016/j.it.2003.08.003
114. Courtney AH, Lo WL, Weiss A (2018) TCR signaling: mechanisms of initiation and propagation. Trends Biochem Sci 43(2):108–123. https://doi.org/10.1016/j.tibs.2017.11.008
115. Proust R, Bertoglio J, Gesbert F (2012) The adaptor protein SAP directly associates with CD3zeta chain and regulates T cell receptor signaling. PLoS One 7(8):e43200. https://doi.org/10.1371/journal.pone.0043200
116. Gil D, Schamel WWA, Montoya M, Sanchez-Madrid F, Alarcon B (2002) Recruitment of Nck by CD3 epsilon reveals a ligand-induced conformational change essential for T cell receptor signaling and synapse formation. Cell 109(7):901–912
117. Kazi JU, Kabir NN, Ronnstrand L (2015) Role of SRC-like adaptor protein (SLAP) in immune and malignant cell signaling. Cell Mol Life Sci 72(13):2535–2544. https://doi.org/10.1007/s00018-015-1882-6
118. Sosinowski T, Pandey A, Dixit VM, Weiss A (2000) Src-like adaptor protein (SLAP) is a negative regulator of T cell receptor signaling. J Exp Med 191:463–473
119. Myers MD, Dragone LL, Weiss A (2005) Src-like adaptor protein down-regulates T cell receptor (TCR)-CD3 expression by targeting TCRzeta for degradation. J Cell Biol 170(2):285–294. https://doi.org/10.1083/jcb.200501164
120. Ksionda O, Saveliev A, Kochl R, Rapley J, Faroudi M, Smith-Garvin JE, Wulfing C, Rittinger K, Carter T, Tybulewicz VLJ (2012) Mechanism and function of Vav1 localisation in TCR signalling. J Cell Sci 125(Pt 22):5302–5314
121. Lewis JB, Scangarello FA, Murphy JM, Eidell KP, Sodipo MO, Ophir MJ, Sargeant R, Seminario M-C, Bunnell SC (2018) ADAP is an upstream regulator that precedes SLP-76 at sites of TCR engagement and stabilizes

signaling microclusters. J Cell Sci 131(21). https://doi.org/10.1242/jcs.215517
122. Yousefi OS, Gunther M, Horner M, Chalupsky J, Wess M, Brandl SM, Smith RW, Fleck C, Kunkel T, Zurbriggen MD, Hofer T, Weber W, Schamel WW (2019) Optogenetic control shows that kinetic proofreading regulates the activity of the T cell receptor. Elife 8. https://doi.org/10.7554/eLife.42475
123. Stefanova I, Dorfman J, Germain R (2002) Self-recognition promotes the foreign antigen sensitivity of naive T lymphocytes. Nature 420(6914):429–434
124. Plas DR, Johnson R, Pingel JT, Matthews RJ, Dalton M, Roy G, Chan AC, Thomas ML (1996) Direct regulation of ZAP-70 by SHP-1 in T cell antigen receptor signaling. Science 272(5265):1173–1176
125. Artyomov MN, Lis M, Devadas S, Davis MM, Chakraborty AK (2010) CD4 and CD8 binding to MHC molecules primarily acts to enhance Lck delivery. Proc Natl Acad Sci U S A 107(39):16916–16921. https://doi.org/10.1073/pnas.1010568107
126. Li Q-J, Dinner AR, Qi S, Irvine DJ, Huppa JB, Davis MM, Chakraborty AK (2004) CD4 enhances T cell sensitivity to antigen by coordinating Lck accumulation at the immunological synapse. Nat Immunol 5(8):791–799
127. Moody AM, Chui D, Reche PA, Priatel JJ, Marth JD, Reinherz EL (2001) Developmentally regulated glycosylation of the CD8alphabeta coreceptor stalk modulates ligand binding. Cell 107(4):501–512
128. Williams CM, Schonnesen AA, Zhang S-Q, Ma K-Y, He C, Yamamoto T, Eckhardt SG, Klebanoff CA, Jiang N (2017) Normalized synergy predicts that CD8 co-receptor contribution to T cell receptor (TCR) and pMHC binding decreases as TCR affinity increases in human viral-specific T cells. Front Immunol 8:894
129. Au-Yeung BB, Levin SE, Zhang C, Hsu L-Y, Cheng DA, Killeen N, Shokat KM, Weiss A (2010) A genetically selective inhibitor demonstrates a function for the kinase Zap70 in regulatory T cells independent of its catalytic activity. Nat Immunol 11(12):1085–1092
130. Comrie WA, Burkhardt JK (2016) Action and traction: cytoskeletal control of receptor triggering at the immunological synapse. Front Immunol 7:68
131. Steinbuck MP, Arakcheeva K, Winandy S (2018) Novel TCR-mediated mechanisms of notch activation and signaling. J Immunol 200(3):997–1007. https://doi.org/10.4049/jimmunol.1700070
132. Wang R, Dillon CP, Shi LZ, Milasta S, Carter R, Finkelstein D, McCormick LL, Fitzgerald P, Chi H, Munger J, Green DR (2011) The transcription factor Myc controls metabolic reprogramming upon T lymphocyte activation. Immunity 35(6):871–882
133. Scholz G, Jandus C, Zhang L, Grandclement C, Lopez-Mejia IC, Soneson C, Delorenzi M, Fajas L, Held W, Dormond O, Romero P (2016) Modulation of mTOR signalling triggers the formation of stem cell-like memory T cells. EBioMedicine 4:50–61. https://doi.org/10.1016/j.ebiom.2016.01.019
134. Macian F (2005) NFAT proteins: key regulators of T-cell development and function. Nat Rev Immunol 5(6):472–484
135. Cockerill PN (2016) Receptor signaling directs global recruitment of pre-existing transcription factors to inducible elements. Yale J Biol Med 89(4):591–596
136. Tan MP, Dolton GM, Gerry AB, Brewer JE, Bennett AD, Pumphrey NJ, Jakobsen BK, Sewell AK (2017) Human leucocyte antigen class I-redirected anti-tumour CD4+ T cells require a higher T cell receptor binding affinity for optimal activity than CD8+ T cells. Clin Exp Immunol 187(1):124–137
137. de la Roche M, Asano Y, Griffiths GM (2016) Origins of the cytolytic synapse. Nat Rev Immunol 16(7):421–432
138. Schuijs MJ, Hammad H, Lambrecht BN (2019) Professional and 'amateur' antigen-presenting cells in type 2 immunity. Trends Immunol 40(1):22–34
139. Lin A, Lore K (2017) Granulocytes: new members of the antigen-presenting cell family. Front Immunol 8:1781
140. Rouhani SJ, Eccles JD, Riccardi P, Peske JD, Tewalt EF, Cohen JN, Liblau R, Makinen T, Engelhard VH (2015) Roles of lymphatic endothelial cells expressing peripheral tissue antigens in CD4 T-cell tolerance induction. Nat Commun 6:6771
141. Kambayashi T, Laufer TM (2014) Atypical MHC class II-expressing antigen-presenting cells: can anything replace a dendritic cell? Nat Rev Immunol 14(11):719–730
142. Verboogen DRJ, Dingjan I, Revelo NH, Visser LJ, ter Beest M, van den Bogaart G (2016) The dendritic cell side of the immunological synapse. Biomol Concepts 7(1):17–28
143. Rodriguez-Fernandez JL, Riol-Blanco L, Delgado-Martin C (2010) What is the function of the dendritic cell side of the immunological synapse? Sci Signal 3(105):re2
144. Fisher PJ, Bulur PA, Vuk-Pavlovic S, Prendergast FG, Dietz AB (2008) Dendritic cell microvilli: a novel membrane structure associated with the multifocal synapse and T-cell clustering. Blood 112(13):5037–5045
145. Ronchese F, Hermans IF (2001) Killing of dendritic cells: a life cut short or a purposeful death? J Exp Med 194(5):F23–F26

146. Brockman JM, Salaita K (2019) Mechanical proofreading: a general mechanism to enhance the fidelity of information transfer between cells. Front Phys 7. https://doi.org/10.3389/fphy.2019.00014
147. Legut M, Cole DK, Sewell AK (2015) The promise of gammadelta T cells and the gammadelta T cell receptor for cancer immunotherapy. Cell Mol Immunol 12(6):656–668. https://doi.org/10.1038/cmi.2015.28
148. Sharma P, Kranz DM (2016) Recent advances in T-cell engineering for use in immunotherapy. F1000Res 5
149. Davenport AJ, Cross RS, Watson KA, Liao Y, Shi W, Prince HM, Beavis PA, Trapani JA, Kershaw MH, Ritchie DS, Darcy PK, Neeson PJ, Jenkins MR (2018) Chimeric antigen receptor T cells form nonclassical and potent immune synapses driving rapid cytotoxicity. Proc Natl Acad Sci U S A 115(9):E2068–E2076
150. van der Merwe PA, Dushek O (2011) Mechanisms for T cell receptor triggering. Nat Rev Immunol 11(1):47–55. https://doi.org/10.1038/nri2887
151. Siller-Farfan JA, Dushek O (2018) Molecular mechanisms of T cell sensitivity to antigen. Immunol Rev 285(1):194–205
152. Boniface JJ, Rabinowitz JD, Wulfing C, Hampl J, Reich Z, Altman JD, Kantor RM, Beeson C, McConnell HM, Davis MM (1998) Initiation of signal transduction through the T cell receptor requires the multivalent engagement of peptide/MHC ligands [corrected]. Immunity 9(4):459–466
153. Grossman Z, Paul WE (2015) Dynamic tuning of lymphocytes: physiological basis, mechanisms, and function. Annu Rev Immunol 33:677–713. https://doi.org/10.1146/annurev-immunol-032712-100027
154. Cho JH, Sprent J (2018) TCR tuning of T cell subsets. Immunol Rev 283(1):129–137. https://doi.org/10.1111/imr.12646
155. Azzam HS, DeJarnette JB, Huang K, Emmons R, Park CS, Sommers CL, El-Khoury D, Shores EW, Love PE (2001) Fine tuning of TCR signaling by CD5. J Immunol 166(9):5464–5472
156. Park J-H, Adoro S, Lucas PJ, Sarafova SD, Alag AS, Doan LL, Erman B, Liu X, Ellmeier W, Bosselut R, Feigenbaum L, Singer A (2007) 'Coreceptor tuning': cytokine signals transcriptionally tailor CD8 coreceptor expression to the self-specificity of the TCR. Nat Immunol 8(10):1049–1059
157. Thomas S, Xue SA, Bangham CR, Jakobsen BK, Morris EC, Stauss HJ (2011) Human T cells expressing affinity-matured TCR display accelerated responses but fail to recognize low density of MHC-peptide antigen. Blood 118(2):319–329. https://doi.org/10.1182/blood-2010-12-326736
158. Junghans V, Santos AM, Lui Y, Davis SJ, Jonsson P (2018) Dimensions and interactions of large T-cell surface proteins. Front Immunol 9:2215
159. Jung Y, Riven I, Feigelson SW, Kartvelishvily E, Tohya K, Miyasaka M, Alon R, Haran G (2016) Three-dimensional localization of T-cell receptors in relation to microvilli using a combination of superresolution microscopies. Proc Natl Acad Sci U S A 113(40):E5916–E5924
160. Sage PT, Varghese LM, Martinelli R, Sciuto TE, Kamei M, Dvorak AM, Springer TA, Sharpe AH, Carman CV (2012) Antigen recognition is facilitated by invadosome-like protrusions formed by memory/effector T cells. J Immunol 188(8):3686–3699
161. Varma R, Campi G, Yokosuka T, Saito T, Dustin ML (2006) T cell receptor-proximal signals are sustained in peripheral microclusters and terminated in the central supramolecular activation cluster. Immunity 25(1):117–127
162. Shaw AS, Dustin ML (1997) Making the T cell receptor go the distance: a topological view of T cell activation. Immunity 6(4):361–369
163. Dustin ML (2014) The immunological synapse. Cancer Immunol Res 2(11): 1023–1033
164. Sadreddini S, Baradaran B, Aghebati-Maleki A, Sadreddini S, Shanehbandi D, Fotouhi A, Aghebati-Maleki L (2019) Immune checkpoint blockade opens a new way to cancer immunotherapy. J Cell Physiol 234(6):8541–8549
165. Parry RV, Chemnitz JM, Frauwirth KA, Lanfranco AR, Braunstein I, Kobayashi SV, Linsley PS, Thompson CB, Riley JL (2005) CTLA-4 and PD-1 receptors inhibit T-cell activation by distinct mechanisms. Mol Cell Biol 25(21):9543–9553. https://doi.org/10.1128/MCB.25.21.9543-9553.2005
166. Hui E, Cheung J, Zhu J, Su X, Taylor MJ, Wallweber HA, Sasmal DK, Huang J, Kim JM, Mellman I, Vale RD (2017) T cell costimulatory receptor CD28 is a primary target for PD-1-mediated inhibition. Science 355(6332):1428–1433. https://doi.org/10.1126/science.aaf1292
167. Walker LSK (2017) PD-1 and CTLA4: two checkpoints, one pathway? Sci Immunol 2(11). https://doi.org/10.1126/sciimmunol.aan3864
168. Walker LSK, Sansom DM (2015) Confusing signals: recent progress in CTLA-4 biology. Trends Immunol 36(2):63–70
169. Hivroz C, Chemin K, Tourret M, Bohineust A (2012) Crosstalk between T lympho-

cytes and dendritic cells. Crit Rev Immunol 32(2):139–155
170. Watchmaker PB, Urban JA, Berk E, Nakamura Y, Mailliard RB, Watkins SC, van Ham SM, Kalinski P (2008) Memory CD8+ T cells protect dendritic cells from CTL killing. J Immunol 180(6):3857–3865
171. Zarour HM (2016) Reversing T-cell dysfunction and exhaustion in cancer. Clin Cancer Res 22(8):1856–1864. https://doi.org/10.1158/1078-0432.CCR-15-1849
172. Grywalska E, Pasiarski M, Gozdz S, Rolinski J (2018) Immune-checkpoint inhibitors for combating T-cell dysfunction in cancer. Onco Targets Ther 11:6505–6524. https://doi.org/10.2147/OTT.S150817
173. Chung AS, Lee J, Ferrara N (2010) Targeting the tumour vasculature: insights from physiological angiogenesis. Nat Rev Cancer 10(7):505–514. https://doi.org/10.1038/nrc2868
174. Aptsiauri N, Cabrera T, Mendez R, Garcia-Lora A, Ruiz-Cabello F, Garrido F (2007) Role of altered expression of HLA class I molecules in cancer progression. Adv Exp Med Biol 601:123–131. https://doi.org/10.1007/978-0-387-72005-0_13
175. Gao D, Jiang L (2018) Exosomes in cancer therapy: a novel experimental strategy. Am J Cancer Res 8(11):2165–2175
176. Andrews MC, Reuben A, Gopalakrishnan V, Wargo JA (2018) Concepts collide: genomic, immune, and microbial influences on the tumor microenvironment and response to cancer therapy. Front Immunol 9:946. https://doi.org/10.3389/fimmu.2018.00946
177. Pageon SV, Tabarin T, Yamamoto Y, Ma Y, Nicovich PR, Bridgeman JS, Cohnen A, Benzing C, Gao Y, Crowther MD, Tungatt K, Dolton G, Sewell AK, Price DA, Acuto O, Parton RG, Gooding JJ, Rossy J, Rossjohn J, Gaus K (2016) Functional role of T-cell receptor nanoclusters in signal initiation and antigen discrimination. Proc Natl Acad Sci U S A 113(37):E5454–E5463. https://doi.org/10.1073/pnas.1607436113
178. Kumar R, Ferez M, Swamy M, Arechaga I, Rejas MT, Valpuesta JM, Schamel WW, Alarcon B, van Santen HM (2011) Increased sensitivity of antigen-experienced T cells through the enrichment of oligomeric T cell receptor complexes. Immunity 35(3):375–387. https://doi.org/10.1016/j.immuni.2011.08.010
179. Sherman E, Barr V, Manley S, Patterson G, Balagopalan L, Akpan I, Regan CK, Merrill RK, Sommers CL, Lippincott-Schwartz J, Samelson LE (2011) Functional nanoscale organization of signaling molecules downstream of the T cell antigen receptor. Immunity 35(5):705–720. https://doi.org/10.1016/j.immuni.2011.10.004
180. Schamel WW, Alarcon B, Hofer T, Minguet S (2017) The allostery model of TCR regulation. J Immunol 198(1):47–52. https://doi.org/10.4049/jimmunol.1601661
181. Martin-Blanco N, Blanco R, Alda-Catalinas C, Bovolenta ER, Oeste CL, Palmer E, Schamel WW, Lythe G, Molina-Paris C, Castro M, Alarcon B (2018) A window of opportunity for cooperativity in the T cell receptor. Nat Commun 9(1):2618. https://doi.org/10.1038/s41467-018-05050-6
182. von Essen MR, Kongsbak M, Geisler C (2012) Mechanisms behind functional avidity maturation in T cells. Clin Dev Immunol 2012:163453
183. Hayes SM, Love PE (2002) Distinct structure and signaling potential of the gamma delta TCR complex. Immunity 16(6):827–838
184. Hayes SM, Love PE (2006) Stoichiometry of the murine gammadelta T cell receptor. J Exp Med 203(1):47–52. https://doi.org/10.1084/jem.20051886
185. Dave VP, Cao Z, Browne C et al (1997) CD3 delta deficiency arrests development of the alpha beta but not the gamma delta T cell lineage. EMBO J 16(6):1360–1370
186. Siegers GM, Swamy M, Fernandez-Malave E, Minguet S, Rathmann S, Guardo AC, Perez-Flores V, Regueiro JR, Alarcon B, Fisch P, Schamel WW (2007) Different composition of the human and the mouse gammadelta T cell receptor explains different phenotypes of CD3gamma and CD3delta immunodeficiencies. J Exp Med 204(11):2537–2544. https://doi.org/10.1084/jem.20070782
187. Cheng HY, Wu R, Gebre AK, Hanna RN, Smith DJ, Parks JS, Ley K, Hedrick CC (2013) Increased cholesterol content in gammadelta (gammadelta) T lymphocytes differentially regulates their activation. PLoS One 8(5):e63746. https://doi.org/10.1371/journal.pone.0063746
188. Haks MC, Lefebvre JM, Lauritsen JPH, Carleton M, Rhodes M, Miyazaki T, Kappes DJ, Wiest DL (2005) Attenuation of gammadeltaTCR signaling efficiently diverts thymocytes to the alphabeta lineage. Immunity 22(5):595–606
189. Nicolas L, Monneret G, Debard AL, Blesius A, Gutowski MC, Salles G, Bienvenu J (2001) Human gammadelta T cells express a higher TCR/CD3 complex density than alphabeta T cells. Clin Immunol 98(3):358–363
190. Hayes SM, Li L, Love PE (2005) TCR signal strength influences alphabeta/gammadelta lineage fate. Immunity 22(5):583–593

191. Dopfer EP, Hartl FA, Oberg HH, Siegers GM, Yousefi OS, Kock S, Fiala GJ, Garcillan B, Sandstrom A, Alarcon B, Regueiro JR, Kabelitz D, Adams EJ, Minguet S, Wesch D, Fisch P, Schamel WWA (2014) The CD3 conformational change in the gammadelta T cell receptor is not triggered by antigens but can be enforced to enhance tumor killing. Cell Rep 7(5):1704–1715. https://doi.org/10.1016/j.celrep.2014.04.049
192. Muro R, Takayanagi H, Nitta T (2019) T cell receptor signaling for gammadeltaT cell development. Inflamm Regen 39:6. https://doi.org/10.1186/s41232-019-0095-z
193. Laird RM, Laky K, Hayes SM (2010) Unexpected role for the B cell-specific Src family kinase B lymphoid kinase in the development of IL-17-producing gammadelta T cells. J Immunol 185(11):6518–6527. https://doi.org/10.4049/jimmunol.1002766
194. Getts D, Hofmeister R, Quintas-Cardama A (2019) Synthetic T cell receptor-based lymphocytes for cancer therapy. Adv Drug Deliv Rev. https://doi.org/10.1016/j.addr.2019.04.002
195. Thomas S, Stauss HJ, Morris EC (2010) Molecular immunology lessons from therapeutic T-cell receptor gene transfer. Immunology 129(2):170–177
196. van Loenen MM, de Boer R, Amir AL, Hagedoorn RS, Volbeda GL, Willemze R, van Rood JJ, Falkenburg JHF, Heemskerk MHM (2010) Mixed T cell receptor dimers harbor potentially harmful neoreactivity. Proc Natl Acad Sci U S A 107(24):10972–10977
197. Voss R-H, Willemsen RA, Kuball J, Grabowski M, Engel R, Intan RS, Guillaume P, Romero P, Huber C, Theobald M (2008) Molecular design of the Calphabeta interface favors specific pairing of introduced TCRalphabeta in human T cells. J Immunol 180(1):391–401
198. Sommermeyer D, Uckert W (2010) Minimal amino acid exchange in human TCR constant regions fosters improved function of TCR gene-modified T cells. J Immunol 184(11):6223–6231
199. Cohen CJ, Zhao Y, Zheng Z, Rosenberg SA, Morgan RA (2006) Enhanced antitumor activity of murine-human hybrid T-cell receptor (TCR) in human lymphocytes is associated with improved pairing and TCR/CD3 stability. Cancer Res 66(17):8878–8886
200. Rosati S, Parkhurst MR, Hong Y et al (2014) A novel murine T-cell receptor targeting NY-ESO-1. J Immunother 37:135–146
201. Legut M, Dolton G, Mian AA, Ottmann OG, Sewell AK (2018) CRISPR-mediated TCR replacement generates superior anticancer transgenic T cells. Blood 131(3): 311–322. https://doi.org/10.1182/blood-2017-05-787598
202. Ahmadi M, King JW, Xue S-A, Voisine C, Holler A, Wright GP, Waxman J, Morris E, Stauss HJ (2011) CD3 limits the efficacy of TCR gene therapy in vivo. Blood 118(13):3528–3537
203. Liddy N, Bossi G, Adams KJ, Lissina A, Mahon TM, Hassan NJ, Gavarret J, Bianchi FC, Pumphrey NJ, Ladell K, Gostick E, Sewell AK, Lissin NM, Harwood NE, Molloy PE, Li Y, Cameron BJ, Sami M, Baston EE, Todorov PT, Paston SJ, Dennis RE, Harper JV, Dunn SM, Ashfield R, Johnson A, McGrath Y, Plesa G, June CH, Kalos M, Price DA, Vuidepot A, Williams DD, Sutton DH, Jakobsen BK (2012) Monoclonal TCR-redirected tumor cell killing. Nat Med 18(6):980–987
204. Rapoport AP, Stadtmauer EA, Binder-Scholl GK, Goloubeva O, Vogl DT, Lacey SF, Badros AZ, Garfall A, Weiss B, Finklestein J, Kulikovskaya I, Sinha SK, Kronsberg S, Gupta M, Bond S, Melchiori L, Brewer JE, Bennett AD, Gerry AB, Pumphrey NJ, Williams D, Tayton-Martin HK, Ribeiro L, Holdich T, Yanovich S, Hardy N, Yared J, Kerr N, Philip S, Westphal S, Siegel DL, Levine BL, Jakobsen BK, Kalos M, June CH (2015) NY-ESO-1-specific TCR-engineered T cells mediate sustained antigen-specific antitumor effects in myeloma. Nat Med 21(8):914–921. https://doi.org/10.1038/nm.3910
205. Kang S, Li Y, Bao Y, Li Y (2019) High-affinity T cell receptors redirect cytokine-activated T cells (CAT) to kill cancer cells. Front Med 13(1):69–82. https://doi.org/10.1007/s11684-018-0677-1
206. Matsuda T, Leisegang M, Park J-H, Ren L, Kato T, Ikeda Y, Harada M, Kiyotani K, Lengyel E, Fleming GF, Nakamura Y (2018) Induction of neoantigen-specific cytotoxic T cells and construction of T-cell receptor-engineered T cells for ovarian cancer. Clin Cancer Res 24(21):5357–5367
207. Kato T, Matsuda T, Ikeda Y, Park J-H, Leisegang M, Yoshimura S, Hikichi T, Harada M, Zewde M, Sato S, Hasegawa K, Kiyotani K, Nakamura Y (2018) Effective screening of T cells recognizing neoantigens and construction of T-cell receptor-engineered T cells. Oncotarget 9(13):11009–11019
208. Gross G, Waks T, Eshhar Z (1989) Expression of immunoglobulin-T-cell receptor chimeric molecules as functional receptors with antibody-type specificity. Proc Natl Acad Sci U S A 86(24):10024–10028
209. Duan H, Huang H, Jing G (2019) An antibody Fab fragment-based chimeric antigen receptor could efficiently eliminate human

thyroid cancer cells. J Cancer 10(8):1890–1895. https://doi.org/10.7150/jca.30163
210. Gross G, Eshhar Z (2016) Therapeutic potential of T cell chimeric antigen receptors (CARs) in cancer treatment: counteracting off-tumor toxicities for safe CAR T cell therapy. Annu Rev Pharmacol Toxicol 56:59–83
211. Ramos CA, Dotti G (2011) Chimeric antigen receptor (CAR)-engineered lymphocytes for cancer therapy. Expert Opin Biol Ther 11(7):855–873. https://doi.org/10.1517/14712598.2011.573476
212. Rodgers DT, Mazagova M, Hampton EN, Cao Y, Ramadoss NS, Hardy IR, Schulman A, Du J, Wang F, Singer O, Ma J, Nunez V, Shen J, Woods AK, Wright TM, Schultz PG, Kim CH, Young TS (2016) Switch-mediated activation and retargeting of CAR-T cells for B-cell malignancies. Proc Natl Acad Sci U S A 113(4):E459–E468. https://doi.org/10.1073/pnas.1524155113
213. Romeo C, Marine A, Seed B (1992) Sequence requirements for induction of cytolysis by the T cell antigen/Fc receptor zeta chain. Cell 68:889–897
214. Lim WA, June CH (2017) The principles of engineering immune cells to treat cancer. Cell 168(4):724–740. https://doi.org/10.1016/j.cell.2017.01.016
215. Dodson LF, Boomer JS, Deppong CM, Shah DD, Sim J, Bricker TL, Russell JH, Green JM (2009) Targeted knock-in mice expressing mutations of CD28 reveal an essential pathway for costimulation. Mol Cell Biol 29(13):3710–3721. https://doi.org/10.1128/MCB.01869-08
216. Boomer JS, Green JM (2010) An enigmatic tail of CD28 signaling. Cold Spring Harb Perspect Biol 2(8):a002436. https://doi.org/10.1101/cshperspect.a002436
217. Wu LX, La Rose J, Chen L, Neale C, Mak T, Okkenhaug K, Wange R, Rottapel R (2005) CD28 regulates the translation of Bcl-xL via the phosphatidylinositol 3-kinase/mammalian target of rapamycin pathway. J Immunol 174(1):180–194. https://doi.org/10.4049/jimmunol.174.1.180
218. Zapata JM, Perez-Chacon G, Carr-Baena P, Martinez-Forero I, Azpilikueta A, Otano I, Melero I (2018) CD137 (4-1BB) signalosome: complexity is a matter of TRAFs. Front Immunol 9:2618. https://doi.org/10.3389/fimmu.2018.02618
219. Nam KO, Kang H, Shin SM, Cho KH, Kwon B, Kwon BS, Kim SJ, Lee HW (2005) Cross-linking of 4-1BB activates TCR-signaling pathways in CD8+ T lymphocytes. J Immunol 174(4):1898–1905. https://doi.org/10.4049/jimmunol.174.4.1898
220. Abate-Daga D, Davila ML (2016) CAR models: next-generation CAR modifications for enhanced T-cell function. Mol Ther Oncolytics 3:16014. https://doi.org/10.1038/mto.2016.14
221. Guedan S, Calderon H, Posey AD Jr, Maus MV (2019) Engineering and design of chimeric antigen receptors. Mol Ther Methods Clin Dev 12:145–156. https://doi.org/10.1016/j.omtm.2018.12.009
222. Weinkove R, George P, Dasyam N, McLellan AD (2019) Selecting costimulatory domains for chimeric antigen receptors: functional and clinical considerations. Clin Transl Immunol 8(5):e1049. https://doi.org/10.1002/cti2.1049
223. Sadelain M, Brentjens R, Riviere I (2013) The basic principles of chimeric antigen receptor design. Cancer Discov 3(4):388–398
224. Shirasu N, Yamada H, Shibaguchi H, Kuroki M, Kuroki M (2015) Corrigendum to "Molecular characterization of a fully human chimeric T-cell antigen receptor for tumor-associated antigen EpCAM". Biomed Res Int 2015:292436
225. Kalaitsidou M, Kueberuwa G, Schutt A, Gilham DE (2015) CAR T-cell therapy: toxicity and the relevance of preclinical models. Immunotherapy 7(5):487–497
226. Hermanson DL, Kaufman DS (2015) Utilizing chimeric antigen receptors to direct natural killer cell activity. Front Immunol 6:195. https://doi.org/10.3389/fimmu.2015.00195
227. Wang D, Aguilar B, Starr R, Alizadeh D, Brito A, Sarkissian A, Ostberg JR, Forman SJ, Brown CE (2018) Glioblastoma-targeted CD4+ CAR T cells mediate superior antitumor activity. JCI Insight 3(10). https://doi.org/10.1172/jci.insight.99048
228. Hong LK, Chen Y, Smith CC, Montgomery SA, Vincent BG, Dotti G, Savoldo B (2018) CD30-redirected chimeric antigen receptor T cells target CD30(+) and CD30(−) embryonal carcinoma via antigen-dependent and Fas/FasL interactions. Cancer Immunol Res 6(10):1274–1287. https://doi.org/10.1158/2326-6066.CIR-18-0065
229. Rabinovich PM, Komarovskaya ME, Wrzesinski SH, Alderman JL, Budak-Alpdogan T, Karpikov A, Guo H, Flavell RA, Cheung NK, Weissman SM, Bahceci E (2009) Chimeric receptor mRNA transfection as a tool to generate antineoplastic lymphocytes. Hum Gene Ther 20(1):51–61. https://doi.org/10.1089/hum.2008.068
230. Du Y, Wei Y (2018) Therapeutic potential of natural killer cells in gastric cancer. Front Immunol 9:3095. https://doi.org/10.3389/fimmu.2018.03095

231. Bollino D, Webb TJ (2017) Chimeric antigen receptor-engineered natural killer and natural killer T cells for cancer immunotherapy. Transl Res 187:32–43. https://doi.org/10.1016/j.trsl.2017.06.003
232. Fisher J, Anderson J (2018) Engineering approaches in human gamma delta T cells for cancer immunotherapy. Front Immunol 9:1409. https://doi.org/10.3389/fimmu.2018.01409
233. Capsomidis A, Benthall G, Van Acker HH, Fisher J, Kramer AM, Abeln Z, Majani Y, Gileadi T, Wallace R, Gustafsson K, Flutter B, Anderson J (2018) Chimeric antigen receptor-engineered human gamma delta T cells: enhanced cytotoxicity with retention of cross presentation. Mol Ther 26(2):354–365. https://doi.org/10.1016/j.ymthe.2017.12.001
234. Kriegsmann K, Kriegsmann M, von Bergwelt-Baildon M, Cremer M, Witzens-Harig M (2018) NKT cells - new players in CAR cell immunotherapy? Eur J Haematol 101(6):750–757. https://doi.org/10.1111/ejh.13170
235. Simon B, Wiesinger M, Marz J, Wistuba-Hamprecht K, Weide B, Schuler-Thurner B, Schuler G, Dorrie J, Uslu U (2018) The generation of CAR-transfected natural killer T cells for the immunotherapy of melanoma. Int J Mol Sci 19(8). https://doi.org/10.3390/ijms19082365
236. Roberts MR, Cooke KS, Tran AC, Smith KA, Lin WY, Wang M, Dull TJ, Farson D, Zsebo KM, Finer MH (1998) Antigen-specific cytolysis by neutrophils and NK cells expressing chimeric immune receptors bearing zeta or gamma signaling domains. J Immunol 161(1):375–384
237. Liu X, Jiang S, Fang C, Yang S, Olalere D, Pequignot EC, Cogdill AP, Li N, Ramones M, Granda B, Zhou L, Loew A, Young RM, June CH, Zhao Y (2015) Affinity-tuned ErbB2 or EGFR chimeric antigen receptor T cells exhibit an increased therapeutic index against tumors in mice. Cancer Res 75(17):3596–3607. https://doi.org/10.1158/0008-5472.CAN-15-0159
238. Park S, Shevlin E, Vedvyas Y, Zaman M, Park S, Hsu YS, Min IM, Jin MM (2017) Micromolar affinity CAR T cells to ICAM-1 achieves rapid tumor elimination while avoiding systemic toxicity. Sci Rep 7(1):14366. https://doi.org/10.1038/s41598-017-14749-3
239. Drent E, Themeli M, Poels R, de Jong-Korlaar R, Yuan H, de Bruijn J, Martens ACM, Zweegman S, van de Donk N, Groen RWJ, Lokhorst HM, Mutis T (2017) A rational strategy for reducing on-target off-tumor effects of CD38-chimeric antigen receptors by affinity optimization. Mol Ther 25(8):1946–1958. https://doi.org/10.1016/j.ymthe.2017.04.024
240. Ma JS, Kim JY, Kazane SA, Choi SH, Yun HY, Kim MS, Rodgers DT, Pugh HM, Singer O, Sun SB, Fonslow BR, Kochenderfer JN, Wright TM, Schultz PG, Young TS, Kim CH, Cao Y (2016) Versatile strategy for controlling the specificity and activity of engineered T cells. Proc Natl Acad Sci U S A 113(4):E450–E458. https://doi.org/10.1073/pnas.1524193113
241. Ying Z, Huang XF, Xiang X, Liu Y, Kang X, Song Y, Guo X, Liu H, Ding N, Zhang T, Duan P, Lin Y, Zheng W, Wang X, Lin N, Tu M, Xie Y, Zhang C, Liu W, Deng L, Gao S, Ping L, Wang X, Zhou N, Zhang J, Wang Y, Lin S, Mamuti M, Yu X, Fang L, Wang S, Song H, Wang G, Jones L, Zhu J, Chen SY (2019) A safe and potent anti-CD19 CAR T cell therapy. Nat Med. https://doi.org/10.1038/s41591-019-0421-7
242. Kumaresan PR, Manuri PR, Albert ND, Maiti S, Singh H, Mi T, Roszik J, Rabinovich B, Olivares S, Krishnamurthy J, Zhang L, Najjar AM, Huls MH, Lee DA, Champlin RE, Kontoyiannis DP, Cooper LJN (2014) Bioengineering T cells to target carbohydrate to treat opportunistic fungal infection. Proc Natl Acad Sci U S A 111(29):10660–10665
243. Townsend MH, Shrestha G, Robison RA, O'Neill KL (2018) The expansion of targetable biomarkers for CAR T cell therapy. J Exp Clin Cancer Res 37(1):163. https://doi.org/10.1186/s13046-018-0817-0
244. Shaffer DR, Savoldo B, Yi Z, Chow KKH, Kakarla S, Spencer DM, Dotti G, Wu M-F, Liu H, Kenney S, Gottschalk S (2011) T cells redirected against CD70 for the immunotherapy of CD70-positive malignancies. Blood 117(16):4304–4314
245. Urbanska K, Lanitis E, Poussin M, Lynn RC, Gavin BP, Kelderman S, Yu J, Scholler N, Powell DJ Jr (2012) A universal strategy for adoptive immunotherapy of cancer through use of a novel T-cell antigen receptor. Cancer Res 72(7):1844–1852
246. Liu K, Liu X, Peng Z, Sun H, Zhang M, Zhang J, Liu S, Hao L, Lu G, Zheng K, Gong X, Wu D, Wang F, Shen L (2015) Retargeted human avidin-CAR T cells for adoptive immunotherapy of EGFRvIII expressing gliomas and their evaluation via optical imaging. Oncotarget 6(27):23735–23747
247. Bridgeman JS, Ladell K, Sheard VE, Miners K, Hawkins RE, Price DA, Gilham DE (2014) CD3zeta-based chimeric antigen receptors mediate T cell activation via cis- and trans-signalling mechanisms: implications for optimization of receptor structure for adoptive

cell therapy. Clin Exp Immunol 175(2):258–267. https://doi.org/10.1111/cei.12216
248. Watanabe K, Terakura S, Martens AC, van Meerten T, Uchiyama S, Imai M, Sakemura R, Goto T, Hanajiri R, Imahashi N, Shimada K, Tomita A, Kiyoi H, Nishida T, Naoe T, Murata M (2015) Target antigen density governs the efficacy of anti-CD20-CD28-CD3 zeta chimeric antigen receptor-modified effector CD8+ T cells. J Immunol 194(3):911–920. https://doi.org/10.4049/jimmunol.1402346
249. Stone JD, Aggen DH, Schietinger A, Schreiber H, Kranz DM (2012) A sensitivity scale for targeting T cells with chimeric antigen receptors (CARs) and bispecific T-cell Engagers (BiTEs). Oncoimmunology 1(6):863–873. https://doi.org/10.4161/onci.20592
250. Long AH, Haso WM, Shern JF, Wanhainen KM, Murgai M, Ingaramo M, Smith JP, Walker AJ, Kohler ME, Venkateshwara VR, Kaplan RN, Patterson GH, Fry TJ, Orentas RJ, Mackall CL (2015) 4-1BB costimulation ameliorates T cell exhaustion induced by tonic signaling of chimeric antigen receptors. Nat Med 21(6):581–590. https://doi.org/10.1038/nm.3838
251. Gomes-Silva D, Mukherjee M, Srinivasan M, Krenciute G, Dakhova O, Zheng Y, Cabral JMS, Rooney CM, Orange JS, Brenner MK, Mamonkin M (2017) Tonic 4-1BB costimulation in chimeric antigen receptors impedes T cell survival and is vector-dependent. Cell Rep 21(1):17–26
252. Ajina A, Maher J (2018) Strategies to address chimeric antigen receptor tonic signaling. Mol Cancer Ther 17(9):1795–1815
253. Watanabe N, Bajgain P, Sukumaran S, Ansari S, Heslop HE, Rooney CM, Brenner MK, Leen AM, Vera JF (2016) Fine-tuning the CAR spacer improves T-cell potency. Oncoimmunology 5(12):e1253656. https://doi.org/10.1080/2162402X.2016.1253656
254. Feucht J, Sun J, Eyquem J, Ho YJ, Zhao Z, Leibold J, Dobrin A, Cabriolu A, Hamieh M, Sadelain M (2019) Calibration of CAR activation potential directs alternative T cell fates and therapeutic potency. Nat Med 25(1):82–88. https://doi.org/10.1038/s41591-018-0290-5
255. Tschumi BO, Dumauthioz N, Marti B, Zhang L, Schneider P, Mach JP, Romero P, Donda A (2018) CART cells are prone to Fas- and DR5-mediated cell death. J Immunother Cancer 6(1):71. https://doi.org/10.1186/s40425-018-0385-z
256. Grupp SA, Kalos M, Barrett D, Aplenc R, Porter DL, Rheingold SR, Teachey DT, Chew A, Hauck B, Wright JF, Milone MC, Levine BL, June CH (2013) Chimeric antigen receptor-modified T cells for acute lymphoid leukemia. N Engl J Med 368(16):1509–1518. https://doi.org/10.1056/NEJMoa1215134
257. Porter DL, Levine BL, Kalos M, Bagg A, June CH (2011) Chimeric antigen receptor-modified T cells in chronic lymphoid leukemia. N Engl J Med 365(8):725–733
258. Teachey DT, Lacey SF, Shaw PA, Melenhorst JJ, Maude SL, Frey N, Pequignot E, Gonzalez VE, Chen F, Finklestein J, Barrett DM, Weiss SL, Fitzgerald JC, Berg RA, Aplenc R, Callahan C, Rheingold SR, Zheng Z, Rose-John S, White JC, Nazimuddin F, Wertheim G, Levine BL, June CH, Porter DL, Grupp SA (2016) Identification of predictive biomarkers for cytokine release syndrome after chimeric antigen receptor T-cell therapy for acute lymphoblastic leukemia. Cancer Discov 6(6):664–679
259. Uttenthal BJ, Chua I, Morris EC, Stauss HJ (2012) Challenges in T cell receptor gene therapy. J Gene Med 14(6):386–399
260. Lee DW, Kochenderfer JN, Stetler-Stevenson M, Cui YK, Delbrook C, Feldman SA, Fry TJ, Orentas R, Sabatino M, Shah NN, Steinberg SM, Stroncek D, Tschernia N, Yuan C, Zhang H, Zhang L, Rosenberg SA, Wayne AS, Mackall CL (2015) T cells expressing CD19 chimeric antigen receptors for acute lymphoblastic leukaemia in children and young adults: a phase 1 dose-escalation trial. Lancet 385(9967):517–528
261. Evans AG, Rothberg PG, Burack WR, Huntington SF, Porter DL, Friedberg JW, Liesveld JL (2015) Evolution to plasmablastic lymphoma evades CD19-directed chimeric antigen receptor T cells. Br J Haematol 171(2):205–209
262. Maude SL, Frey N, Shaw PA, Aplenc R, Barrett DM, Bunin NJ, Chew A, Gonzalez VE, Zheng Z, Lacey SF, Mahnke YD, Melenhorst JJ, Rheingold SR, Shen A, Teachey DT, Levine BL, June CH, Porter DL, Grupp SA (2014) Chimeric antigen receptor T cells for sustained remissions in leukemia. N Engl J Med 371(16):1507–1517
263. Cho JH, Collins JJ, Wong WW (2018) Universal chimeric antigen receptors for multiplexed and logical control of T cell responses. Cell 173(6):1426–1438.e1411. https://doi.org/10.1016/j.cell.2018.03.038
264. Walseng E, Koksal H, Sektioglu IM, Fane A, Skorstad G, Kvalheim G, Gaudernack G, Inderberg EM, Walchli S (2017) A TCR-based chimeric antigen receptor. Sci Rep 7(1):10713. https://doi.org/10.1038/s41598-017-11126-y

265. Helsen CW, Hammill JA, Lau VWC, Mwawasi KA, Afsahi A, Bezverbnaya K, Newhook L, Hayes DL, Aarts C, Bojovic B, Denisova GF, Kwiecien JM, Brain I, Derocher H, Milne K, Nelson BH, Bramson JL (2018) The chimeric TAC receptor co-opts the T cell receptor yielding robust anti-tumor activity without toxicity. Nat Commun 9(1):3049
266. Baeuerle PA, Ding J, Patel E, Thorausch N, Horton H, Gierut J, Scarfo I, Choudhary R, Kiner O, Krishnamurthy J, Le B, Morath A, Baldeviano GC, Quinn J, Tavares P, Wei Q, Weiler S, Maus MV, Getts D, Schamel WW, Hofmeister R (2019) Synthetic TRuC receptors engaging the complete T cell receptor for potent anti-tumor response. Nat Commun 10(1):2087. https://doi.org/10.1038/s41467-019-10097-0
267. DeFord-Watts LM, Young JA, Pitcher LA, van Oers NSC (2007) The membrane-proximal portion of CD3 epsilon associates with the serine/threonine kinase GRK2. J Biol Chem 282(22):16126–16134
268. Yamazaki T, Hamano Y, Tashiro H, Itoh K, Nakano H, Miyatake S, Saito T (1999) CAST, a novel CD3epsilon-binding protein transducing activation signal for interleukin-2 production in T cells. J Biol Chem 274(26):18173–18180
269. Deford-Watts LM, Tassin TC, Becker AM, Medeiros JJ, Albanesi JP, Love PE, Wulfing C, van Oers NSC (2009) The cytoplasmic tail of the T cell receptor CD3 epsilon subunit contains a phospholipid-binding motif that regulates T cell functions. J Immunol 183(2):1055–1064
270. Kagoya Y, Tanaka S, Guo T, Anczurowski M, Wang CH, Saso K, Butler MO, Minden MD, Hirano N (2018) A novel chimeric antigen receptor containing a JAK-STAT signaling domain mediates superior antitumor effects. Nat Med 24(3):352–359. https://doi.org/10.1038/nm.4478
271. Roybal KT, Rupp LJ, Morsut L, Walker WJ, McNally KA, Park JS, Lim WA (2016) Precision tumor recognition by T cells with combinatorial antigen-sensing circuits. Cell 164(4):770–779
272. Shah NN, Maatman T, Hari P, Johnson B (2019) Multi targeted CAR-T cell therapies for B-cell malignancies. Front Oncol 9:146. https://doi.org/10.3389/fonc.2019.00146
273. D'Aloia MM, Zizzari IG, Sacchetti B, Pierelli L, Alimandi M (2018) CAR-T cells: the long and winding road to solid tumors. Cell Death Dis 9(3):282. https://doi.org/10.1038/s41419-018-0278-6
274. Bagashev A, Sotillo E, Tang C-HA, Black KL, Perazzelli J, Seeholzer SH, Argon Y, Barrett DM, Grupp SA, Hu C-CA, Thomas-Tikhonenko A (2018) CD19 alterations emerging after CD19-directed immunotherapy cause retention of the misfolded protein in the endoplasmic reticulum. Mol Cell Biol 38(21):e00383–e00318
275. Schneider D, Xiong Y, Wu D, Nölle V, Schmitz S, Haso W, Kaiser A, Dropulic B, Orentas RJ (2017) A tandem CD19/CD20 CAR lentiviral vector drives on-target and off-target antigen modulation in leukemia cell lines. J Immunother Cancer 5:42
276. Hegde M, Mukherjee M, Grada Z, Pignata A, Landi D, Navai SA, Wakefield A, Fousek K, Bielamowicz K, Chow KK, Brawley VS, Byrd TT, Krebs S, Gottschalk S, Wels WS, Baker ML, Dotti G, Mamonkin M, Brenner MK, Orange JS, Ahmed N (2016) Tandem CAR T cells targeting HER2 and IL13Ralpha2 mitigate tumor antigen escape. J Clin Invest 126(8):3036–3052. https://doi.org/10.1172/JCI83416
277. Kloss CC, Condomines M, Cartellieri M, Bachmann M, Sadelain M (2013) Combinatorial antigen recognition with balanced signaling promotes selective tumor eradication by engineered T cells. Nat Biotechnol 31(1):71–75
278. Fedorov VD, Themeli M, Sadelain M (2013) PD-1- and CTLA-4-based inhibitory chimeric antigen receptors (iCARs) divert off-target immunotherapy responses. Sci Transl Med 5(215):215ra172. https://doi.org/10.1126/scitranslmed.3006597
279. Liu X, Ranganathan R, Jiang S, Fang C, Sun J, Kim S, Newick K, Lo A, June CH, Zhao Y, Moon EK (2016) A chimeric switch-receptor targeting PD1 augments the efficacy of second-generation CAR T cells in advanced solid tumors. Cancer Res 76(6):1578–1590
280. Ankri C, Shamalov K, Horovitz-Fried M, Mauer S, Cohen CJ (2013) Human T cells engineered to express a programmed death 1/28 costimulatory retargeting molecule display enhanced antitumor activity. J Immunol 191(8):4121–4129
281. Kobold S, Grassmann S, Chaloupka M, Lampert C, Wenk S, Kraus F, Rapp M, Duwell P, Zeng Y, Schmollinger JC, Schnurr M, Endres S, RothenfuSser S (2015) Impact of a new fusion receptor on PD-1-mediated immunosuppression in adoptive T cell therapy. J Natl Cancer Inst 107(8):djv146
282. Prosser ME, Brown CE, Shami AF, Forman SJ, Jensen MC (2012) Tumor PD-L1 co-stimulates primary human CD8(+) cytotoxic T cells modified to express a PD1:CD28 chimeric receptor. Mol Immunol 51(3–4):263–272. https://doi.org/10.1016/j.molimm.2012.03.023

283. Chmielewski M, Abken H (2015) TRUCKs: the fourth generation of CARs. Expert Opin Biol Ther 15(8):1145–1154. https://doi.org/10.1517/14712598.2015.1046430
284. Chmielewski M, Kopecky C, Hombach AA, Abken H (2011) IL-12 release by engineered T cells expressing chimeric antigen receptors can effectively Muster an antigen-independent macrophage response on tumor cells that have shut down tumor antigen expression. Cancer Res 71(17):5697–5706
285. Yeku OO, Brentjens RJ (2016) Armored CAR T-cells: utilizing cytokines and pro-inflammatory ligands to enhance CAR T-cell anti-tumour efficacy. Biochem Soc Trans 44(2):412–418
286. Avanzi MP, Yeku O, Li X, Wijewarnasuriya DP, van Leeuwen DG, Cheung K, Park H, Purdon TJ, Daniyan AF, Spitzer MH, Brentjens RJ (2018) Engineered tumor-targeted T cells mediate enhanced anti-tumor efficacy both directly and through activation of the endogenous immune system. Cell Rep 23(7):2130–2141
287. Adachi K, Kano Y, Nagai T, Okuyama N, Sakoda Y, Tamada K (2018) IL-7 and CCL19 expression in CAR-T cells improves immune cell infiltration and CAR-T cell survival in the tumor. Nat Biotechnol 36(4):346–351. https://doi.org/10.1038/nbt.4086
288. Shum T, Kruse RL, Rooney CM (2018) Strategies for enhancing adoptive T-cell immunotherapy against solid tumors using engineered cytokine signaling and other modalities. Expert Opin Biol Ther 18(6):653–664
289. Curran KJ, Seinstra BA, Nikhamin Y, Yeh R, Usachenko Y, van Leeuwen DG, Purdon T, Pegram HJ, Brentjens RJ (2015) Enhancing antitumor efficacy of chimeric antigen receptor T cells through constitutive CD40L expression. Mol Ther 23(4):769–778
290. Rafiq S, Yeku OO, Jackson HJ, Purdon TJ, van Leeuwen DG, Drakes DJ, Song M, Miele MM, Li Z, Wang P, Yan S, Xiang J, Ma X, Seshan VE, Hendrickson RC, Liu C, Brentjens RJ (2018) Targeted delivery of a PD-1-blocking scFv by CAR-T cells enhances anti-tumor efficacy in vivo. Nat Biotechnol 36(9):847–856
291. Di Stasi A, Tey S-K, Dotti G, Fujita Y, Kennedy-Nasser A, Martinez C, Straathof K, Liu E, Durett AG, Grilley B, Liu H, Cruz CR, Savoldo B, Gee AP, Schindler J, Krance RA, Heslop HE, Spencer DM, Rooney CM, Brenner MK (2011) Inducible apoptosis as a safety switch for adoptive cell therapy. N Engl J Med 365(18):1673–1683
292. Roybal KT, Williams JZ, Morsut L, Rupp LJ, Kolinko I, Choe JH, Walker WJ, McNally KA, Lim WA (2016) Engineering T cells with customized therapeutic response programs using synthetic notch receptors. Cell 167(2):419–432.e416
293. Kamiya T, Wong D, Png YT, Campana D (2018) A novel method to generate T-cell receptor-deficient chimeric antigen receptor T cells. Blood Adv 2(5):517–528
294. Legrand F, Driss V, Woerly G, Loiseau S, Hermann E, Fournie JJ, Heliot L, Mattot V, Soncin F, Gougeon ML, Dombrowicz D, Capron M (2009) A functional gammadeltaTCR/CD3 complex distinct from gammadeltaT cells is expressed by human eosinophils. PLoS One 4(6):e5926. https://doi.org/10.1371/journal.pone.0005926
295. Beham AW, Puellmann K, Laird R, Fuchs T, Streich R, Breysach C, Raddatz D, Oniga S, Peccerella T, Findeisen P, Kzhyshkowska J, Gratchev A, Schweyer S, Saunders B, Wessels JT, Mobius W, Keane J, Becker H, Ganser A, Neumaier M, Kaminski WE (2011) A TNF-regulated recombinatorial macrophage immune receptor implicated in granuloma formation in tuberculosis. PLoS Pathog 7(11):e1002375. https://doi.org/10.1371/journal.ppat.1002375
296. Kaminski WE, Beham AW, Kzhyshkowska J, Gratchev A, Puellmann K (2013) On the horizon: flexible immune recognition outside lymphocytes. Immunobiology 218(3):418–426. https://doi.org/10.1016/j.imbio.2012.05.024
297. Mensali N, Dillard P, Hebeisen M, Lorenz S, Theodossiou T, Myhre MR, Fane A, Gaudernack G, Kvalheim G, Myklebust JH, Inderberg EM, Walchli S (2019) NK cells specifically TCR-dressed to kill cancer cells. EBioMedicine 40:106–117. https://doi.org/10.1016/j.ebiom.2019.01.031
298. Hu W, Wang G, Huang D, Sui M, Xu Y (2019) Cancer immunotherapy based on natural killer cells: current progress and new opportunities. Front Immunol 10:1205
299. Heczey A, Liu D, Tian G, Courtney AN, Wei J, Marinova E, Gao X, Guo L, Yvon E, Hicks J, Liu H, Dotti G, Metelitsa LS (2014) Invariant NKT cells with chimeric antigen receptor provide a novel platform for safe and effective cancer immunotherapy. Blood 124(18):2824–2833. https://doi.org/10.1182/blood-2013-11-541235
300. Tosolini M, Pont F, Poupot M, Vergez F, Nicolau-Travers ML, Vermijlen D, Sarry JE, Dieli F, Fournie JJ (2017) Assessment of tumor-infiltrating TCRVgamma9Vdelta2 gammadelta lymphocyte abundance by deconvolution of human cancers microarrays. Oncoimmunology 6(3):e1284723.

https://doi.org/10.1080/2162402X.2017.1284723
301. Deniger DC, Moyes JS, Cooper LJ (2014) Clinical applications of gamma delta T cells with multivalent immunity. Front Immunol 5:636. https://doi.org/10.3389/fimmu.2014.00636
302. Morita CT, Jin C, Sarikonda G, Wang H (2007) Nonpeptide antigens, presentation mechanisms, and immunological memory of human Vgamma2Vdelta2 T cells: discriminating friend from foe through the recognition of prenyl pyrophosphate antigens. Immunol Rev 215:59–76
303. Wu YL, Ding YP, Tanaka Y, Shen LW, Wei CH, Minato N, Zhang W (2014) gammadelta T cells and their potential for immunotherapy. Int J Biol Sci 10(2):119–135. https://doi.org/10.7150/ijbs.7823
304. Caccamo N, Todaro M, Sireci G, Meraviglia S, Stassi G, Dieli F (2013) Mechanisms underlying lineage commitment and plasticity of human gammadelta T cells. Cell Mol Immunol 10(1):30–34. https://doi.org/10.1038/cmi.2012.42
305. Cordova A, Toia F, La Mendola C, Orlando V, Meraviglia S, Rinaldi G, Todaro M, Cicero G, Zichichi L, Donni PL, Caccamo N, Stassi G, Dieli F, Moschella F (2012) Characterization of human gammadelta T lymphocytes infiltrating primary malignant melanomas. PLoS One 7(11):e49878. https://doi.org/10.1371/journal.pone.0049878
306. Pauza CD, Liou ML, Lahusen T, Xiao L, Lapidus RG, Cairo C, Li H (2018) Gamma delta T cell therapy for cancer: it is good to be local. Front Immunol 9:1305. https://doi.org/10.3389/fimmu.2018.01305
307. Sheridan BS, Romagnoli PA, Pham QM, Fu HH, Alonzo F III, Schubert WD, Freitag NE, Lefrancois L (2013) gammadelta T cells exhibit multifunctional and protective memory in intestinal tissues. Immunity 39(1):184–195. https://doi.org/10.1016/j.immuni.2013.06.015
308. Zhao Y, Niu C, Cui J (2018) Gamma-delta (gammadelta) T cells: friend or foe in cancer development? J Transl Med 16(1):3. https://doi.org/10.1186/s12967-017-1378-2
309. Lawand M, Dechanet-Merville J, Dieu-Nosjean MC (2017) Key features of gamma-delta T-cell subsets in human diseases and their immunotherapeutic implications. Front Immunol 8:761. https://doi.org/10.3389/fimmu.2017.00761
310. Handgretinger R, Schilbach K (2018) The potential role of gammadelta T cells after allogeneic HCT for leukemia. Blood 131(10):1063–1072. https://doi.org/10.1182/blood-2017-08-752162
311. van der Veken LT, Coccoris M, Swart E, Falkenburg JH, Schumacher TN, Heemskerk MH (2009) Alpha beta T cell receptor transfer to gamma delta T cells generates functional effector cells without mixed TCR dimers in vivo. J Immunol 182(1):164–170. https://doi.org/10.4049/jimmunol.182.1.164
312. Hiasa A, Nishikawa H, Hirayama M, Kitano S, Okamoto S, Chono H, Yu SS, Mineno J, Tanaka Y, Minato N, Kato I, Shiku H (2009) Rapid alphabeta TCR-mediated responses in gammadelta T cells transduced with cancer-specific TCR genes. Gene Ther 16(5):620–628. https://doi.org/10.1038/gt.2009.6
313. Rischer M, Pscherer S, Duwe S, Vormoor J, Jurgens H, Rossig C (2004) Human gammadelta T cells as mediators of chimaeric-receptor redirected anti-tumour immunity. Br J Haematol 126(4):583–592
314. Deniger DC, Switzer K, Mi T, Maiti S, Hurton L, Singh H, Huls H, Olivares S, Lee DA, Champlin RE, Cooper LJ (2013) Bispecific T-cells expressing polyclonal repertoire of endogenous gammadelta T-cell receptors and introduced CD19-specific chimeric antigen receptor. Mol Ther 21(3):638–647. https://doi.org/10.1038/mt.2012.267
315. Fisher J, Abramowski P, Wisidagamage Don ND, Flutter B, Capsomidis A, Cheung GW, Gustafsson K, Anderson J (2017) Avoidance of on-target off-tumor activation using a co-stimulation-only chimeric antigen receptor. Mol Ther 25(5):1234–1247. https://doi.org/10.1016/j.ymthe.2017.03.002
316. van Willigen WW, Bloemendal M, Gerritsen WR, Schreibelt G, de Vries IJM, Bol KF (2018) Dendritic cell cancer therapy: vaccinating the right patient at the right time. Front Immunol 9:2265. https://doi.org/10.3389/fimmu.2018.02265
317. Braza MS, Klein B (2013) Anti-tumour immunotherapy with Vgamma9Vdelta2 T lymphocytes: from the bench to the bedside. Br J Haematol 160(2):123–132
318. Hotblack A, Holler A, Piapi A, Ward S, Stauss HJ, Bennett CL (2018) Tumor-resident dendritic cells and macrophages modulate the accumulation of TCR-engineered T cells in melanoma. Mol Ther 26(6):1471–1481. https://doi.org/10.1016/j.ymthe.2018.03.011

Chapter 2

Identification of Cell Surface Targets for CAR T Cell Immunotherapy

Diana C. DeLucia and John K. Lee

Abstract

Immunotherapy has become a prominent approach for the treatment of cancer. Targeted killing of malignant cells by adoptive transfer of chimeric antigen receptor (CAR) T cells is a promising immunotherapy technique in oncology. However, the identification of cell surface antigens unique to tumor cells against which CAR T cells can be engineered has historically been challenging and not well documented in solid tumors. Here, we describe a generalized method to construct a cell subtype-specific surface antigen profile (i.e., surfaceome) from cell lines and identify high-confidence antigens as effective targets for CAR T cell therapy by integrating transcriptomics and cell surface proteomics. This method is widely applicable to all therapies utilizing CAR T cells, such as cancer, as well as infectious and autoimmune diseases.

Key words Cell surface antigens, Chimeric antigen receptor, CAR T cell, RNA sequencing, Transcriptomics, Proteomics, Mass spectrometry

1 Introduction

Chimeric antigen receptor (CAR) T cells are T lymphocytes engineered to express a unique receptor that recognizes a specific cell surface antigen and mediates cytotoxic elimination of such cells. Typically, T cells recognize antigens in the form of peptides presented on the surface of target cells in association with the major histocompatibility complex (MHC). Usually, MHC molecules are responsible for recognizing nonself or foreign antigens, such as those derived from viral or bacterial proteins, but can also recognize and bind self-antigens [1]. The interaction of the T cell receptor (TCR) on T cells with the MHC on target cells during antigen presentation is essential for the induction of T cell effector functions, such as target cell killing. CARs are unique compared to endogenous TCRs in that they contain an extracellular single-chain variable antibody fragment linked to a single TCR transmembrane domain and intracellular signaling domain [2]. The variable region

Samuel G. Katz and Peter M. Rabinovich (eds.), *Cell Reprogramming for Immunotherapy: Methods and Protocols*, Methods in Molecular Biology, vol. 2097, https://doi.org/10.1007/978-1-0716-0203-4_2,

of the CAR antibody fragment is designed to recognize antigens in the form of intact protein on the surface of target cells in an MHC-independent manner. This allows CAR T cells to perform targeted killing of cells with downregulated MHC surface expression, a mechanism of immune escape observed in several tumors [3, 4] and utilized by many pathogens, such as HIV-1 [5, 6]. CAR T cell immunotherapy has had promising antitumor effects against hematological and solid tumors in preclinical studies, and numerous constructs have been assessed in clinical trials [7, 8]. Unlike the foreign pathogen-specific proteins targeted by CAR T cells in infectious disease studies, the majority of CAR T cell targets against tumors are self-antigens that are inherently expressed on various cell lineages and tissues throughout the body, which can be tolerated or cause deleterious off-target side effects [9]. To increase the likelihood of efficient tumor-specific cytotoxicity and minimal damage to normal tissue, CAR T cell antigens should ideally be high-density neoantigens or self-antigens that are exclusively expressed on tumor cells.

Historically, the comparison of healthy and malignant cell transcriptomes has been the primary method of identifying CAR protein targets against cancer. This method relies on the assumption that mRNA expression equates to and properly represents protein expression on the cell surface. However, the quantitative relationship between mRNA and surface protein expression is often unpredictable due to the number of variables affecting surface expression of proteins following gene transcription, including mRNA translation, and posttranslational protein regulation and trafficking [10]. Here, we explain an integrated method to identify high-confidence protein antigens for CAR T cell therapy using transcriptomic and proteomic data [11]. The method is broken into three parts: (1) RNA sequencing and differential gene expression analysis, (2) generation of cell-specific surface protein profiles, and (3) integrated cell surface transcriptome and proteome analysis. Importantly, this is a generalized method that can be applied to CAR antigen identification in the context of any available cell type or disease of interest for which the appropriate proteomic and transcriptomic datasets are available.

2 Materials

2.1 RNA Extraction and Library Preparation

1. miRNeasy Mini Kit (Qiagen).
2. Chloroform.
3. Diethyl pyrocarbonate (DEPC)-treated water. Add 0.1% volume/volume DEPC to distilled water. Mix well, and incubate at 37 °C for 8–12 h. Autoclave for at least 15 min to inactivate DEPC. Alternatively, DEPC-treated water may be purchased.
4. TruSeq RNA Library Prep Kit v2 (Illumina).

2.2 Cell Surface Protein Biotinylation and Cell Lysis

1. Pierce Cell Surface Protein Isolation Kit (Thermo Fisher Scientific).
2. Cell scrapers.
3. Lysis buffer: 8 M urea, 2% SDS, 100 mM Tris–HCl, pH 8.0. Weigh 24 g of high-purity urea and 1 g of SDS, and transfer to a 50 ml polypropylene conical centrifuge tube (*see* **Note 1**). Add 5 ml of 1 M Tris–HCl and molecular biology-grade water to a total volume of 50 ml. Add one tablet of the cOmplete Protease Inhibitor Cocktail (Roche). Mix on a tube revolver for at least 4 h, and ensure that the urea is completely dissolved. Adjust pH with HCl to 8.0 (*see* **Note 2**). Store at −20 °C until use.
4. Protein LoBind conical bottom polypropylene 2.0 ml centrifuge tubes (Eppendorf).
5. Benzonase Nuclease (Sigma). Store at −20 °C, and avoid repeated freeze thaw cycles.
6. Pierce BCA Protein Assay Kit (Thermo Fisher Scientific).

2.3 Immunoaffinity Purification of Cell Surface Proteins

1. Pierce High-Capacity Streptavidin Agarose (Thermo Fisher Scientific).
2. Reductive Alkylation Kit from Hampton Research (*see* **Note 3**).
3. GELoader 0.5–20 μl tips from Eppendorf.
4. Lysis buffer, as above.
5. 1 M Dimethylamine borane complex.
6. 1 M Methanol-free formaldehyde.
7. Wash buffer A: 8 M urea, 2% SDS, 100 mM Tris–HCl, 500 mM NaCl, pH 8.0. Weigh 24 g of urea and 1 g of SDS, and transfer to a 50 ml polypropylene conical centrifuge tube. Add 5 ml of 1 M Tris–HCl, 5 ml of 5 M NaCl, and molecular biology-grade water to a total volume of 50 ml. Mix on a tube revolver for at least 4 h, and ensure that the solutes are completely dissolved. Adjust pH with HCl to 8.0. Store at −20 °C until use.
8. Wash buffer B: 8 M urea, 0.2% SDS, 100 mM Tris–HCl, 500 mM NaCl, 5% ethanol, 5% isopropanol, pH 8.0. Weigh 24 g of urea and 100 mg of SDS, and transfer to a 50 ml polypropylene conical centrifuge tube. Add 5 ml of 1 M Tris–HCl, 5 ml of 5 M NaCl, 2.5 ml absolute ethanol, 2.5 ml isopropanol, and molecular biology-grade water to a total volume of 50 ml. Mix on a tube revolver for at least 4 h, and ensure that the solutes are completely dissolved. Adjust pH with HCl to 8.0. Store at −20 °C until use.
9. Digestion buffer: 8 M urea, 100 mM Tris–HCl, pH 8.5. Weigh 24 g of high-purity urea and 100 mg of SDS, and transfer to a 50 ml polypropylene conical centrifuge tube. Add 5 ml of 1 M Tris–HCl and molecular biology-grade water to a total volume of 50 ml (*see* **Note 4**). Mix on a tube revolver for at

least 4 h, and ensure that the solutes have completely dissolved. Adjust pH with HCl to 8.5. Store at −20 °C until use.

2.4 Tryptic Digestion of Purified Proteins

1. 200 mM Tris(carboxyethyl)phosphine (TCEP). Add 2.293 g of TCEP to a 50 ml polypropylene conical centrifuge tube. Add 35 ml of cold molecular biology-grade water to the tube, and dissolve the TCEP (*see* **Note 5**). Adjust pH with NaOH to 7.0. Add molecular biology-grade water to a final volume of 40 ml. Aliquot and store at −20 °C.
2. 500 mM Iodoacetamide. Add 0.046 g iodoacetamide to 500 μl of water (*see* **Note 6**).
3. 0.1 μg/μl Lys-C. Reconstitute 5 μg lyophilized, sequencing-grade Lys-C in 50 μl molecular biology-grade water. Aliquot and store at −20 °C.
4. 100 mM Tris–HCl, pH 8.5. Prepare solution using high-purity, sterile 1 M Tris–HCl, pH 8.5, and molecular biology-grade water.
5. 100 mM $CaCl_2$. Dissolve 1.47 g $CaCl_2 \cdot 2H_2O$ in 100 ml of molecular biology-grade water. Aliquot and store at 4 °C.
6. 0.4 μg/μl Trypsin. Reconstitute 20 μg lyophilized, mass spectrometry-grade trypsin in 50 μl 50 mM acetic acid. Aliquot and store at −80 °C.
7. 98% Formic acid.

3 Methods

3.1 High-Throughput RNA Sequencing (RNA-Seq)

1. Expand cell lines of interest in 100 mm tissue culture dishes.
2. Follow the miRNeasy Mini Kit protocol (Qiagen) to isolate RNA from the cell lines (*see* **Note 7**). Elute RNA in RNase-free DEPC-treated water.
3. Submit aliquots of cell line RNA samples for analysis on a TapeStation Automated Electrophoretic System (*see* **Note 8**).
4. Store RNA samples at −80 °C.
5. Perform library preparation according to the TruSeq RNA Library Prep Kit v2 protocol (Illumina) (*see* **Note 9**).
6. Submit libraries for high-throughput sequencing on an Illumina sequencing system (*see* **Note 10**).

3.2 Differential Expression Analysis of Genes Encoding Putative Cell Surface Proteins

1. Generate a set of genes encoding cell surface proteins. The human cell surfaceome characterized by mass spectrometry is described in multiple manuscripts and publicly available databases including the Cell Surface Protein Atlas [12].
2. After mapping reads and calculating gene-level counts, perform differential gene expression analysis between cell line

sample classes based on an appropriate statistical cutoff (*see* **Note 11**).

3. From the output of differentially expressed genes, extract and sort genes encoding cell surface proteins identified in **step 1** based on decreasing log-fold change.

3.3 Cell Surface Protein Isolation

1. Expand cell lines of interest in T75 cm^2 vented cap tissue culture flasks. Determine the cell count when 90–95% confluent.
2. Grow cell lines in a sufficient number of T75 cm^2 tissue culture flasks for a goal of 4×10^7 cells (*see* **Note 12**).
3. Follow the Pierce Cell Surface Protein Isolation Kit protocol through Part A: Biotinylation to perform biotinylation of cell surface proteins, quenching of the reactions, and cell washing.
4. Perform cell lysis by adding 2 ml of lysis buffer with protease inhibitor cocktail to each cell pellet, and transfer each lysate to an Eppendorf Protein LoBind 2.0 ml centrifuge tube.
5. Add 1 μl Benzonase Nuclease to each sample, and place on a microcentrifuge rotator at room temperature for 30 min (*see* **Note 13**).
6. Centrifuge the samples at $20{,}000 \times g$ for 15 min at room temperature.
7. Collect the supernatants using a 1 ml micropipette, avoiding the lipid layer on top of the supernatant, and transfer to new Eppendorf Protein LoBind 2.0 ml centrifuge tubes.
8. Perform total protein quantitation of each sample using the Pierce BCA Protein Assay Kit.
9. Normalize the concentration of the protein lysates by adding appropriate volumes of lysis buffer to the samples (*see* **Note 14**).

3.4 Affinity Purification and Digestion of Isolated Proteins for Mass Spectrometry

1. Transfer 1 ml of Pierce High-Capacity Streptavidin Agarose into a 1.5 ml centrifuge tube. Pellet agarose beads in a microcentrifuge at $100 \times g$ for 2 min.
2. Remove supernatant without removing agarose beads. Add 1 ml of ice-cold phosphate-buffered saline (PBS). Pellet agarose beads in a microcentrifuge at $100 \times g$ for 2 min. Repeat a total of five times to equilibrate the agarose beads.
3. After the final spin, add 1 ml of ice-cold PBS, and proceed to reductive methylation of the streptavidin agarose beads below in **steps 4** through 7.
4. Add 20 μl 1 M dimethylamine borane complex and 40 μl 1 M methanol-free formaldehyde to the microcentrifuge tube (*see* **Note 15**). Rotate at 4 °C for 2 h.

5. Add an additional 20 µl 1 M dimethylamine borane complex and 40 µl 1 M methanol-free formaldehyde to the microcentrifuge tube. Rotate at 4 °C for 2 h.
6. Add an additional 10 µl 1 M dimethylamine borane complex to the microcentrifuge tube. Rotate at 4 °C overnight.
7. Add 125 µl 1 M glycine, pH 8.6, and 125 µl 50 mM DTT to the microcentrifuge tube. Rotate at 4 °C for 2 h.
8. Wash the reductively methylated streptavidin agarose beads by adding 1 ml of ice-cold PBS, pelleting the agarose beads in a microcentrifuge at 100 × *g* for 2 min, and removing the supernatant without removing agarose beads. Repeat a total of ten times.
9. Add 400 µl of ice-cold PBS, and store at 4 °C.
10. Transfer 125 µl of reductively methylated streptavidin agarose beads to an Eppendorf Protein LoBind 2.0 ml centrifuge tube for each sample.
11. Add 1 ml lysis buffer to each tube, rotate tubes for 3 min at room temperature, pellet the agarose beads in a microcentrifuge at 100 × *g* for 2 min, and remove supernatant. Repeat a total of three times to equilibrate the agarose beads.
12. Load 2 ml of lysates normalized for total protein concentration onto the agarose beads. Rotate tubes for 2 h at room temperature.
13. Pellet the agarose beads in a microcentrifuge at 100 × *g* for 2 min, and remove supernatant.
14. Add 1 ml of wash buffer A, rotate the tube for 5 min at room temperature, pellet the agarose beads in a microcentrifuge at 100 × *g* for 2 min, and remove supernatant. Repeat two more times.
15. Add 1 ml of wash buffer B, rotate the tube for 5 min at room temperature, pellet the agarose beads in a microcentrifuge at 100 × *g* for 2 min, and remove supernatant. Repeat two more times.
16. Add 1 ml of digestion buffer, rotate the tube for 5 min at room temperature, pellet the agarose beads in a microcentrifuge at 100 × *g* for 2 min, and remove supernatant. Repeat two more times.
17. Add 1 ml digestion buffer to the agarose beads, and transfer to new Eppendorf Protein LoBind 2.0 ml centrifuge tubes.
18. Perform three washes with digestion buffer as in **step 16**.
19. Pellet the agarose beads in a microcentrifuge at 100 × *g* for 5 min. Carefully aspirate as much liquid as possible using Eppendorf GELoader tips without removing agarose beads.
20. Resuspend the agarose beads in 50 µl digestion buffer (*see* **Note 16**).

21. Add 1.25 μl of 200 mM TCEP to each tube. Shake at 1300 rpm at room temperature for 20 min.
22. Add 1.2 μl of 500 mM iodoacetamide to each tube. Shake in the dark at 1300 rpm at room temperature for 20 min.
23. Add 2.5 μl of 0.1 μg/μl Lys-C to each tube. Shake in the dark at 1300 rpm at 37 °C for 4 h.
24. Add 2 μl of 100 mM $CaCl_2$ to each tube.
25. Add 4 μl of 0.4 μg/μl trypsin. Shake in the dark at 1300 rpm at 37 °C for 16 h.
26. Add 10.33 μl of 98% formic acid. Shake in the dark at 1300 rpm at room temperature for 5 min.
27. Spin tubes at 20,800 × *g* at room temperature for 10 min. Collect and transfer the supernatants to new Eppendorf Protein LoBind 1.5 ml centrifuge tubes (*see* **Note 17**).
28. Repeat **step 27**.
29. Split supernatant into two new Eppendorf Protein LoBind 1.5 ml centrifuge tubes. Store at −80 °C.
30. Submit samples to your local proteomics core facility for desalting and tandem mass spectrometry.

3.5 Integrative Cell Surface Transcriptome and Proteome Analysis

1. Use protein quantitation data analyzed from the mass spectrometry data to perform differential expression analysis on identified proteins between cell line classes based on an appropriate statistical cutoff (*see* **Note 18**).
2. From the output of differentially expressed proteins, extract and sort only cell surface proteins from Subheading 3.2, **step 1** based on decreasing log-fold change.
3. Perform Rank-Rank Hypergeometric Overlap analysis [13] as described in the user guide using the online tool (http://systems.crump.ucla.edu/rankrank/rankranksimple.php) with ranked differentially expressed genes (rank 1) and ranked differentially expressed proteins (rank 2) between cell line classes (*see* **Note 19**). Examine the output of genes/proteins within regions of maximal statistical overlap for high-confidence cell line class-specific cell surface proteins.

4 Notes

1. Wear a mask and gloves when measuring SDS.
2. Concentrated HCl (10 N) may be used initially followed by HCl with lower ionic strength (5 N and 1 N) to avoid overshooting the target pH.
3. Reagents found in the Reductive Alkylation Kit may be purchased individually at reduced cost.

4. Use high-purity, sterile 1 M Tris–HCl and pH 8.5 which can be purchased from Alfa Aesar or Teknova.
5. When TCEP is dissolved in water, the solution is highly acidic with a pH of about 2.5.
6. Use HPLC-grade iodoacetamide. Always prepare this solution fresh.
7. Collect cell lines when equally 90–95% confluent. Cell lines lysed in Qiazol Lysis Reagent may be stored at −80 °C for several weeks without significant RNA degradation.
8. The optimal RNA integrity number (RIN) from TapeStation analysis should be >8 to proceed with library preparation and high-throughput RNA sequencing. If the RIN is <8, recommend repeating the extraction of RNA from cell line samples in culture.
9. Ensure that each sample is prepared with a unique indexed adapter to enable multiplexing of multiple libraries in lanes of the flow cell.
10. Generally, we have used an Illumina HiSeq 3000 system with paired-end 150 bp reads and four libraries multiplexed on each lane of the flow cell. Using this approach, we obtain between 60 and 80 million mapped reads per sample. However, single-end short read length protocols are usually sufficient for gene expression analysis.
11. We have used the Toil RNA-seq Pipeline [14] to perform RNA-seq expression quantification. Rounded, estimated RSEM gene-level counts are analyzed in DESeq2 [15] for differential gene expression analysis. Typically, we use a *q*-value cutoff of ≤ 0.05.
12. Approximately 4–6 T75 cm^2 tissue culture flasks with cells grown to 90–95% confluence will yield 4×10^7 cells.
13. Benzonase Nuclease is added to enzymatically degrade nucleic acids from the sample. The viscosity of the cell lysate should be significantly reduced after this step. If not, repeat this step with fresh Benzonase Nuclease. Alternatively, a handheld sonicator can be used at low power to disrupt the cells on ice using multiple bursts of 1 s each.
14. Normalized protein lysates may be stored at −80 °C for several weeks.
15. We recommend using Parafilm to seal the cap of the microcentrifuge tube to prevent any leakage of contents while on the rotator.
16. Consult with your local proteomic core facilities as most will accept protein samples and perform or assist with on-bead digestion followed by desalting for mass spectrometry.

17. Avoid transferring agarose beads with the supernatant.
18. We have used log-transformed normalized abundance (NSAFe5) to perform differential protein analysis between cell line classes.
19. The tab-delimited text input for Rank-Rank Hypergeometric Overlap analysis is stringent. Carefully follow the formatting instructions. Each row representing a specific gene/protein must include both a differential gene rank (rank 1) and a differential protein rank (rank 2). If either of these values is missing due to not meeting statistical criteria, the gene/protein should be removed from the analysis.

Acknowledgments

We thank the UCLA Technology Center for Genomics & Bioinformatics and the UCLA Proteomics Research Center for providing assistance with the workflows. This work was supported by the Department of Defense Prostate Cancer Research Program Physician Research Award (W81XWH-17-1-0129) and a Prostate Cancer Young Investigator Award to J.K.L.

References

1. Chaplin DD (2006) 1. Overview of the human immune response. J Allergy Clin Immunol 117(2 Suppl Mini-Primer):S430–S435. https://doi.org/10.1016/j.jaci.2005.09.034
2. Sadelain M, Brentjens R, Riviere I (2013) The basic principles of chimeric antigen receptor design. Cancer Discov 3(4):388–398. https://doi.org/10.1158/2159-8290.CD-12-0548
3. Marincola FM, Jaffee EM, Hicklin DJ, Ferrone S (2000) Escape of human solid tumors from T-cell recognition: molecular mechanisms and functional significance. Adv Immunol 74:181–273
4. Seliger B, Cabrera T, Garrido F, Ferrone S (2002) HLA class I antigen abnormalities and immune escape by malignant cells. Semin Cancer Biol 12(1):3–13. https://doi.org/10.1006/scbi.2001.0404
5. Hale M, Mesojednik T, Romano Ibarra GS, Sahni J, Bernard A, Sommer K, Scharenberg AM, Rawlings DJ, Wagner TA (2017) Engineering HIV-resistant, anti-HIV chimeric antigen receptor T cells. Mol Ther 25(3):570–579. https://doi.org/10.1016/j.ymthe.2016.12.023
6. Collins KL, Chen BK, Kalams SA, Walker BD, Baltimore D (1998) HIV-1 Nef protein protects infected primary cells against killing by cytotoxic T lymphocytes. Nature 391(6665):397–401. https://doi.org/10.1038/34929
7. Lipowska-Bhalla G, Gilham DE, Hawkins RE, Rothwell DG (2012) Targeted immunotherapy of cancer with CAR T cells: achievements and challenges. Cancer Immunol Immunother 61(7):953–962. https://doi.org/10.1007/s00262-012-1254-0
8. Fesnak AD, June CH, Levine BL (2016) Engineered T cells: the promise and challenges of cancer immunotherapy. Nat Rev Cancer 16(9):566–581. https://doi.org/10.1038/nrc.2016.97
9. Bonifant CL, Jackson HJ, Brentjens RJ, Curran KJ (2016) Toxicity and management in CAR T-cell therapy. Mol Ther Oncolytics 3:16011. https://doi.org/10.1038/mto.2016.11
10. Vogel C, Marcotte EM (2012) Insights into the regulation of protein abundance from proteomic and transcriptomic analyses. Nat Rev Genet 13(4):227–232. https://doi.org/10.1038/nrg3185
11. Lee JK, Bangayan NJ, Chai T, Smith BA, Pariva TE, Yun S, Vashisht A, Zhang Q, Park JW, Corey E, Huang J, Graeber TG, Wohlschlegel J, Witte ON (2018) Systemic

surfaceome profiling identifies target antigens for immune-based therapy in subtypes of advanced prostate cancer. Proc Natl Acad Sci U S A 115(19):E4473–E4482. https://doi.org/10.1073/pnas.1802354115

12. Bausch-Fluck D, Hofmann A, Bock T, Frei AP, Cerciello F, Jacobs A, Moest H, Omasits U, Gundry RL, Yoon C, Schiess R, Schmidt A, Mirkowska P, Hartlova A, Van Eyk JE, Bourquin JP, Aebersold R, Boheler KR, Zandstra P, Wollscheid B (2015) A mass spectrometric-derived cell surface protein atlas. PLoS One 10(3):e0121314. https://doi.org/10.1371/journal.pone.0121314
13. Plaisier SB, Taschereau R, Wong JA, Graeber TG (2010) Rank-rank hypergeometric overlap: identification of statistically significant overlap between gene-expression signatures. Nucleic Acids Res 38(17):e169. https://doi.org/10.1093/nar/gkq636
14. Vivian J, Rao AA, Nothaft FA, Ketchum C, Armstrong J, Novak A, Pfeil J, Narkizian J, Deran AD, Musselman-Brown A, Schmidt H, Amstutz P, Craft B, Goldman M, Rosenbloom K, Cline M, O'Connor B, Hanna M, Birger C, Kent WJ, Patterson DA, Joseph AD, Zhu J, Zaranek S, Getz G, Haussler D, Paten B (2017) Toil enables reproducible, open source, big biomedical data analyses. Nat Biotechnol 35(4):314–316. https://doi.org/10.1038/nbt.3772
15. Love MI, Huber W, Anders S (2014) Moderated estimation of fold change and dispersion for RNA-seq data with DESeq2. Genome Biol 15(12):550. https://doi.org/10.1186/s13059-014-0550-8

Chapter 3

Microfluidic Approach for Highly Efficient Viral Transduction

Reginald Tran and Wilbur A. Lam

Abstract

Lentiviral vectors enable gene transfer into target cells, but manufacturing is complex, scale-limited, and costly. Here, we describe the use of microfluidic devices for efficient ex vivo gene transfer. Up to four- to fivefold reductions in viral vector usage and two- to fourfold reductions in transduction times can be obtained by using this method.

Key words Microfluidics, Gene therapy, Cell therapy, Transduction, Viral vectors, Lentiviral vectors, Retroviral vectors

1 Introduction

Gene transfer using viral vectors is used in a variety of applications, from generating animal models in the research setting [1, 2] to the genetic modification of human cells for therapeutic benefit in gene therapy [3–5] and chimeric antigen receptor (CAR) T cell immunotherapy [6–8] in the clinical setting. Various products and methods have been developed to enhance transduction efficiency, though many issues still persist and require excessive amounts of viral vector in addition to lengthy transduction times. For example, cationic polymers such as polybrene [9] are added in many transduction protocols to reduce charge shielding between cell membranes and viral vector envelope proteins. However, high concentrations of polybrene have been shown to negatively impact cell expansion [10]. Spinoculation/spinfection is another commonly used method that relies on centrifugation to concentrate viral vectors and cells [11], though the mechanism of action may be due to physical changes in the cell membrane during centrifugation [12]. In all cases, traditional cell culture vessels such as well plates, flasks, and gas-permeable bags are the largest con-

Samuel G. Katz and Peter M. Rabinovich (eds.), *Cell Reprogramming for Immunotherapy: Methods and Protocols*, Methods in Molecular Biology, vol. 2097, https://doi.org/10.1007/978-1-0716-0203-4_3,

tributor to excessive viral vector usage. Due to the large working volumes of these vessels and the short half-life of viral vectors at 37 °C, only a small percentage of active viral vectors will diffuse to target cells for transduction [13, 14]. Therefore, excessive amounts of viral vector are routinely added so that enough cells are transduced. As a result, significant amounts of viral vector are wasted, which is problematic due to the high cost and lengthy manufacturing times.

Using microfluidics as a transduction platform circumvents this issue in addition to being compatible with existing transduction-enhancing strategies [15]. By reducing the distances that viral vectors must travel to reach cells, viral vector usage per cell target can be reduced by four- to fivefold, and transduction can be shortened by two- to eightfold over standard transduction protocols depending on the target cell type and viral vector used.

2 Materials

1. Silicon wafer.
2. 150 mm Polystyrene petri dish.
3. Sigmacote.
4. Sylgard 184 Silicone Encapsulant Kit (Dow Corning, MI, USA).
5. Scalpel.
6. 1 mm Biopsy punch with plunger (Harris Uni-Core, Miltex, or similar, Ted Pella, Inc., CA, USA).
7. Plasma cleaner (PDC-32, Harrick Plasma, NY, USA) or corona gun (BD-20AC, Electro-Technic Products, IL, USA).
8. Oven.
9. Viral vector stock titered on easily transduced cell lines such as Jurkats or 293T cells. When ready to use, thaw viral vector stock on ice. Avoid multiple freeze/thaw cycles.
10. RetroNectin solution (optional). Based on the dimensions of your microfluidic device, make a stock solution of RetroNectin diluted in phosphate-buffered saline (PBS) such that the surface concentration of RetroNectin is ≥1 μg/cm². Refer to the RetroNectin product sheet for more details.
11. Blocking solution: make a 2% (w/v) solution of bovine serum albumin (BSA) dissolved in PBS.
12. Tweezers.
13. 15 cm Section of 0.020″ × 0.060″ OD Tygon tubing (per device).
14. 5 mL Syringe with a Luer lock tip.
15. 20 Gauge × 1/2″ pink blunt tip needle.
16. 15 mL Conical tube.

3 Methods

This protocol only provides a detailed explanation for using microfluidic devices for ex vivo cell transduction. Detailed protocols for microfluidic mold design and fabrication using photolithography can be found elsewhere [16, 17]. Molds may also be designed and purchased through various vendors (*see* **Note 1**). An example design is shown in Tran et al. [15].

Due to the physical mechanism of microfluidic transduction, the height of the device channels should not exceed 200 μm to ensure that the distance required for a viral vector to reach a cell is minimized while inside the device. Working within this range will minimize the amount of viral vector particles that are unable to diffuse to target cells before decaying. The titer of virus used and number of cells targeted should also be considered when designing microfluidic devices for transduction. We have outlined the standard soft lithography process for making microfluidic devices in Subheadings 3.1–3.5. Alternatively, microfluidic devices may be made using double-sided silicone adhesives (*see* **Note 2**).

Two modes of microfluidic transduction are described below for suspension cells (*see* Fig. 1 for workflow summary). First, the Standard Protocol refers to a basic microfluidic transduction requiring only cells and virus. In the second optional protocol, RetroNectin can be employed if greater efficiency of transduction is required. This reagent can increase viral adhesion to cells, but

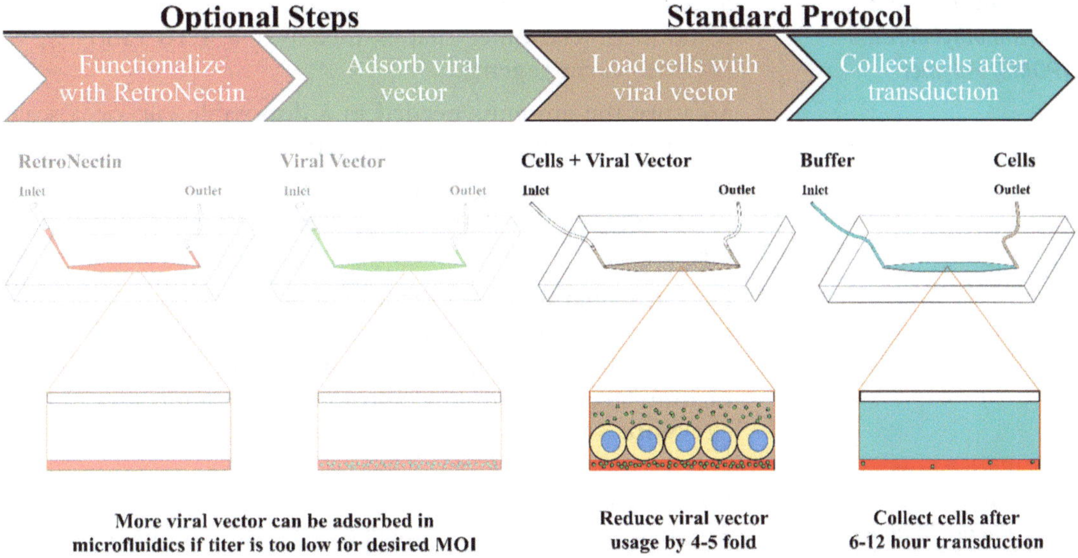

Fig. 1 Workflow diagram of microfluidic transduction. Cells are premixed with viral vector and loaded into a microfluidic device for 6–12 h. Afterward, cells can be collected by flushing the device with buffer. Optionally, microfluidics may be coated with RetroNectin prior to cell loading to adsorb additional viral vector or to immobilize non-adherent cell types

actual efficiency of such treatment should be tested experimentally. For this, optional steps are provided to coat microfluidic channels with RetroNectin for adsorption of viral particles. For adherent cells, adjust the protocol accordingly to allow for a static cell-seeding step for initial adhesion before starting perfusion and grow to confluence under perfusion culture [18, 19]. Once the cells have grown to confluence, switch to perfusion of viral vector-containing media. Extracellular matrix (ECM) proteins such as fibronectin or collagen are generally used for microfluidic adherent cell culture, but refer to cell specification sheets to determine the optimal ECM for growth. Perform under aseptic conditions and use sterile reagents.

3.1 Silicon Microfluidic Mold Preparation

1. Carefully tape the silicon wafer mold to the bottom of a 150 mm polystyrene petri dish to prevent movement. Avoid taping within 1 cm of any of the channel features on the mold to prevent damage or deformities that would transfer to all subsequent microfluidic devices.
2. In a fume hood, silanize the surface of the silicon wafer mold to facilitate de-molding of polydimethylsiloxane (PDMS), and preserve the longevity of the mold. Drop 1 mL of Sigmacote directly on top of the wafer, cap the petri dish, and swirl around to ensure complete coating of the surface. The solvent should evaporate within a few minutes. Remove the lid, and allow the wafer to air out.

3.2 Uncured Polydimethylsiloxane (PDMS) Polymer

1. Sylgard 184 Silicone Encapsulant Kit: this kit comes as a two-component elastomer that is mixed at a 1:10 ratio (w/w) (PDMS/curing agent).
2. Measure the amounts needed in a disposable cup on a scale. Add 1 g of curing agent for every 10 g of PDMS polymer used.
3. Mix vigorously with a disposable scoopula or plastic stick. The appearance should change from clear to white due to bubble formation.
4. Place the mixture into a vacuum desiccator to degas the bubbles out. When the mixture becomes clear again, the PDMS is ready for pouring. The time required will vary depending on vacuum strength but ranges from several minutes to an hour (*see* **Note 3**).

3.3 PDMS Microfluidic Channel

1. Pour uncured PDMS onto the silicon microfluidic mold. Approximately 100 g of uncured PDMS is required to fill a 150 mm petri dish to a height of 5 mm (*see* **Note 4**). If bubbles form during pouring, they will dissipate within a few minutes or can be quickly removed with a sharp sterile needle (*see* **Note 5**).
2. After covering the mold with uncured PDMS to the desired height, place in an oven set to 60 °C for at least 1 h to allow

the PDMS to fully cure. Faster cure times can be achieved by increasing the temperature, but temperatures should not exceed 90 °C to avoid damaging the petri dishes or photoresist on the molds (*see* **Note 6**).

3. Cut the microfluidic channel out using a knife or scalpel (*see* **Note 7**). Leave at least 1 cm around the periphery of the channels, or cut out the entire mold to avoid accidental damage to the microfluidic channel features.
4. If necessary, use a razor blade to trim away the edges of the microfluidic channel so that the bonding surface surrounding the channels is completely flat (*see* **Note 8**).
5. Use a 1 mm biopsy punch to create inlet and outlet ports at the desired locations (*see* **Note 9**).

3.4 Bare PDMS Sheet for Bonding

1. Pour uncured PDMS into a clean 150 mm petri dish (*see* **Note 10**). Approximately 20 g of uncured PDMS is required to fill the petri dish to a height of 1 mm.
2. Store in a 60 °C oven until ready for covalent bonding of the two PDMS layers (*see* Subheading 3.5, **step 2**), which will create a closed microfluidic channel. It will take approximately 30–60 min for PDMS to fully cure at this temperature.
3. Once cured and ready for bonding, use a pair of tweezers or a knife/scalpel to lift the flat PDMS sheet from the edge and remove the entire sheet from the dish.
4. Cut out a section of flat PDMS for bonding that is slightly larger than the microfluidic device.

3.5 Assembled PDMS Microfluidic Device

1. Take components from Subheading 2, **items 3** and **4** and use Scotch Magic tape to remove debris from the bonding surfaces (*see* **Note 11**).
2. In a plasma cleaner, expose the bonding surfaces of the components from Subheadings 2, **items 3** and **4** to oxygen plasma for 30 s. Alternatively, use a corona gun, and treat the bonding surfaces for 1 min with several passes to ensure even treatment (*see* **Note 12**).
3. Bring the treated surfaces of the two PDMS components in physical contact so that the microfluidic channels are sealed by the flat PDMS sheet (*see* **Note 13**).
4. Place the bonded microfluidic device in an oven until ready for use (*see* **Note 14**).
5. Insert a >5 cm section of 0.020″ × 0.060″ OD Tygon tubing into the outlet port.
6. Before use, sterilize by flushing with a 70% ethanol solution in deionized water followed by sterile phosphate-buffered saline (PBS).

3.6 RetroNectin Coating of Microfluidic Devices (Optional, See Note 15)

1. Calculate the volume of the microfluidic channel based on the surface area and channel height. This is the working volume of the microfluidic device.
2. Fill the entire microfluidic channel with RetroNectin solution by pipetting directly into the inlet port. Leave a large droplet of fluid over the inlet and outlet ports to avoid evaporation (*see* **Note 16**).
3. Incubate at 4 °C overnight.
4. When ready to use, flush the RetroNectin solution out with PBS.
5. Fill the microfluidic channel with a 2% BSA blocking solution (*see* Subheading 2, **item 11**). Incubate at room temperature for 30 min.
6. Flush with PBS. The device is now ready for use.

3.7 Cell Loading and Transduction Without RetroNectin (Standard Protocol)

1. Calculate the volume of the microfluidic channel based on the surface area and channel height. This is the working volume of the microfluidic device.
2. Microfluidic transductions should be performed at a surface density of 2.1×10^5 cells/cm^2. Calculate the amount of cells required for each device. Optimization of surface density may need to be performed depending on cell type, MOI desired, and titer of the viral vector stock.
3. Calculate the volume of viral vector to be used for each microfluidic device based on titer and MOI desired. As a general rule of thumb, viral vector volume should not exceed 30% of the total volume to preserve cell viability (*see* **Note 17**). Refer to Subheading 3.8 for more advanced techniques that allow for more viral vector to be delivered.
4. Once the appropriate number of cells is determined, centrifuge the cells at $200 \times g$ for 10 min.
5. While the cells are spinning down, prepare a mixture of the cell culture media with the appropriate amount of viral vector stock for the desired MOI or viral vector concentration.
6. Aspirate the supernatant, and resuspend the cells in the viral vector mixture. Prepare in bulk for multiple devices to avoid error from small volumes.
7. Based on the working volume of the microfluidic device, pipet the appropriate volume of the cell/viral vector mixture, and directly load into the inlet port.
8. Using a pair of sterilized tweezers, insert a 15 cm section of 0.020″ × 0.060″ OD Tygon tubing to the inlet port.

9. Place the loaded microfluidic device in a petri dish with a moistened Kimwipe, and transfer to an incubator set to 37 °C/5% CO_2.
10. Transduction times will need to be optimized depending on cell type and viral vector, but 6–12 h generally allows for effective gene transfer. If necessary, longer transductions can be performed (*see* **Note 18**).

3.8 Adsorption of Viral Vector onto RetroNectin (Optional, See Note 15)

1. Viral vector may be adsorbed onto RetroNectin-coated channels if higher transduction efficiency is required.
2. Thaw viral vector on ice. Based on the working volume of the microfluidic channel, make a viral vector stock solution diluted in cell culture media to adsorb the desired number of viral particles onto the RetroNectin-coated channels.
3. Fill the RetroNectin-coated microfluidic channel with the desired amount of viral vector stock solution. Dilute in media if needed (*see* **Note 19**).
4. Incubate at 37 °C for 2 h.
5. Gently flush unabsorbed viral vector from the microfluidic channel with cell culture media or buffer.
6. The device is now ready for use. Proceed with the steps listed in Subheading 3.7 for cell loading and transduction.

3.9 Cell Removal

1. Once the cells have been incubated with viral vector in the microfluidic device for the desired transduction time, connect a ≥ 5 mL syringe with PBS to the inlet tubing.
2. Place the section of outlet tubing into the end of a 15 mL conical tube so that the cells can be flushed directly into the tube.
3. If using RetroNectin or lightly adherent cells, first flush the device with PBS to remove medium. Then, switch to a nonenzymatic cell dissociation buffer, and incubate for 5–10 min. Adherent cells may require trypsin or other digestive enzymes. Otherwise, ignore and continue to **step 4**.
4. Gently tap the bottom surface of the microfluidic channel to dislodge any lightly adhering cells. Be careful not to let the outlet tubing fall out of the 15 mL conical tube.
5. Carefully push on the plunger of the syringe so that PBS is gently flushed through the device. The majority of the cells should come out in the first few droplets. Continue at a steady rate over the course of 5 or more minutes until all of the fluid has been pushed through the device.
6. Top off the collected cell suspension with additional PBS to a volume of 15 mL. Pipet mix or vortex to ensure complete washing.

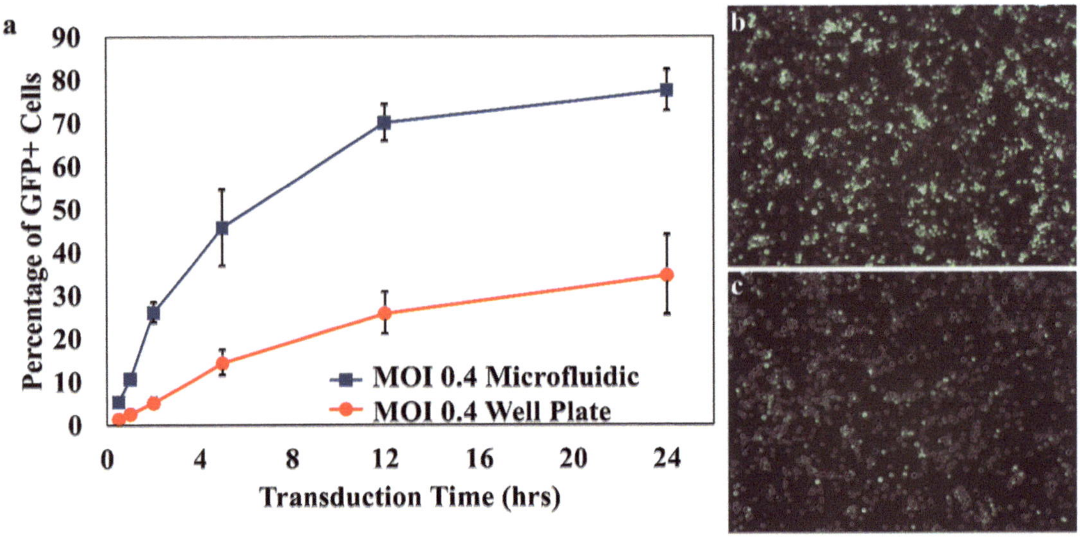

Fig. 2 Example of microfluidic transduction comparison with static well plate transductions using the Standard Protocol. (**a**) Using the same amount of viral vector, Jurkat cells are more efficiently transduced in microfluidics, even for short transduction times. Fluorescent images of Jurkat cells transduced with a GFP-encoding lentiviral vector in (**b**) microfluidics and (**c**) well plates

7. Centrifuge at 200 × *g* for 10 min to pellet the cells.
8. Aspirate the supernatant, and resuspend in culture media.
9. Replate the cells at the recommended seeding density depending on the cell type.
10. Culture the cells in a 37 °C/5% CO_2 incubator for at least 3 days before assessing transduction efficiency for stable gene expression (*see* Fig. 2 for example transduction results).

4 Notes

1. Silicon wafer molds for microfluidics can be purchased from companies such as FlowJEM or Darwin Microfluidics.
2. Microfluidic channels can be made without a clean room by using a craft cutter and double-sided silicone adhesive. The height of the channel is determined by the thickness of the adhesive, which can be as thin as 50 μm. The channel pattern can be cut into the adhesive with the craft cutter, which can then be sandwiched between two flat sheets of PDMS. Make sure to punch inlet and outlet ports on the top layer before affixing the double-sided adhesive.
3. During the first few seconds of degassing, the PDMS level quickly rises before settling. Make sure to monitor over the first minute to quickly release the vacuum to prevent overflowing. Uncured PDMS may be cleaned by wiping with acetone and then isopropyl alcohol.

4. Once the mold has been poured and cut out for the first time, the amount of PDMS used should be less since the majority of the dead volume has already been filled with cured PDMS.
5. Pouring the uncured PDMS as close as possible to the silicon wafer will minimize the chance of bubble formation. A quick blast of nitrogen can also be used to pop small bubbles on the surface.
6. Once the curing agent has been mixed with the PDMS, cross-linking will occur in a temperature-dependent manner. PDMS will cure at room temperature over the course of several hours or within minutes at a high enough temperature. Although PDMS is generally considered to be nontoxic, the cross-linker is volatile, and care should be taken to avoid inhalation. Refer to the material safety data sheet (MSDS) for proper precautions.
7. Make sure the knife or scalpel is sharp; otherwise, the PDMS may become ripped instead of cut. Apply pressure so that the blade just makes contact with the wafer surface, and then cut around the whole device in a uniform stroke.
8. Cutting out the microfluidic device with a knife may result in uneven edges that are hard to see, but using a razor blade to trim away the perimeter will ensure that these are removed.
9. Inspect the punch before use to ensure that the tip is still sharp and not deformed. Dull or deformed punches may rip PDMS while forming the inlets/outlets, which cause leakages.
10. Any size of petri dish or bioassay dish can be used. Be careful not to scratch the surface, as this will transfer the patterns to the PDMS, which will affect bonding.
11. Scotch Magic tape does not leave behind any sticky residue, but any other similar tape should work. Typically, these kinds of tape will loosely adhere. Be sure to test on other surfaces before use.
12. The color of the oxygen plasma should be pink or bluish/purple. If using the corona gun, make sure to perform this step in a fume hood. Ozone is generated as a by-product and may cause respiratory irritation if exposed for long periods of time.
13. Once the two components have been treated, the surfaces will temporarily be activated and become hydrophilic. They should be brought in contact as soon as possible (within a few minutes) before loss of surface activation occurs.
14. PDMS can be stored for several weeks under dehydrated conditions but will become brittle over time. For best results, use freshly cured PDMS within a few days.
15. For low-titer viral vector stock, microfluidic channels can be coated with RetroNectin to adsorb additional viral vector in the

microfluidic channels and to immobilize suspension cells. Cell immobilization on RetroNectin also allows for perfusion culture/transduction. A syringe pump can be connected to continually deliver more cell media and viral vector if necessary.

16. When loading new microfluidic devices with fluid, issues of incomplete loading such as the appearance of bubbles may arise. If this is a recurrent problem, placing microfluidic devices in a vacuum desiccator for at least 30 min prior to use will help facilitate more uniform loading. Once the channel is primed, be careful not to introduce additional bubbles when exchanging fluids.
17. Viral vector stock solutions can contain contaminants such as producer cell line debris.
18. Transductions exceeding 12 h may require cell immobilization and perfusion of media to preserve cell viability. RetroNectin is suggested because it is able to bind both viral vectors and cells, but other adhesive proteins may be used.
19. Optimization of the amount of viral vector pre-adsorbed on RetroNectin should be performed. Filling the entire microfluidic channel with viral vector will maximize the amount of viral vector adsorbed on the RetroNectin but may result in undesired loss of excess unadsorbed viral vector.

Acknowledgments

This work was supported by NIH (R01-HL129141 to W.A.L.) and a research partnership between Children's Healthcare of Atlanta and the Georgia Institute of Technology. This work was performed in part at the Georgia Tech Institute for Electronics and Nanotechnology, a member of the National Nanotechnology Coordinated Infrastructure, which is supported by the National Science Foundation (grant ECCS-1542174).

References

1. Nguyen TK, Morse SJ, Fleischman AG (2016) Transduction-transplantation mouse model of myeloproliferative neoplasm. J Vis Exp (118):54624
2. Wernig G et al (2006) Expression of Jak2V617F causes a polycythemia vera-like disease with associated myelofibrosis in a murine bone marrow transplant model. Blood 107(11):4274–4281
3. Cartier N et al (2009) Hematopoietic stem cell gene therapy with a lentiviral vector in X-linked adrenoleukodystrophy. Science 326(5954):818–823
4. Cavazzana-Calvo M et al (2010) Transfusion independence and HMGA2 activation after gene therapy of human β-thalassaemia. Nature 467(7313):318
5. Boztug K et al (2010) Stem-cell gene therapy for the Wiskott–Aldrich syndrome. N Engl J Med 363(20):1918–1927
6. Maude SL et al (2014) Chimeric antigen receptor T cells for sustained remissions in leukemia. N Engl J Med 371(16):1507–1517
7. Lee DW et al (2015) T cells expressing CD19 chimeric antigen receptors for acute lymphoblastic leukaemia in children and young

adults: a phase 1 dose-escalation trial. Lancet 385(9967):517–528

8. Kochenderfer JN et al (2015) Chemotherapy-refractory diffuse large B-cell lymphoma and indolent B-cell malignancies can be effectively treated with autologous T cells expressing an anti-CD19 chimeric antigen receptor. J Clin Oncol 33(6):540
9. Davis HE, Morgan JR, Yarmush ML (2002) Polybrene increases retrovirus gene transfer efficiency by enhancing receptor-independent virus adsorption on target cell membranes. Biophys Chem 97(2–3):159–172
10. Flasshove M et al (1995) Ex vivo expansion and selection of human CD34+ peripheral blood progenitor cells after introduction of a mutated dihydrofolate reductase cDNA via retroviral gene transfer. Blood 85(2):566–574
11. O'Doherty U, Swiggard WJ, Malim MH (2000) Human immunodeficiency virus type 1 spinoculation enhances infection through virus binding. J Virol 74(21):10074–10080
12. Guo J et al (2011) Spinoculation triggers dynamic actin and cofilin activity facilitating HIV-1 infection of transformed and resting CD4 T cells. J Virol 2011:JVI-05170
13. Chuck AS, Clarke MF, Palsson BO (1996) Retroviral infection is limited by Brownian motion. Hum Gene Ther 7(13):1527–1534
14. Andreadis S et al (2000) Toward a more accurate quantitation of the activity of recombinant retroviruses: alternatives to titer and multiplicity of infection. J Virol 74(3):1258–1266
15. Tran R et al (2017) Microfluidic transduction harnesses mass transport principles to enhance gene transfer efficiency. Mol Ther 25(10):2372–2382
16. McDonald JC, Whitesides GM et al (2002) Acc Chem Res 35(7):491–499
17. Myers DR et al (2012) Endothelialized microfluidics for studying microvascular interactions in hematologic diseases. J Vis Exp (64). https://doi.org/10.3791/3958
18. Kim L et al (2007) A practical guide to microfluidic perfusion culture of adherent mammalian cells. Lab Chip 7(6):681–694
19. Barbulovic-Nad I, Au SH, Wheeler AR (2010) A microfluidic platform for complete mammalian cell culture. Lab Chip 10(12):1536–1542

Chapter 4

Single-Cell Cytokine Analysis to Characterize CAR-T Cell Activation

Amanda Finck and Rong Fan

Abstract

With the increase in the implementation of adoptive transfer anti-CD19 chimeric antigen receptor (CAR) T cell therapy, we introduce a novel platform to study their heterogeneous cytokine response at the single-cell level following activation. Here, we describe an activation platform which incubates single CAR-T cells with single-target CD19 antigen-presenting cells (APCs) and measures the resulting cytokine secretions. This can be compared to the more traditional bulk activation modalities. Both pipelines of activation are measured using our high-throughput, multiplexed single-cell secretomic microchip composed of an antibody barcode array and subnanoliter microchambers, capable of detecting up to 42 unique proteins. This platform has also been adapted to cell lines as well as primary cells.

Key words Single-cell omics, Single-cell cytokine profiling, Immunophenotyping, Functional heterogeneity, High throughput, Multiplexed, Cytokine secretion, Activation, Anti-CD19 CAR-T cells

1 Introduction

In recent years, there have been a rise in the efficacy of adoptive transfer anti-CD19 chimeric antigen receptor (CAR) T cell therapy in hematological malignancies such as chronic lymphocytic leukemia (CLL), B cell acute lymphoblastic leukemia (B-ALL), and pediatric acute lymphoblastic leukemia (ALL). With this, there remains a need for more effective platforms to fully understand their functional heterogeneity following activation [1–3]. Common platforms used for cytokine analysis include fluorescent-activated cell sorting (FACS), enzyme-linked immunosorbent assays (ELISA), and ELISpot/FluoroSpot, each with inherent limitations. FACS uses surface markers and intracellular proteins to investigate cellular characteristics. However, this does not directly measure the actual secreted proteins and can lead to the overestimation of cytokine-secreting cells [4]. Further, only a limited number of intracellular cytokines and chemokines can be measured, thereby preventing a

Samuel G. Katz and Peter M. Rabinovich (eds.), *Cell Reprogramming for Immunotherapy: Methods and Protocols*, Methods in Molecular Biology, vol. 2097, https://doi.org/10.1007/978-1-0716-0203-4_4,

comprehensive view of CAR-T polyfunctionality [5]. ELISA, a microtiter plate immunoassay, has a limited multiplexing capability and sensitivity. Additionally, this platform requires a large sample volume and analysis time [6, 7]. Alternatively, the ELISpot/FluoroSpot method allows for the direct analysis of cytokine secretions but again falters in its multiplexing capabilities [8]. Given the need for a higher-throughput and highly multiplexed immunophenotyping technology, we have developed a single-cell cytokine secretion microchip that integrates an antibody barcode array and subnanoliter microchambers to detect up to 42 unique proteins [9, 10]. This modality allows for the deep functional profiling of anti-CD19 CAR-T cells, imperative in fully understanding their heterogeneity and polyfunctionality [5].

Not only do multiple platforms exist for the measurement of cytokine secretion, but also multiple methodologies exist for the activation of anti-CD19 CAR-T cells. One method uses soluble ligands, such as transforming growth factor beta (TGF-β), CD19 ectodomain, and GFP variants [11]. However, these only allow for an assessment of cytokine secretion at the populational level. Another platform utilizes CD19-coated beads to stimulate CD19–BB-3z CAR-T cells for subsequent measurement of cytokine output, but these beads are not commercially available [5]. Finally, the use of CD19 engineered antigen-presenting cells (APCs) for anti-CD19 CAR-T cell activation has been widely implemented. This strategy has facilitated the correlative study between stimulated pre-infusion CAR-T cell polyfunctionality, CAR-T cell expansion in vivo, and patient response [12]. Our platform uses these CD19 APCs to stimulate anti-CD19 CAR-T cells in two ways (Fig. 1). The first methodology investigates their cell–cell interactions following co-incubation by magnetically separating the CAR-T cells and APCs and performing single-cell cytokine analysis of the CAR-T cells. The latter, novel activation platform incubates a single CAR-T cell with a single-target APC within the same microchamber and measures the resulting cytokines. Here, we outline these activation platforms in tandem with our protein secretion methodology. For the purpose of this protocol, we focus on CAR-T cells. However, this platform has the potential to be adapted to other immunomodulatory therapies such as tumor-infiltrating lymphocytes.

2 Materials

2.1 CAR-T Cell Preparation and Activation

2.1.1 CAR-T Cell Culture

1. Anti-CD19 scFv–CD28–4-1BB–CD3-ζ T cells (Promab).
2. Complete X-VIVO 10 media.
3. IL-2 (10 ng/mL).
4. Anti-CD4 FITC.
5. Anti-CD8 Alexa Fluor 647.

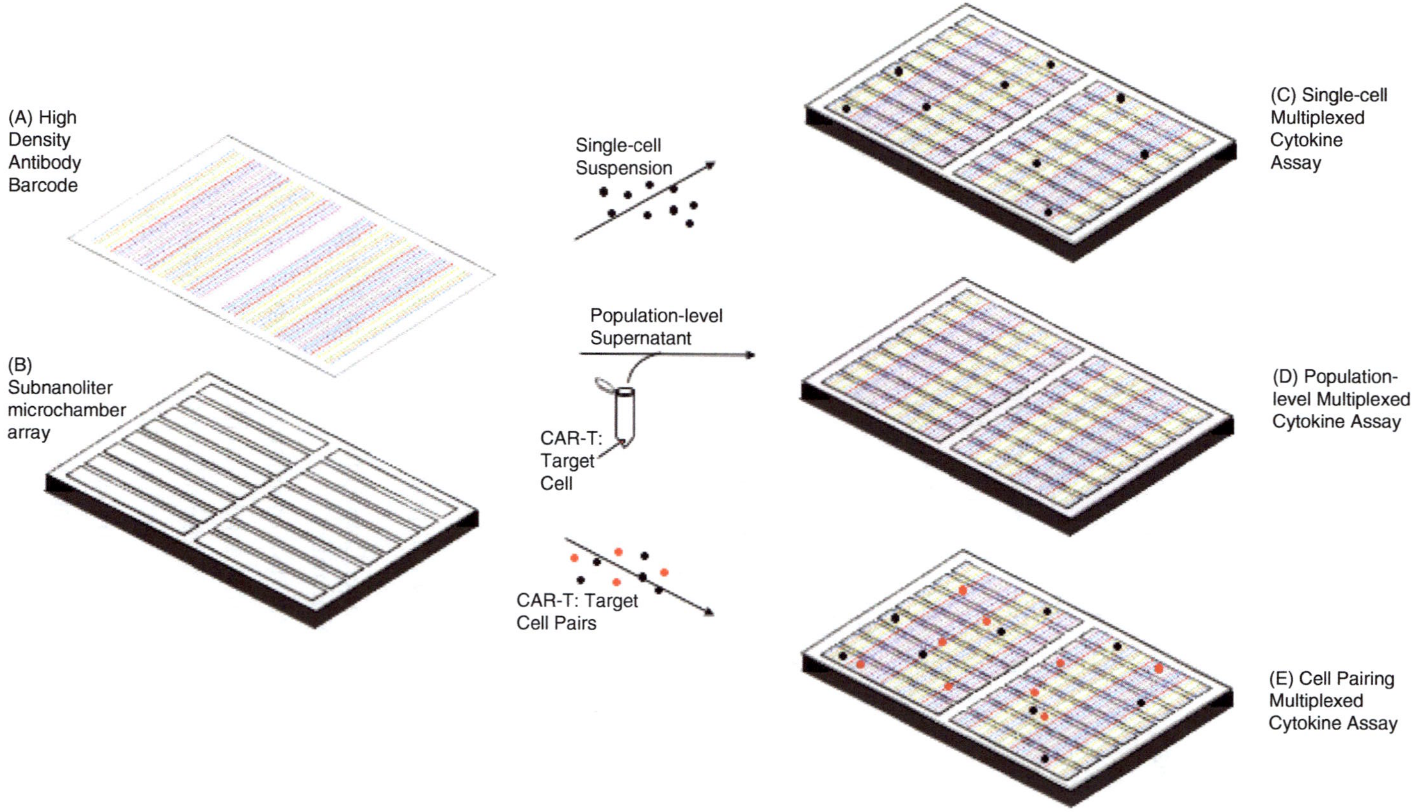

Fig. 1 Multiplexed cytokine analysis for CAR-T cell therapy. Illustration of the (**a**) high-density antibody barcode chip and the (**b**) subnanoliter microchamber array chip for a high-throughput multiplexed cytokine secretion assay at the (**c**) single-cell, (**d**) population, and (**e**) cell-pair levels

6. Phosphate-buffered solution (PBS).
7. RPMI medium supplemented with 10% FBS (*see* **Note 1**).

2.1.2 Target Cell Culture (See **Note 2***)*

1. Vybrant DiD.
2. RPMI medium.
3. RPMI medium supplemented with 10% fetal bovine serum (FBS).

2.2 Device Preparation for Cytokine Analysis

2.2.1 Silicon Master Mold

1. Patterned photomask (*see* **Note 3**).
2. 4 in. p-Type CZ silicon wafer (Silicon Quest Int'l).
3. SPR 220 i-line 1.5 (Dow).
4. MF-319 Microposit developer (Shipley).
5. Chlorotrimethylsilane (TMCS) (Sigma-Aldrich).
6. SU-8 negative photoresist (MicroChem).
7. SU-8 developer (MicroChem).
8. 100% Isopropanol.

2.2.2 Barcode Flow Patterning and Cell Capture Microchamber Chip Fabrication

1. Silicon master mold (barcode flow patterning and cell capture microchamber).
2. Aluminum foil.
3. Polydimethylsiloxane (PDMS) RTV615 A (siloxane/potting agent), RTV615 B (cross-linking agent) (Momentive) (*see* **Note 4**).
4. Desiccator.
5. 70% Ethanol.
6. 100% Isopropanol.
7. DI water.
8. PDMS hole puncher.
9. Magic Tape 810 (*see* **Note 5**).

2.3 Barcode Chip Fabrication

1. SuperChip microarray poly-L-lysine glass slide (Thermo Scientific).
2. PDMS flow-patterning chip.
3. 23G Metal pins, blunt ends.
4. Microbore tubing 0.02 in. (Tygon).
5. 23G Luer-Stub Adapter (Becton Dickinson).
6. Nitrogen gas.
7. Purified capture monoclonal antibodies, ELISA compatible.
8. Phosphate-buffered solution (PBS).
9. Blocking solution: 3% bovine serum albumin (BSA) in PBS (*see* **Note 6**).
10. DI water.

2.4 Cytokine Assay

1. PDMS cell capture microchamber array chip.
2. PDMS flow-patterning chip (barcode chip).
3. Culture media (*see* **Note 7**).
4. Cell suspension.
5. Plexiglass clamps.
6. 18-8 Stainless steel socket head cap screw 4-40 thread, 5/8″ length.
7. Ballpoint L-key 3/32″ Hex.
8. Blocking solution: 3% BSA in PBS.
9. Biotinylated antibody, ELISA compatible (*see* **Note 8**).
10. Streptavidin–fluorophores.
11. Microscope with automatic image stitching (*see* **Note 9**).
12. Genepix array scanner (Molecular Devices).

3 Methods

All procedures were conducted at room temperature unless otherwise specified.

3.1 CAR-T Cell Preparation and Activation

3.1.1 CAR-T Cell Culture

1. Culture human CAR-T cells at 37 °C; 5% CO_2 in complete X-VIVO 10 media supplemented with IL-2 at a concentration of 0.5×10^6 cells/mL.
2. Label CD4/CD8 subsets using anti-CD4 FITC and anti-CD8 Alexa Fluor 647 for 15 min, rinse 2× with PBS and 1× in target cell medium (*see* **Notes 2** and **10**).

3.1.2 Target Cell Culture

1. Culture target cells following manufacturing instructions in respective media (*see* **Note 2**).
2. To label, resuspend cells at a final concentration of 1×10^6 cells/mL in serum-free media with Vybrant DiD (5 μL/mL of cells) for 20 min at 37 °C; 5% CO_2 (*see* **Note 11**).
3. Wash labeled cells three times in respective medium (*see* **Note 2**).

3.2 Silicon Master Mold Fabrication

This procedure should be performed in the cleanroom.

3.2.1 Fabrication of Flow-Patterning Master Mold

1. Heat hot plates to 120 °C.
2. Bake a new silicon wafer on the preheated hot plate for 5 min.
3. Place wafer at the center of the photoresist spinner's wafer holder. Fix its position with the vacuum pressure of the wafer holder.

4. Use nitrogen air to remove any dust or particles.
5. For the photoresist coat, apply photoresist SPR 220 i-line 1.5 (*see* **Note 12**) on the wafer (4 in.), and spin-coat at 900 rpm for 5 s and then at 3000 g for 1 min (*see* **Notes 13** and **14**).
6. Place coated wafer on 100 °C hot plate for 3 min to soft bake.
7. Decrease the hot plate temperature to 55 °C at a rate of 5 °C/min (*see* **Note 15**).
8. Expose wafer to 150 mJ/cm^2 UV light using EVG 620 contact/proximity mask aligner (*see* **Note 16**).
9. Bake wafer on a 120 °C hotplate for 3 min or more, and gradually cool down to room temperature at a rate of 5 °C/min (*see* **Note 15**).
10. Develop wafer with Microposit MF-319 developer for approximately 1 min.
11. Use dry reactive-ion etching (DRIE) for wafer pattern etching.
12. Following DRIE, treat with TMCS via vapor deposition in a vacuum chamber for 20 min.

3.2.2 Fabrication of Single-Cell Capture Microchamber Master Mold

1. Heat hot plates to 120 °C.
2. Bake a new silicon wafer on the preheated hot plate for 5 min.
3. Place wafer at the center of the photoresist spinner's wafer holder. Fix its position with the vacuum pressure of the wafer holder.
4. Use nitrogen air to remove any dust or particles.
5. For the photoresist coat, apply SU-8 negative photoresist on the wafer, and spin-coat at 500 rpm for 20 s and then 2000 g for 60 s (*see* **Notes 13** and **14**).
6. To soft bake, place coated wafer on a 65 °C hot plate, and increase the temperature to 95 °C at a rate of 5 °C/min. Then bake for 15 min at 95 °C.
7. Decrease the hot plate temperature to 55 °C at a rate of 5 °C/min (*see* **Note 15**).
8. Expose the wafer employing the EVG 620 contact/proximity mast aligner using the following parameters: soft contact/exposure time = 3 s/interval = 20 s/separation = 10 cycles.
9. Repeat **steps** 7 and **8**.
10. Develop wafer in a glass dish filled of SU-8 developer for 7–8 min with constant, gentle shaking.
11. Clean the wafer with IPA, and air-dry to remove excess liquid on the wafer surface.

3.3 Barcode Flow-Patterning Chip Device Preparation (See Note 17)

1. Vigorously mix PDMS siloxane/potting agent with the PDMS cross-linking agent in a 10:1 ratio (*see* **Note 18**).
2. Remove any dust or particles from the flow-patterning silicon master mold using filtered air (*see* **Note 19**).
3. Wrap the wafer in an aluminum foil boat (*see* **Note 20**).
4. Pour PDMS mix (50 g) over the flow-patterning silicon master mold. Degas in a desiccator for 1 h to remove all air bubbles from the PDMS.
5. Transfer the mold to a 60 °C oven, and bake for 4 h (*see* **Note 21**).
6. Remove the mold, and allow to cool to room temperature.
7. Separate the cured PDMS from silicon wafer, and keep it clean and dust free (*see* **Note 22**).
8. Cut the patterned area out using a razor blade, ensuring the final product is proportionate to the poly-L-lysine glass slide.
9. Using the PDMS hole puncher, cut inlet and outlet holes for the antibodies (*see* **Note 23**).
10. Sonicate the PDMS chip in 70% ethanol for 20 min. Repeat.
11. Sonicate the PDMS chip in 100% isopropanol for 20 min.
12. Dry the surface of the chip with filtered air, and place in an 80 °C oven for 2 h to remove all liquids (*see* **Note 24**).
13. Cover the patterned area with Scotch tape. After applying firm pressure, remove tape, and repeat until the PDMS is free of dust and particles (*see* **Notes 3** and **25**).
14. Remove any dust or particles from the poly-L-lysine-coated glass slide using filtered air, and attach to the PDMS barcode flow-patterning chip (*see* **Note 26**).
15. Using a standard microscope, analyze channels section by section. Check for debris, dust, cross talk, air bubbles, deformities, or blockages (*see* **Note 27**).
16. Thermal bond the glass slide to the PDMS chip in a 75 °C oven for 2 h, and allow to cool to room temperature (*see* **Note 28**). This bonded PDMS and glass slide will be referenced as the *Flowpattern Assembly* in the proceeding steps.

3.4 Cell Capture Microchamber Chip Device Preparation

1. Follow **steps 1–7** from Subheading 3.2, now using the Cell Capture Microchamber Chip Silicon Master Mold.
2. Cut the patterned area out using a razor blade, ensuring the final product is proportionate to the barcode area of the flow-patterning chip.
3. Store microchamber chip in a dust-free container until ready to utilize.

3.5 Flow Patterning of Barcode Chip

1. With Luer-Stub Adapters attached to tubing, insert hollow pins at the alternate end.
2. Load 2 μL of primary antibody into the pin, and insert the pin into the inlet hole of the *Flowpattern Assembly* (*see* **Note 29**).
3. Attach the Luer-Stub Adapter to pressure-regulated nitrogen gas.
4. Repeat for all antibodies required for the cytokine analysis.
5. Using the pressure regulator, slowly open nitrogen flow to 2 PSI (*see* **Note 30**).
6. Incubate chip overnight in a clean enclosed dark space (*see* **Note 31**).
7. Turn off the nitrogen flow, and remove pins from the chip (*see* **Note 32**).
8. Submerge *Flowpattern Assembly* in 5% BSA, and carefully remove the PDMS chip from the glass barcode chip. Set PDMS chip aside to clean for later reuse (*see* **Note 33**).
9. Remove the glass barcode chip from the solution, and wash 3× with 1 mL of 5% BSA.
10. Block the surface of the barcode chip with ~1 mL of 5% BSA in a clean enclosed dark space for 1 h (*see* **Note 31**).
11. Wash with PBS, followed by (1:1) PBS/DI water, and then DI water (*see* **Note 32**).
12. Using filtered compressed air, blow the surface of the barcode chip dry (*see* **Note 34**).
13. Store at 4 °C until use (*see* **Note 35**).

3.6 Cytokine Assay

3.6.1 Population-Level Multiplexed Cytokine Assay of CAR-T: Target Cell Pairings

1. Co-incubate CAR-T cells with target cells in a 1:1 ratio and CAR-T cells and target cells alone, each at a 100,000 cells/ mL concentration (*see* **Note 36**).
2. Incubate for 10 h at 37 °C; 5% CO_2.
3. Following incubation, pellet cells at 200 × *g*, and save the supernatant.
4. Sonicate Cell Capture Microchamber Chip in 5% BSA for 20 min.
5. Oxygen-plasma treat the Cell Capture Microchamber Chip, and top with culture media (*see* **Note 37**).
6. Place Cell Capture Microchamber Chip on the bottom clamp plate.
7. Pipette 40 μL of supernatant onto cell capture area evenly, and let cells settle for at least 2 min (*see* **Note 38**).
8. Add ~40 μL of media along the top of the Cell Capture Microchamber Chip (*see* **Note 37**).

9. With the antibody barcode facing the microchambers, place the glass barcode chip lengthwise onto the Cell Capture Microchamber Chip (*see* **Note 39**).
10. Top with the clamp plate and screw device together evenly (*see* **Note 40**).
11. Image chip with a microscope with automatic image stitching (*see* **Note 41**), and incubate at 37 °C and 5% CO_2 for 12–24 h.
12. Submerge assembly into 5% BSA, disassemble the clamp plates, and carefully remove the barcode chip from the Cell Capture Microchamber Chip (*see* **Note 42**).
13. Rinse chip 5× with 1 mL of 5% BSA (*see* **Note 43**).
14. In a dark flat environment, add 400 μL of secondary antibody (1:200 dilution in 5% BSA), and incubate for 40 min (*see* **Note 31**).
15. Following incubation, repeat **step 10**.
16. In a dark flat environment, add 400 μL of streptavidin–fluorophore (1:100 dilution in 5% BSA), and incubate for 30 min (*see* **Note 31**).
17. Following incubation, repeat **step 10**.
18. In a dark flat environment, block chip with 5% BSA, and incubate for 30 min (*see* **Note 31**).
19. Wash with PBS, followed by (1:1) PBS/DI water, and then DI water (*see* **Note 32**).
20. Using filtered compressed air, blow the surface of the barcode chip dry (*see* **Note 34**).
21. Scan chip using Genepix array scanner (*see* **Note 44**).
22. Combine Genepix array scanner with the positions of the microchambers and cell counts.

3.6.2 Single-Cell Multiplexed Cytokine Assay

1. Sonicate Cell Capture Microchamber Chip in 5% BSA for 20 min.
2. Oxygen-plasma treat the Cell Capture Microchamber Chip, and top with culture media (*see* **Note 37**).
3. Place Cell Capture Microchamber Chip on the bottom clamp plate.
4. Pipette 200 μL of 0.3 million cells/mL cell suspension onto cell capture area evenly, and let cells settle for at least 2 min (*see* **Note 38**).
5. Add ~40 μL of media along the top of the Cell Capture Microchamber Chip (*see* **Note 37**).
6. With the antibody barcode facing the microchambers, place the glass barcode chip lengthwise onto the Cell Capture Microchamber Chip (*see* **Note 39**).

7. Top with the clamp plate and screw device together evenly (*see* **Note 40**).
8. Image chip with a microscope with automatic image stitching (*see* **Note 41**), and incubate at 37 °C for 12–24 h.
9. Submerge assembly into 5% BSA, disassemble the clamp plates, and carefully remove the barcode chip from the Cell Capture Microchamber Chip (*see* **Note 42**).
10. Rinse chip 5× with 1 mL of 5% BSA (*see* **Note 43**).
11. In a dark flat environment, add 400 μL of secondary antibody (1:200 dilution in 5% BSA), and incubate for 40 min (*see* **Note 31**).
12. Following incubation, repeat **step 10**.
13. In a dark flat environment, add 400 μL of streptavidin–fluorophore (1:100 dilution in 5% BSA), and incubate for 30 min (*see* **Note 31**).
14. Following incubation, repeat **step 10**.
15. In a dark flat environment, block chip with 5% BSA, and incubate for 30 min (*see* **Note 31**).
16. Wash with PBS, followed by (1:1) PBS/DI water, and then DI water (*see* **Note 32**).
17. Using filtered compressed air, blow the surface of the barcode chip dry (*see* **Note 34**).
18. Scan chip using Genepix array scanner (*see* **Note 44**).
19. Combine Genepix array scanner with the positions of the microchambers and cell counts (Fig. 2).

3.6.3 Cell-Pairing Multiplexed Cytokine Assay

1. After labeling CAR and cells with membrane dye, incubate cells in a 1:1 ratio for 15 min in culture medium (*see* **Note 7**).
2. Sonicate Cell Capture Microchamber Chip in 5% BSA for 20 min.
3. Oxygen-plasma treat the Cell Capture Microchamber Chip, and top with culture media (*see* **Note 37**).
4. Place Cell Capture Microchamber Chip (Fig. 2) on the bottom clamp plate.
5. Pipette 200 μL of 0.3 million cells/mL cell suspension onto cell capture area evenly, and let cells settle for at least 2 min (*see* **Note 38**).
6. Add ~40 μL of media along the top of the Cell Capture Microchamber Chip (*see* **Note 37**).
7. With the antibody barcode facing the microchambers, place the glass barcode chip lengthwise onto the Cell Capture Microchamber Chip (*see* **Note 39**).

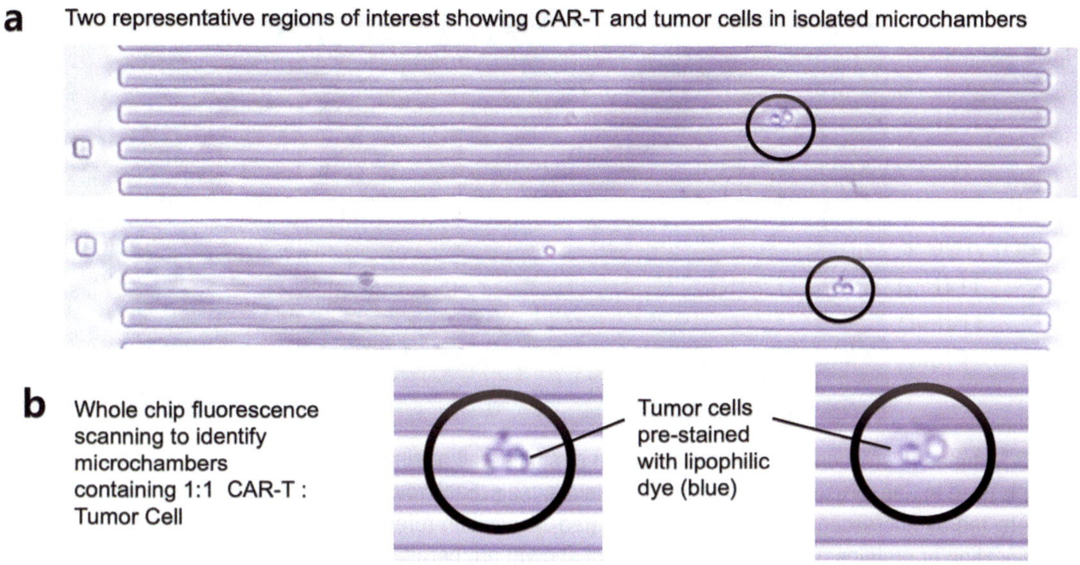

Fig. 2 Loading of CAR-T and tumor cells in microchambers for multiplex cytokine secretion assay. (**a**) Representative regions of the device showing the loading of CAR-T and/or tumor cells in isolated microchambers. (**b**) Enlarged view of 1:1 CAR-T:tumor cell pairs in microchambers. Tumor cells can be readily identified by prestaining with lipophilic dye (blue)

8. Top with the clamp plate and screw device together evenly (*see* **Note 40**).
9. Image chip with a microscope with automatic image stitching, and incubate at 37 °C for 14 h (*see* **Note 41**).
10. Submerge assembly into 5% BSA, disassemble the clamp plates, and carefully remove the barcode chip from the Cell Capture Microchamber Chip (*see* **Note 42**).
11. Rinse chip 5× with 1 mL of 5% BSA (*see* **Note 43**).
12. In a dark flat environment, add 400 μL of secondary antibody (1:200 dilution in 5% BSA), and incubate for 40 min (*see* **Note 31**).
13. Following incubation, repeat **step 10**.
14. In a dark flat environment, add 400 μL of streptavidin–fluorophore (1:100 dilution in 5% BSA), and incubate for 30 min (*see* **Note 31**).
15. Following incubation, repeat **step 10**.
16. In a dark flat environment, block chip with 5% BSA, and incubate for 30 min (*see* **Note 31**).
17. Wash with PBS, followed by (1:1) PBS/DI water, and then DI water (*see* **Note 32**).
18. Using filtered compressed air, blow the surface of the barcode chip dry (*see* **Note 34**).

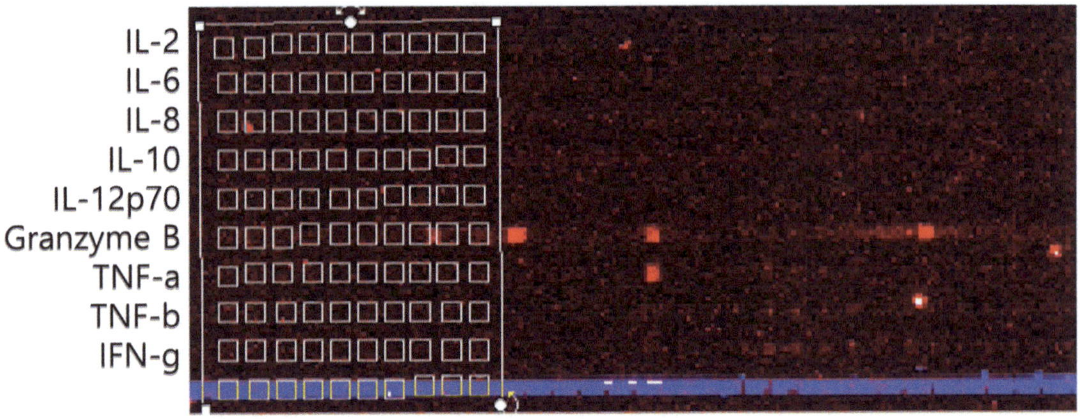

Fig. 3 Scanned fluorescence image for detection of single-cell or single-cell-pair cytokine sections. A square array is used in Genepix software to quantify protein signals. Each row represents a protein of interest and can be detected by fluorescent detection antibodies (red). A position reference (blue) is patterned to facilitate the identification of proteins. The microchambers (not shown) are perpendicular to the antibody lines

19. Scan chip using Genepix array scanner (Fig. 3) (*see* **Note 44**).
20. Combine Genepix array scanner with the positions of the microchambers and cell counts (Fig. 2).

4 Notes

1. This medium is dependent on the target cell culture media. For K562 and Raji cell lines, this is complete RPMI supplemented with 10% FBS.
2. For K562 and Raji cell lines, this is complete RPMI/RPMI supplemented with 10% FBS.
3. Obtained via Rong Fan Lab or CAD/Art Services, Inc., Engineering and Imaging Support Team.
4. PDMS Sylgard 184 (Dow Corning) can also be used.
5. Scotch brand magic tape is the only tape that does not leave a residue.
6. Blocking solution is best when prepared fresh.
7. Use the media that the cells were cultured with.
8. These should be different clones than those of the capture antibodies in order to prevent nonspecific signals.
9. We use the Nikon Eclipse Ti microscope.
10. The final culture medium of the CAR-T cells will be dictated by the target cell culture media.
11. For K562 and Raji cell lines, this step would be RPMI only (no FBS).

12. Quantity should be enough to cover approximately 3/4 of the wafer. Use the positive photoresist when making the flow-patterning master mold. Use the negative photoresist when making the cell capture microchamber mold.
13. If there are bubbles in the photoresist when pouring, the resulting pattern will have defects. Therefore, ensure there are no bubbles present.
14. Varying the spin speed and size of the wafer (i.e., 4″ in this protocol) changes the thickness of the photoresist. For technical specifications, check the photoresist manufacturer's spin speed curves (available here—http://www.microchem.com/PDFs_Dow/SPR%20220%20DATA%20SHEET%20R%26H.pdf).
15. Steady cool down is required to avoid cracking the photoresist.
16. To calculate the proper exposure time, use the mask aligner to measure lamp intensity, and divide this value by 150 mJ/cm^2.
17. It is best to fabricate an extra barcode flow-patterning chip in order to avoid delays in experiments.
18. Make sure the two agents are thoroughly mixed (at least 30 s) so that they cross-link and cure evenly.
19. Compressed air can be filtered by attaching a one-way filter to the outlet. Dust can be detrimental to the channels because it can cause the PDMS channels to connect, making them unusable.
20. Ensure boat is as close as possible to the wafer in order to discourage leakage and an uneven device.
21. To make sure that the chip has a uniform thickness, ensure that the oven surface is level. PDMS can also be left in overnight.
22. The silicon master mold is extremely delicate; therefore, be cautious. Wrapping the PDMS in plastic wrap while not in use helps keep it clean.
23. The holes should be created at the 20 inlet and 20 outlet templates in the pattern.
24. Without the complete removal of the liquid, the PDMS and glass slide will not be able to bond in subsequent steps.
25. Tweezers can also be used to apply force to the tape and PDMS.
26. If possible, perform this step in a clean room or hood to ensure this step is dust-free.
27. It is okay if there is debris in the channel; however, if the debris augments the shape, then it is not usable. Write down the

channels with deformities so that they're not used during flow patterning.

28. Make sure you mark the flow-patterned area for reference when attaching the single cell capture microchamber.
29. When loading antibodies, touch the pipette tip to the bottom of the inlet, and then, release the liquid. Otherwise, air bubbles can be introduced which can slow down the flow-patterning speed. Note the channel each antibody was placed in.
30. The PDMS bound to the glass slide can be delaminated if pressure is released too fast.
31. Petri dish covered in aluminum foil works well for this.
32. Remove these carefully as to avoid delaminating the chip.
33. Take care to avoid smearing antibody channels. Clean PDMS channels in a container of 1:1 IPA/DI water, and sonicate for 20–30 min. Discard IPA/DI water, refill with DI water, and sonicate again for 20–30 min. Prepare PDMS according to Subheading 3.3, beginning at **steps 10** and **11**.
34. Tweezers are helpful to avoid making contact with the barcode.
35. The chip can be stored for up to 2 weeks.
36. Target cells include Raji cells, K562, NIH 3T3, and ELM cell lines expressing CD19. Run this portion in triplicate, and incubate CAR-T and target cell controls in the same u-bottom 96-well plate.
37. Oxygen-plasma treatment decreases the PDMS hydrophobicity. Use the media that the cells are cultured with.
38. Ensure there is just a thin layer of media before adding cells. Following incubation, check that the cells have settled into the wells.
39. When laying the barcode chip down, align starting at the edge you placed the media along. Make sure that the chip is straight. However, once you have placed the barcode chip down, do not alter its position.
40. It is best to insert and tighten the screws in a diagonal fashion to ensure that an even amount of pressure is applied across the device.
41. We use the Nikon Eclipse Ti microscope.
42. Take care to avoid smearing antibody channels.
43. Dry the bottom of the chip with a Kimwipe, and place in a new petri dish.
44. Afterward, chip can be stored in a dark desiccator.

References

1. Maude SL, Frey N, Shaw PA, Aplenc R, Barrett DM, Bunin NJ, Chew A, Gonzalez VE, Zheng Z, Lacey SF (2014) Chimeric antigen receptor T cells for sustained remissions in leukemia. N Engl J Med 371(16):1507–1517
2. Porter DL, Hwang W-T, Frey NV, Lacey SF, Shaw PA, Loren AW, Bagg A, Marcucci KT, Shen A, Gonzalez V (2015) Chimeric antigen receptor T cells persist and induce sustained remissions in relapsed refractory chronic lymphocytic leukemia. Sci Transl Med 7(303):303ra139
3. Neelapu SS, Locke FL, Bartlett NL, Lekakis LJ, Miklos DB, Jacobson CA, Braunschweig I, Oluwole OO, Siddiqi T, Lin Y (2017) Axicabtagene ciloleucel CAR T-cell therapy in refractory large B-cell lymphoma. N Engl J Med 377(26):2531–2544
4. Bendall SC, Nolan GP, Roederer M, Chattopadhyay PK (2012) A deep profiler's guide to cytometry. Trends Immunol 33(7):323–332
5. Xue Q, Bettini E, Paczkowski P, Ng C, Kaiser A, McConnell T, Kodrasi O, Quigley MF, Heath J, Fan R (2017) Single-cell multiplexed cytokine profiling of CD19 CAR-T cells reveals a diverse landscape of polyfunctional antigen-specific response. J Immunother Cancer 5(1):85
6. Woehrling E, Hill E, Torr E, Coleman M (2011) Single-cell ELISA and flow cytometry as methods for highlighting potential neuronal and astrocytic toxicant specificity. Neurotox Res 19(3):472–483
7. Edwards BS, Oprea T, Prossnitz ER, Sklar LA (2004) Flow cytometry for high-throughput, high-content screening. Curr Opin Chem Biol 8(4):392–398
8. Rebhahn JA, Bishop C, Divekar AA, Jiminez-Garcia K, Kobie JJ, Lee FE-H, Maupin GM, Snyder-Cappione JE, Zaiss DM, Mosmann TR (2008) Automated analysis of two-and three-color fluorescent Elispot (Fluorospot) assays for cytokine secretion. Comput Methods Prog Biomed 92(1):54–65
9. Lu Y, Xue Q, Eisele MR, Sulistijo ES, Brower K, Han L, Amir e-AD, Pe'er D, Miller-Jensen K, Fan R (2015) Highly multiplexed profiling of single-cell effector functions reveals deep functional heterogeneity in response to pathogenic ligands. Proc Natl Acad Sci U S A 112(7):E607–E615
10. Lu Y, Chen JJ, Mu L, Xue Q, Wu Y, Wu P-H, Li J, Vortmeyer AO, Miller-Jensen K, Wirtz D (2013) High-throughput secretomic analysis of single cells to assess functional cellular heterogeneity. Anal Chem 85(4):2548–2556
11. Chang ZL, Lorenzini MH, Chen X, Tran U, Bangayan NJ, Chen YY (2018) Rewiring T-cell responses to soluble factors with chimeric antigen receptors. Nat Chem Biol 14(3):317
12. Rossi J, Paczkowski P, Shen Y-W, Morse K, Flynn B, Kaiser A, Ng C, Gallatin K, Cain T, Fan R, Mackay S, Heath JR, Rosenberg SA, Kochenderfer JN, Zhou J, Bot A (2017) Preinfusion polyfunctional anti-CD19 chimeric antigen receptor T cells associate with clinical outcomes in NHL, Blood. https://doi.org/10.1182/blood-2018-01-828343

Chapter 5

Testing the Specificity of Compounds Designed to Inhibit CPT1A in T Cells

Roddy S. O'Connor and Michael C. Milone

Abstract

In response to antigen and costimulation, T cells undergo a series of metabolic transitions that fulfill the biosynthetic demands of clonal expansion, differentiation, and effector function. Following antigen clearance, the oxidation of long-chain fatty acids (LCFAO) has been implicated in the transition from effector to central memory T cells. However, studies demonstrating a role for LCFAO in memory T-cell development have largely relied on the use of etomoxir (ETO), a small molecule inhibitor of the long-chain fatty acid transporter CPT1A. Understanding how the depletion of nutrients including LCFA that might occur in tumor microenvironments affects T-cell proliferation, differentiation, and function has important implications for tumor immunotherapy. Here, we combine the analysis of posttranscriptional gene silencing with extracellular flux assays to determine if etomoxir exerts nonspecific effects on oxidative metabolism. The off-target effects of ETO that we describe highlight the challenges of using pharmacologic inhibitors in loss-of-function approaches in T cells.

Key words Etomoxir, CPT1A, shRNA

1 Introduction

Following antigen stimulation, T cells follow an ordered differentiation program involving activation, proliferation, and differentiation. T-cell activation induces a series of metabolic transitions that support the entire proliferation and differentiation processes. A metabolic shift to glycolysis in activated T cells supports energy generation, macromolecular biosynthesis, and redox homeostasis. Since it was widely accepted that etomoxir inhibits CPT1A, the rate-limiting enzyme in LCFAO, long-chain fatty acids have been implicated in the formation of memory T cells following antigen clearance [1–3].

Concerns about the specificity of etomoxir have been raised in recent studies on immune cells. We showed that etomoxir concentrations at commonly used concentrations elicit nonspecific effects on

Samuel G. Katz and Peter M. Rabinovich (eds.), *Cell Reprogramming for Immunotherapy: Methods and Protocols*, Methods in Molecular Biology, vol. 2097, https://doi.org/10.1007/978-1-0716-0203-4_5,

oxidative metabolism, culminating in ROS production, mitochondrial matrix swelling, and GSH depletion in primary human T cells undergoing high rates of clonal expansion [4]. Complementary studies provided genetic evidence that ablating CPT1A in T cells had minimal consequence on either memory T-cell formation or regulatory T-cell development [5]. In studies on macrophage metabolism, etomoxir treatment limited the polarization of macrophages toward an M2 lineage [6]. In contrast to T cells, there was no phenotype in macrophages lacking either CPT1a [7] or CPT2 [8] despite their inability to oxidize LCFA. In isolated mitochondrial preparations, etomoxir concentrations above 10 μM inhibited the mitochondrial adenine nucleotide transporter as well as complex 1 of the electron transport chain [9]. Here, we developed an assay to test the specificity of etomoxir in cells genetically engineered to lack CPT1A expression. T cells expressing lentiviral shRNA against CPT1A were seeded on matrix-coated XF "Seahorse" cell culture dishes. Their metabolic parameters, including rates of oxygen consumption and glycolysis, were assessed in response to varying doses of etomoxir treatment.

2 Materials

2.1 Cell Culture

1. Primary human T cells.
2. T-cell growth medium: RPMI 1640 supplemented with 10% FBS, 10 mM HEPES, 2 mM L-glutamine, 100 U/ml penicillin G, and 100 μg/ml streptomycin.
3. 4.5 μm Dynabeads containing immobilized antihuman CD3 and antihuman CD28.
4. Etomoxir (sterile) diluted in QH_2O and sterile filtered with a 2 μM filter.
5. Lentiviral supernatants with shRNA targeting CPT1A or scramble shRNA against the human β-actin gene as described previously in [10].

2.2 XF Flux Analyzer

1. XF96 analyzer with WAVE software.
2. XF96 cell culture microplate and XF sensor cartridge.
3. 1.36 mg/ml Cell-Tak (Corning).
4. 0.1 M sodium bicarbonate, pH 8.0 (sterile filtered).
5. 1 N NaOH.
6. Etomoxir.
7. XF assay medium: Non-buffered RPMI 1640 containing 5.5 mM glucose, 2 mM L-glutamine, and 5 mM HEPES. Adjust the pH to 7.4 with 0.1 N NaOH.
8. Sterile filter assay medium with a 0.2 μM filter following pH adjustment.

2.3 Immunoblotting

1. RIPA-2 lysis buffer (50 mM Tris-HCl, pH 8.0, 150 mM NaCl, 1% NP-40, 0.5% deoxycholic acid, 0.1% SDS).
2. Protease inhibitor (Mini Complete, Roche).
3. Phosphatase inhibitor (PhosSTOP, Roche).
4. PDVF membrane.
5. Running buffer (0.1% SDS, Tris/Glycine solution).
6. Transfer buffer (Tris/Glycine).
7. Tris-buffered saline (TBS).
8. TBS containing 0.1% Tween-20 (TBS-T).
9. Blocking solution: 5% nonfat dry milk in TBS-T.
10. 7.5% separating gel (solutions used: 1.5 M Tris–HCl, pH 8.8, 10% SDS, 30% acrylamide bis 37.5:1, 10% ammonium persulfate, TEMED). *See* **Note 1** for details.
11. 4% stacking gel (solutions used: 0.5 M Tris–HCl, pH 6.8, 10% SDS, 30% acrylamide bis 37.5:1, 10% ammonium persulfate, TEMED). *See* **Note 2** for details.
12. Sample buffer (solutions used: 0.5 M Tris-HCl, pH 6.8, 50% glycerol in QH_2O, 10% SDS, 2-mercaptoethanol, pinch of bromophenol blue). *See* **Note 3** for details.

3 Methods

3.1 T-Cell Activation

1. Day 0. Activate 6e6 primary human T cells with anti-CD3/anti-CD28 Dynabeads at a ratio of 3 beads to 1 cell.
2. Day 1. Activated T cells are infected with either scramble or shRNA lentivirus against CPT1A.
3. Day 3. Determine infection efficiencies by flow cytometry. Count and refeed T cells at a concentration of $0.8–1.0 \times 10^6$ cells/ml.
4. Day 5. Lyse 1.0×10^6 scramble vs. shRNA CPT1A cells in RIPA-2 containing protease and phosphatase inhibitors.

3.2 XF Flux Analysis (See Fig. 1)

1. Day 0 (7 days prior to assay). Prepare the extracellular matrix adhesive (Cell-Tak) by combining 3 ml sodium bicarbonate with 50 μl of Cell-Tak solution and 20 μl NaOH.
2. Coat all wells of the XF cell microplate with 12 μl Cell-Tak.
3. Incubate the microplate overnight at 37 °C.
4. Rinse three times in sterile QH_2O, air-dry, and then store at 4 °C until use.
5. Day 6 (1 day prior to assay). Add 200 μl of the calibrant solution to hydrate the XF sensor cartridge. Leave overnight at room temperature.

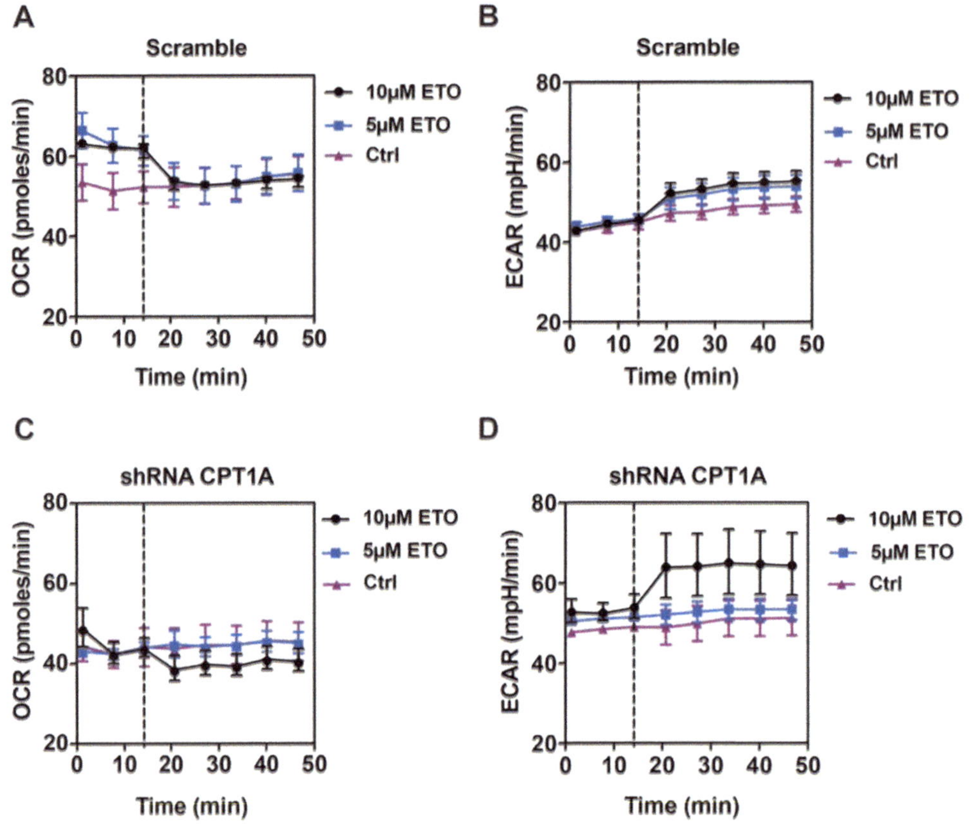

Fig. 1 The specificity of etomoxir for CPT1A is lost at concentrations above 5 μM. Primary human T cells expressing either control or shRNA against CPT1A were restimulated with Dynabeads and expanded for 5 days. Oxygen consumption rates (OCR) in control (**a**) vs. shRNA cells (**c**) are shown. OCR was measured under basal conditions and after the introduction of etomoxir (dotted lines). The corresponding glycolytic rates (ECAR) in control (**b**) and shRNA cells (**d**) are shown. (Figure adapted from R. S. O'Connor et al., The CPT1a inhibitor, etomoxir induces severe oxidative stress at commonly used concentrations. *Sci Rep* **8**, 6289 (2018))

6. Day 7 (day of assay). Transfer the hydrated XF sensor cartridge to a CO_2-free, 37 °C incubator for 5–6 h (*see* **Note 4**).
7. Prepare etomoxir in assay medium at a concentration of 40 vs. 400 μM. Store in a 37 °C water bath until use.
8. Design the experiment using the WAVE software; include triplicate measurements to ensure steady-state measurement.
9. Centrifuge 2.7 × 10^6 T cells per condition at 1200 × *g* for 5 min.
10. Aspirate, and wash with 1× PBS.
11. Resuspend the cell pellets in 1.575 ml of XF assay medium prepared in Subheading 2.2 **step** 7.
12. Mix and seed T cells at 175 μl/well for a total of 8 technical replicates.

13. Centrifuge the XF microplate at 1000 × *g* for 5 min and transfer to a CO_2-free, 37 °C incubator for 30 min.
14. Add 25 μl of etomoxir to injection port A of the sensor cartridge.
15. Insert the hydrated XF sensor cartridge to calibrate the instrument.
16. Following instrument calibration, replace the calibration plate with the cell culture microplate.
17. Run the XF assay to measure cellular oxygen consumption (OCR) and extracellular acidification (ECAR) under basal conditions and in response to either 5 μM or 50 μM etomoxir (*see* **Note 5**).

3.3 Immunoblotting

1. Apply the protein/sample buffer mixture to the SDS-PAGE gel matrix.
2. Perform electrophoresis for 1.5 h at 150 V.
3. Transfer the proteins from the gel to the PDVF membrane using a current of 250 mA for 110 min.
4. Incubate the membrane in 5% nonfat dry milk in TBS containing 0.1% Tween-20 (TBS-T) for 1 h.
5. With a razor blade, cut the membranes into two segments: 250–60 kDa, 60–30 kDa.
6. Probe the 250–60 kDa membrane overnight at 4 °C with a 1:500 dilution of anti-CPT1A antibody (Cell Signaling Technology) in 0.5% nonfat milk in TBS-T.
7. After a series of three washes (1× quick, 1 × 10 min, 3 × 5 min) in TBS-T, incubate the membrane with an HRP-conjugated goat anti-rabbit IgG (cell signaling), diluted 1:10,000 in the blocking solution.
8. Wash in TBS-T as follows: 1× quick, 1 × 10 min, 3 × 5 min washes.
9. Perform a final rinse in TBS.
10. Probe the 60–30 kDa segment with a 1:2000 dilution of anti-beta actin antibody (Cell Signaling Technology) in 0.5% nonfat milk in TBS-T.
11. Wash as described above, and incubate the membrane with an HRP-conjugated sheep anti-mouse IgG (Amersham), diluted 1:5000 in the blocking solution.
12. Mix equal volumes of the West Femto SuperSignal Chemiluminescent reagent A and B prior to use. Add to the blot for 5 min.
13. Wrap on saran wrap, expose the ECL Hyperfilm for 1 min in a cassette, and develop as necessary.

4 Notes

1. Separating gel preparation.

7.5% separating gel	Volume
QH_2O	38.8 ml
1.5 M Tris–HCl, pH 8.8	20.0 ml
10% SDS	800 μl
30% acrylamide/bis solution	20 ml
Vacuum 5 min	
10% ammonium persulfate	400 μl
TEMED	40 μl
Total	80 ml

2. Stacking gel preparation.

4% stacking gel	Volume
QH_2O	12.2 ml
0.5 M Tris–HCl, pH 6.8	5.0 ml
10% SDS	200 μl
30% acrylamide/bis solution	2.66 ml
Vacuum 5 min	
10% ammonium persulfate	200 μl
TEMED	20 μl
Total	20 ml

3. Sample buffer.

5x sample buffer	Volume
QH_2O	175 μl
0.5 M Tris–HCl, pH 6.8	125 μl
50% glycerol in QH_2O	500 μl
10% SDS	200 μl
2-Mercaptoethanol	50 μl
Bromophenol blue	Pinch
Final volume	1 ml

4. Previous protocols recommended overnight hydration of the XF sensor cartridge in a CO_2-free, 37 °C incubator. As this extended duration results in dehydration, it is preferable to add 200 μl of the calibration solution and leave overnight on the bench at room temperature. On the morning of the assay, transfer to a CO_2-free, 37 °C incubator for 5 h.
5. The inherent metabolic flexibility of T cells ensures they can use fuel sources interchangeably. Inhibiting long-chain fatty acid oxidation will promote a corresponding shift to glycolysis (increase in ECAR measurements). In cells expressing shRNA against CPT1A, there should be no further increase in ECAR at doses specific to the CPT1A target.

Acknowledgments

This work was supported by grants from the University of Pennsylvania-Novartis Alliance.

References

1. van der Windt GJ, O'Sullivan D, Everts B, Huang SC, Buck MD, Curtis JD, Chang CH, Smith AM, Ai T, Faubert B, Jones RG, Pearce EJ, Pearce EL (2013) CD8 memory T cells have a bioenergetic advantage that underlies their rapid recall ability. Proc Natl Acad Sci U S A 110(35):14336–14341. https://doi.org/10.1073/pnas.1221740110
2. O'Sullivan D, van der Windt GJ, Huang SC, Curtis JD, Chang CH, Buck MD, Qiu J, Smith AM, Lam WY, Di Plato LM, Hsu FF, Birnbaum MJ, Pearce EJ, Pearce EL (2014) Memory CD8(+) T cells use cell-intrinsic lipolysis to support the metabolic programming necessary for development. Immunity 41(1):75–88. https://doi.org/10.1016/j.immuni.2014.06.005
3. Pearce EL, Walsh MC, Cejas PJ, Harms GM, Shen H, Wang LS, Jones RG, Choi Y (2009) Enhancing CD8 T-cell memory by modulating fatty acid metabolism. Nature 460(7251):103–107. https://doi.org/10.1038/nature08097
4. O'Connor RS, Guo L, Ghassemi S, Snyder NW, Worth AJ, Weng L, Kam Y, Philipson B, Trefely S, Nunez-Cruz S, Blair IA, June CH, Milone MC (2018) The CPT1a inhibitor, etomoxir induces severe oxidative stress at commonly used concentrations. Sci Rep 8(1):6289. https://doi.org/10.1038/s41598-018-24676-6
5. Raud B, Roy DG, Divakaruni AS, Tarasenko TN, Franke R, Ma EH, Samborska B, Hsieh WY, Wong AH, Stuve P, Arnold-Schrauf C, Guderian M, Lochner M, Rampertaap S, Romito K, Monsale J, Bronstrup M, Bensinger SJ, Murphy AN, McGuire PJ, Jones RG, Sparwasser T, Berod L (2018) Etomoxir actions on regulatory and memory T cells are independent of Cpt1a-mediated fatty acid oxidation. Cell Metab 28(3):504–515. https://doi.org/10.1016/j.cmet.2018.06.002
6. Huang SC, Everts B, Ivanova Y, O'Sullivan D, Nascimento M, Smith AM, Beatty W, Love-Gregory L, Lam WY, O'Neill CM, Yan C, Du H, Abumrad NA, Urban JF Jr, Artyomov MN, Pearce EL, Pearce EJ (2014) Cell-intrinsic lysosomal lipolysis is essential for alternative activation of macrophages. Nat Immunol 15(9):846–855. https://doi.org/10.1038/ni.2956
7. Divakaruni AS, Hsieh WY, Minarrieta L, Duong TN, Kim KKO, Desousa BR, Andreyev AY, Bowman CE, Caradonna K, Dranka BP, Ferrick DA, Liesa M, Stiles L, Rogers GW, Braas D, Ciaraldi TP, Wolfgang MJ, Sparwasser T, Berod L, Bensinger SJ, Murphy AN (2018) Etomoxir inhibits macrophage polarization by disrupting CoA homeostasis. Cell Metab 28(3):490–503. https://doi.org/10.1016/j.cmet.2018.06.001

8. Nomura M, Liu J, Rovira II, Gonzalez-Hurtado E, Lee J, Wolfgang MJ, Finkel T (2016) Fatty acid oxidation in macrophage polarization. Nat Immunol 17(3):216–217. https://doi.org/10.1038/ni.3366
9. Divakaruni AS, Rogers GW, Andreyev AY, Murphy AN (2016) 12.03: The CPT inhibitor etomoxir has an off-target effect on the nucleotide translocase and respiratory complex I. Biochim Biophys Acta 1857:e118
10. Milone MC, Fish JD, Carpenito C, Carroll RG, Binder GK, Teachey D, Samanta M, Lakhal M, Gloss B, Danet-Desnoyers G, Campana D, Riley JL, Grupp SA, June CH (2009) Chimeric receptors containing CD137 signal transduction domains mediate enhanced survival of T cells and increased antileukemic efficacy in vivo. Mol Ther 17(8):1453–1464. https://doi.org/10.1038/mt.2009.83

Chapter 6

Engineering of Natural Killer Cells for Clinical Application

Noriko Shimasaki and Dario Campana

Abstract

The clinical success of chimeric antigen receptor-directed T cells in leukemia and lymphoma has boosted the interest in cellular therapy of cancer. It has been known for nearly half a century that a subset of lymphocytes called natural killer (NK) cells can recognize and kill cancer cells, but their clinical potential as therapeutics has not yet been fully explored. Progress in methods to expand and genetically modify human NK cells has resulted in technologies that allow the production of large numbers of highly potent cells with specific anticancer activity. Here, we describe clinically applicable protocols for NK cell engineering, including expansion of NK cells and genetic modification using electroporation of messenger RNA.

Key words Natural killer cells, Cell therapy, Cell expansion, Electroporation, GMP compliance

1 Introduction

Natural killer (NK) cells can identify virally infected cells and tumor cells because these express an altered profile of ligands for NK activating and inhibitory receptors [1–5]. Thus, underexpression of human leukocyte antigen (HLA) class I molecules in tumor cells reduces activity of inhibitory receptors, such as killer immunoglobulin-like receptors (KIRs) KIR2DL1 and KIR3DL1. Conversely, tumor cells may overexpress stress-induced ligands, such as MHC class I chain-related A (MICA) and B (MICB) and UL-16-binding proteins, which stimulate NK activation by binding the NKG2D activating receptor [1–5]. An additional mechanism by which NK cells can identify and kill tumor cells is mediated by antibodies directed to surface molecules expressed by tumor cells, i.e., antibody-dependent cell-mediated cytotoxicity (ADCC), which occurs through the engagement of the low-affinity Fc receptor (CD16) by the Fc portion of immunoglobulins [1, 5, 6].

Activated NK cells can kill tumor cells by multiple mechanisms. The most prominent relies on the release of cytolytic granules containing granzyme B and perforin, which results in activation of

Samuel G. Katz and Peter M. Rabinovich (eds.), *Cell Reprogramming for Immunotherapy: Methods and Protocols*, Methods in Molecular Biology, vol. 2097, https://doi.org/10.1007/978-1-0716-0203-4_6,

apoptosis [7]. Apoptosis of target cells can also result from binding of the tumor necrosis factor (TNF) family member Fas ligand and/or TNF-related apoptosis-inducing ligand (TRAIL) to their corresponding receptors on tumor cells [8]. Besides direct tumor cell killing, NK cell recognition of target cells and the resulting secretion of cytokines, such as interferon gamma (IFNγ), stimulates other immune cells, including T cells, dendritic cells, and macrophages, therefore magnifying antitumor responses [2].

One of the factors that is likely to be critical for the antitumor effect of NK cell infusions is the number of NK cells that can reach the tumor site [9]. Because the number of NK cells that can be obtained from a peripheral blood draw or a leukapheresis is relatively small, there has been considerable interest in developing methods for ex vivo NK cell expansion. Attempts to expand NK cells include culture with interleukin (IL)-2 and anti-CD3 antibody [10] or anti-CD3 and anti-CD52 antibodies [11]. Other cytokines, such as IL-15, IL-12, IL-18, and IL-21, have also been used [12]. Cocultures with irradiated peripheral blood mononucleated cells [13], EBV-transformed lymphoblastoid cells [14], and Wilms tumor-derived cell line HFWT [15, 16] have also been used. We established a genetically modified variant of the K562 leukemia cell line by expressing two NK stimulatory molecules: membrane-bound IL-15 and 4-1BB ligand. Coculture of peripheral blood mononucleated cells with irradiated K562-mb15-41BBL greatly stimulates NK cell expansion (Fig. 1a, b) [17–19]. Other investigators have used similar K562 variants to expand NK cells [20, 21]. The K562-mb15-41BBL culture system also stimulates propagation of NK cells from patients with acute leukemia in remission [18], from patients with multiple myeloma [22], and from patients with breast or gastric cancer (Fig. 1b). Cord blood can also be used as a source of NK cells for expansion in culture with K562-mb15-41BBL cells, or other K562 variants can induce proliferation of NK cells from cord blood (Shimasaki, Campana, unpublished results) [23]. Finally, NK cells derived from induced pluripotent stem cells can be propagated ex vivo [24]. In general, culture of NK cells with cytokines and/or stimulatory cells enhances their cytotoxicity and production of cytokines. For example, NK cells cocultured with K562-mb15-41BBL exerted superior cytotoxicity against several malignancies including acute myeloid leukemia [18], multiple myeloma [25], sarcoma [26], hepatocellular carcinoma [27], and gastric cancer [28].

Activated NK cells have higher antitumor potential than non-activated NK cells, and ex vivo expansion allows the infusion of larger cell products, possibly leading to higher effector-to-target ratios. The antitumor potential of NK cells can be further increased by genetic modification. Thus, expression of chimeric antigen receptors (CAR) consisting of a single-chain variable fragment (scFv) against a tumor marker, a costimulatory molecule

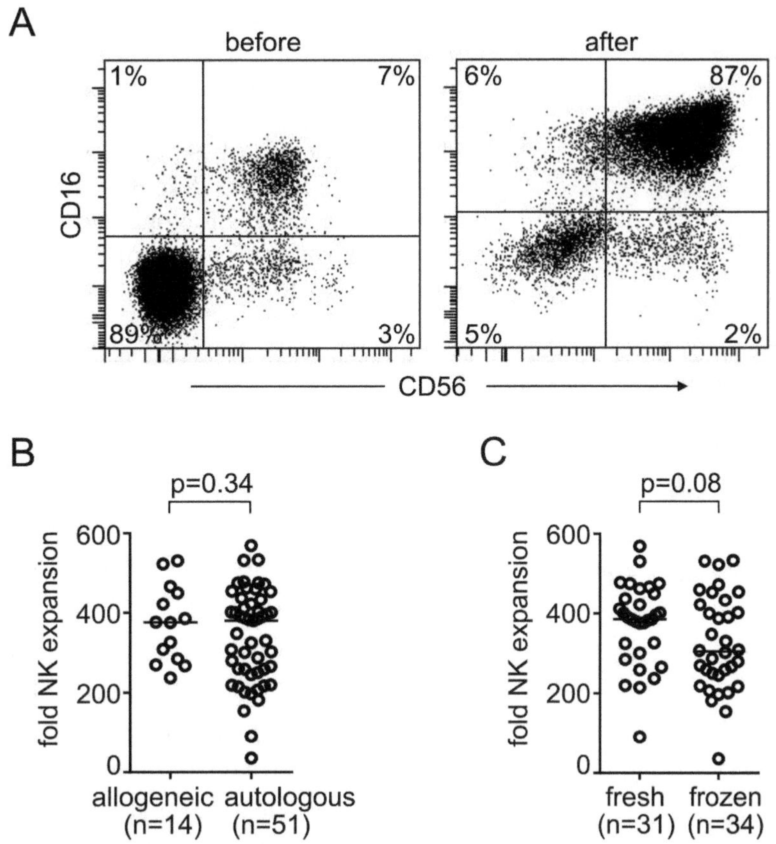

Fig. 1 Expansion of NK cells. (**a**) Flow cytometric dot plots illustrate expression of CD56 and CD16 on cells before (left) and after (after) coculture with irradiated K562-mb15-41BBL for 10 days. (**b**) Fold expansion of NK cells after 10 days of coculture from healthy allogeneic donors (***n*** = 14) or autologous patients (***n*** = 51). Bars represent the median of NK cell expansion (***P*** = 0.34 by ***t***-test). (**c**) Fold expansion of NK cells from fresh (***n*** = 31) or cryopreserved (***n*** = 34) mononuclear cells. Bars represent the median of NK expansion (***P*** = 0.08 by ***t***-test)

(e.g., 4-1BB and CD28), and a signal molecule (CD3ζ) can redirect NK cells to specific target cells and stimulate cytotoxicity and cytokine release upon ligation [17, 29–33]. Tumor cell killing by activated NK cells can also be augmented by increasing the expression of NK activating receptors. To this end, we expressed a chimeric activating receptor including NKG2D, its co-receptor DAP10, and CD3ζ which significantly increased antitumor activity against multiple tumor types [27, 34]. Another approach to increasing NK activity in vivo is to prolong their survival. For this purpose, we transduced the IL-15 gene in its membrane-bound form and found that it stimulated NK cell expansion in vivo and prolonged NK cell survival [35]. One method to genetically modify NK cells is viral transduction [36]. In our experience with retroviral transduction, a transfection efficiency of 70–80% can be routinely achieved after expanding NK cells with K562-mb15-41BBL cells. A nonviral transfection method which is also very effective is electroporation of mRNA [31, 36, 37], although it produces a transient instead of permanent gene expression. Other methods, such as lipofection [38] and trogocytosis [39], have also been reported.

In this chapter, we describe methods for large-scale NK cells expansion, depletion of allogeneic T cells, and genetic modification by electroporation of mRNA in compliance with current Good Manufacturing Practices (cGMP) regulations.

2 Materials (*See* Note 1)

2.1 Expansion of NK Cells

2.1.1 Irradiation of K562-mb15- 41BBL

1. K562 cell line retrovirally transduced with the genes encoding 4-1BB ligand and membrane-bound IL-15 (K562-mb15-41BBL) plus green fluorescent protein (GFP) (*see* **Note 2**).
2. RPMI-1640/10% fetal bovine serum (FBS) medium.
3. Dimethyl sulfoxide (DMSO).
4. G-Rex100M (Wilson Wolf, New Brighton, MN).
5. Transfer bag (T-600, Terumo, Tokyo, Japan or equivalent).
6. Cryobag (CryoMACS Freezing Bag, Miltenyi Biotec, Bergisch Gladbach, Germany or equivalent).
7. Linear particle accelerator (Elekta Synergy, Stockholm, Sweden or equivalent) or Cs-137 irradiator.
8. Control rate freezer.
9. Vapor-phase liquid nitrogen storage tank.

2.1.2 Isolation of Peripheral Blood Mononuclear Cells (PBMC)

1. Leukapheresis product.
2. Saline (0.9% sodium chloride in water) containing 5% human serum albumin or Human Albumin Grifols 5% (Grifols, Barcelona, Spain or equivalent).
3. Ficoll.
4. SCGM media (CellGenix, Freiburg, Germany or equivalent).
5. DMSO.
6. Cryobag.
7. Control rate freezer.
8. Vapor-phase liquid nitrogen storage tank.

2.1.3 Expansion of NK Cells by Coculture with K562-mb15- 41BBL

1. Fresh or cryopreserved peripheral blood mononucleated cells (PBMC).
2. Fresh or cryopreserved irradiated K562-mb15-41BBL.
3. SCGM media.

4. Recombinant human interleukin-2 (IL-2): aldesleukin (Proleukin, Novartis, Basel, Switzerland or IL-2 IS, Miltenyi Biotech) or other sources (*see* **Note 3**).
5. G-Rex100 (Wilson Wolf).

2.1.4 Harvest of Expanded NK Cells

1. SCGM media.
2. Saline (0.9% sodium chloride in water) containing 5% human serum albumin, Human Albumin Grifols 5%, or equivalent.
3. Transfer bag.

2.1.5 T Lymphocytes Depletion

1. CliniMACS CD3 Reagent (Miltenyi Biotec).
2. CliniMACS PBS/EDTA Buffer (phosphate-buffered saline, pH 7.2, plus 1 mM EDTA, Miltenyi Biotec) containing 0.5% human serum albumin (0.5% HSA/PBS/EDTA buffer).
3. CliniMACS Depletion Tubing Set (Miltenyi Biotec).
4. Pre-System Filter (Miltenyi Biotec).
5. Transfer bag.
6. CliniMACS Plus Instrument (Miltenyi Biotec).
7. Heat sealer.

2.2 Electroporation of mRNA

2.2.1 Stimulation of NK Cells

1. NK cells.
2. SCGM media.
3. IL-2.
4. G-Rex100.

2.2.2 Electroporation Procedure

1. NK cells.
2. mRNA encoding the gene of interest.
3. Electroporation buffer (MaxCyte, Gaithersburg, ML).
4. SCGM media.
5. IL-2.
6. CL-2 bag (MaxCyte).
7. G-Rex100.
8. Electroporator (MaxCyte GT or equivalent).

3 Methods

3.1 Expansion of NK Cells

3.1.1 Irradiation of K562-mb15- 41BBL

1. Thaw cryopreserved K562-mb15-41BBL (*see* **Note 4**).
2. Wash the K562-mb15-41BBL cells once and resuspend 1.0×10^6 cells in 50 ml of RPMI-1640/10% FBS medium.
3. Transfer the cells to G-Rex100M (*see* **Note 5**).

4. Add 950 ml of RPMI-1640/10% FBS medium to the vessel.
5. Incubate at 37 °C in 5% CO_2.
6. Seven days after culture, remove 800 ml of the cell culture supernatant leaving cells at the bottom of the vessel undisturbed.
7. Add the same volume of RPMI-1640/10% FBS medium.
8. Incubate at 37 °C in 5% CO_2 for three more days.
9. Remove 800 ml of the cell culture supernatant.
10. Transfer the cells in the vessel into a transfer bag.
11. Irradiate the cells in the transfer bag at 120 Gy (*see* **Note 6**).
12. Centrifuge at 300 × *g* for 10 min and remove the supernatant.
13. Resuspend the cell pellet in RPMI-1640/10% FBS medium at 6–8 × 10^6 cells/ml.
14. Transfer 40 ml of the irradiated cells into a cryobag (*see* **Notes 7** and **8**).
15. Add 20 ml of 30% DMSO to the cryobag (*see* **Note 9**).
16. Collect a sample from the cryobag for assay (*see* **Note 10**).
17. Freeze the cells using a control rate freezer (*see* **Note 11**).
18. Store the frozen cells in a vapor-phase liquid nitrogen storage tank until use.

3.1.2 Isolation of PBMC

1. Transfer leukapheresis product into a flask (*see* **Notes 12** and **13**).
2. Add the same volume of saline with 5% human serum albumin (*see* **Note 14**).
3. Transfer 15 ml of Ficoll to a 50 ml centrifuge tube.
4. Layer 30 ml of diluted leukapheresis product on Ficoll in the tube.
5. Centrifuge at 400 × *g* for 45 min. Do not use a break to stop the spin.
6. Transfer mononuclear cells in new centrifuge tubes.
7. Wash the cells twice and resuspend the cell pellet in SCGM medium.
8. Collect some aliquot of cells for assay (*see* **Note 15**).
9. Dilute the cells with SCGM medium to get 1–3 × 10^7 cells/ml.
10. Transfer 40 ml of the cells into a cryobag (*see* **Note 16**).
11. Add 20 ml of 30% DMSO to the cryobag (*see* **Note 9**).
12. Collect some aliquot from the cryobag for assay (*see* **Note 17**).

13. Freeze cells using a control rate freezer (*see* **Note 11**).
14. Store the frozen cells in a vapor-phase liquid nitrogen tank until use.

3.1.3 Expansion of NK Cells by Coculture with K562-mb15- 41BBL

1. Thaw cryopreserved irradiated K562-mb15-41BBL.
2. Wash cells twice and resuspend the cell pellet in SCGM medium.
3. Transfer 3×10^7 cells in 50 ml of SCGM medium to G-Rex100 (*see* **Notes 18** and **19**).
4. Add 150 ml of SCGM medium.
5. Incubate cells at 37 °C in 5% CO_2 (*see* **Note 20**).
6. Prepare PBMC containing 3×10^6 NK cells in 50 ml of SCGM medium (*see* **Notes 21** and **22**).
7. Transfer PBMC into G-Rex100 containing irradiated K562-mb15-41BBL (*see* **Note 23**).
8. Add 150 ml of SCGM medium.
9. Add IL-2 at a final concentration of 40 IU/ml (*see* **Note 3**).
10. Incubate cells at 37 °C in 5% CO_2.
11. Every 2 or 3 days, add IL-2 at a final concentration of 40 IU/ml to the cell culture (*see* **Notes 24–26**).

3.1.4 Harvest of Expanded NK Cells

1. Ten days after coculture, remove 200 ml of cell culture supernatant.
2. Mix cells well and transfer the cells to centrifuge tubes.
3. Centrifuge at 300 × *g* for 15 min.
4. Wash the cells with saline containing human serum albumin twice.
5. Resuspend the cell pellet with saline containing human serum albumin.
6. Transfer cells to a transfer bag.
7. Collect an aliquot for the assay (*see* **Note 27**).
8. Adjust the volume for the infusion (*see* **Notes 28** and **29**).

3.1.5 T Lymphocytes Depletion

1. Ten days after coculture, remove 200 ml of cell culture supernatant.
2. Mix cells well and transfer the cells to centrifuge tubes.
3. Centrifuge at 300 × *g* for 15 min.
4. Remove the supernatant, and resuspend the cell pellet with 0.5% HSA/PBS/EDTA buffer (*see* **Note 30**).
5. Centrifuge at 300 × *g* for 15 min.
6. Remove the supernatant.

7. Resuspend the cell pellet with 95 ml of 0.5% HSA/PBS/EDTA buffer in a 225 ml centrifuge tube.
8. Add 7.5 ml of CliniMACS CD3 Reagent.
9. Incubate cells with the reagent for 30 min, swaying the tube periodically.
10. Add 100 ml of 0.5% HSA/PBS/EDTA buffer to the tube.
11. Centrifuge at 300 × *g* for 15 min.
12. Remove the supernatant, and resuspend the cell pellet with 150 ml of 0.5% HSA/PBS/EDTA buffer.
13. Transfer cells into a transfer bag and seal the bag by a heat sealer.
14. Clamp the tubing of CliniMACS Depletion Tubing Set.
15. Connect the transfer bag of cells with CliniMACS Depletion Tubing Set through Pre-System Filter.
16. Connect a bag of 0.5% HSA/PBS/EDTA buffer with the tubing set.
17. Install the tubing set to CliniMACS Plus Instrument.
18. Select "Depletion 3.1".
19. Release the clamps and start separation.
20. After separation, close the tubing of the collection bag (*see* **Note 31**).

3.2 Electroporation of mRNA

3.2.1 Stimulation of NK Cells

1. Seed expanded NK cells with 400 ml of SCGM medium in G-Rex100 (*see* **Notes 32** and **33**).
2. Add IL-2 at a final concentration of 1000 IU/ml (*see* **Note 34**).
3. Incubate cells at 37 °C in 5% CO_2 overnight.

3.2.2 Electroporation Procedure

1. Add the appropriate volume of SCGM medium to G-Rex100 (*see* **Notes 35** and **36**).
2. Incubate SCGM medium in G-Rex100 at 37 °C in 5% CO_2.
3. Warm up 80 ml of SCGM medium at 37 °C.
4. Harvest stimulated NK cells to centrifuge tubes (*see* **Note 37**).
5. Centrifuge at 200 × *g* for 10 min (*see* **Note 38**).
6. Remove the supernatant and resuspend the cell pellet with electroporation buffer (*see* **Note 39**).
7. Centrifuge at 200 × *g* for 10 min.
8. Remove the supernatant completely.
9. Resuspend the cell pellet with electroporation buffer at 2×10^8 cells/ml.

10. Transfer the cells for electroporation to a new tube.
11. Add the appropriate amount of mRNA to the tube (*see* **Notes 40** and **41**).
12. Transfer the cells with mRNA into CL-2 bag.
13. Place the CL-2 bag in MaxCyte GT and electroporate cells under MaxCyte Program flow #2 (*see* **Note 42**).
14. Incubate electroporated cells in the CL-2 bag at 37 °C in 5% CO_2 for 10 min (*see* **Note 43**).
15. Infuse 40 ml of warm SCGM medium to the CL-2 bag gently, swaying the bag.
16. Transfer cells in the CL-2 bag to the G-Rex100 containing warm SCGM medium.
17. Infuse 40 ml of warm SCGM medium to the CL-2 bag.
18. Transfer remaining cells in the CL-2 bag to the G-Rex100.
19. Incubate the cells at 37 °C in 5% CO_2 (*see* **Note 44**).

4 Notes

1. All materials listed in the material section should be cGMP or pharmaceutical grade, whenever available.
2. A master cell bank of K562-mb15-41BBL cells should be prepared under cGMP conditions. The cells from the master cell bank should be used as the starting material of the irradiation procedure.
3. Two types of recombinant IL-2 are widely available: IL-2 with a wild-type sequence and IL-2 with a substitution of cysteine-125 by serine. We use the IL-2 with the substitution, Proleukin (Novartis). The potency of Proleukin at 40 IU/ml is comparable with that of the wild-type IL-2 at 10 IU/ml [40].
4. K562-mb15-41BBL can be expanded in other cell culture media without FBS, such as SCGM media.
5. Cells are cultured in G-Rex100M which can hold a large quantity of cell culture medium. Oxygen and carbon dioxide are exchanged through the gas permeable membrane at the bottom of the vessel. The cells can also be expanded in a T-flask or a bioreactor system such as WAVE Bioreactor (GE Healthcare, Chicago, IL). Figure 2a shows the outline of cell expansion.
6. We irradiate the cells with X-rays generated from linear particle accelerator (Elekta Synergy). A gamma irradiator can be used instead of an X-ray irradiator.
7. The cell volume depends on the size of cryobag.

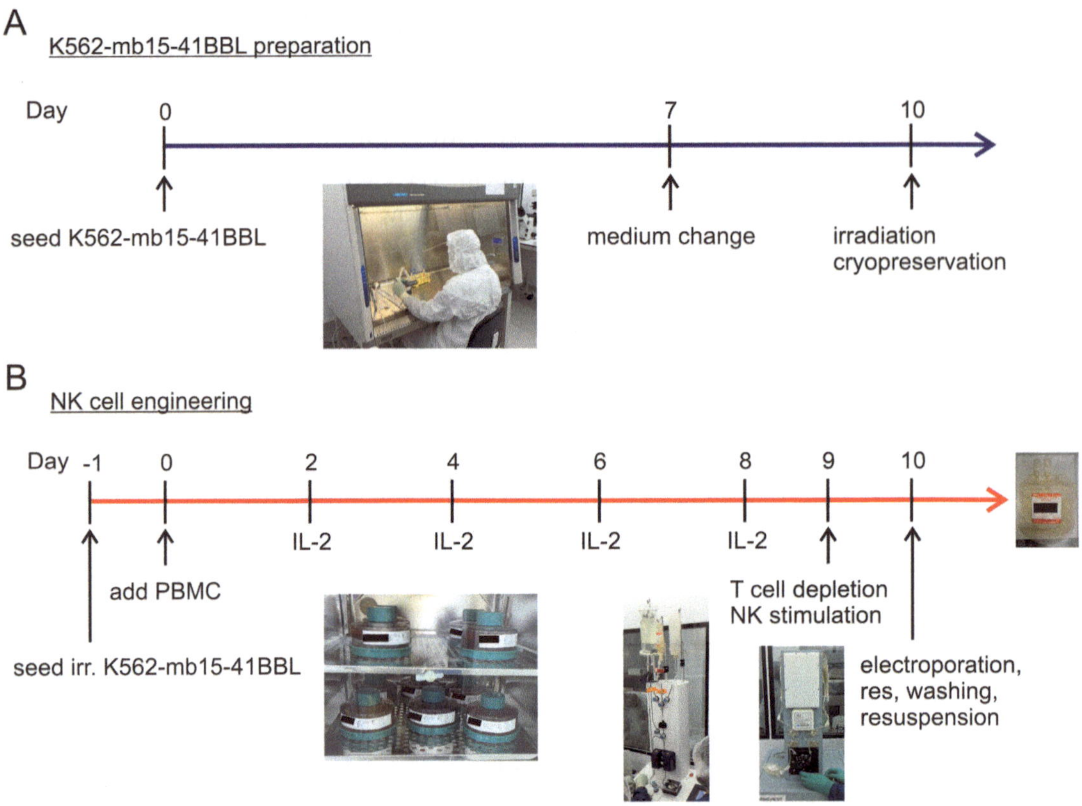

Fig. 2 Overview of cell processing. (a) Preparation of irradiated K562-mb15-41BBL. (**b**) NK cell engineering including expansion, T lymphocytes depletion, and electroporation of mRNA

8. Irradiated cells can be used directly for NK cell expansion without freezing. In our laboratory, we cryopreserve the irradiated cells and validate the cells before coculture with NK cells.
9. The cells are frozen in 10% DMSO. Another freezing buffer can also be used.
10. We routinely examine the cell line for sterility (bacteria, fungi, mycoplasma, and endotoxin), as well as the expression of IL-15 and 4-1BBL on the cell surface. After irradiation, we test the cell line for proliferation over 4 weeks. We also confirm the absence of cells which enter S phase by staining with 5-ethynyl-2-deoxyuridine (EdU) (Click-iT Assay, Thermo Fisher Scientific, Waltham, MA) and FxCycle Violet Stain (Thermo Fisher Scientific) [41].
11. Cooling rate is set at −1.0 °C per minute.
12. We start with a leukapheresis to obtain a sufficient number of PBMC for NK expansion.

13. PBMC can be isolated in a closed system using Sepax2 S-100 (GE Healthcare).
14. The leukapheresis product is diluted 2–3 times with saline containing human serum albumin before density gradient centrifugation.
15. We determine the percentage of NK cells in PBMC by staining cells with anti-CD45, anti-CD56, and anti-CD3 antibodies for flow cytometry.
16. Isolated PBMC can be used directly for NK cell expansion without freezing. We cryopreserve PBMC for multiple infusions of NK cells. NK cells can be expanded from cryopreserved PBMC as well as fresh PBMC (Fig. 1c).
17. The sterility (bacteria and fungi) of the cryopreserved cells is examined using Trypsin Soy Broth and Fluid Thioglycollate Medium.
18. Addition of FBS or plate lysate to SCGM media enhances NK cell expansion. NK cells can also be expanded in another medium such as RPMI with 10% FBS and NK MACS medium (Miltenyi Biotec).
19. A T-flask or a culture bag can be used instead of G-Rex100.
20. Irradiated K562-mb15-41BBL are placed in the G-Rex100 chambers either 1 h before coculture or later up to 24 h (Fig. 2b). This depends on the time of the arrival of the leukapheresis to the cGMP laboratory. Within this time range, we observed no significant difference in NK cell proliferation and their killing potency after culture.
21. Cryopreserved PBMC are washed in SCGM media twice before coculture with K562-mb15-41BBL.
22. The percentage of NK cells in PBMC is variable (1–25%). The number of PBMC containing 3×10^6 NK cells is typically between 1×10^7 and 3×10^8 cells.
23. NK cells are cocultured with irradiated K562-mb15-41BBL at 1:10.
24. A parallel coculture of NK cells and irradiated K562-mb15-41BBL, with one log less cells than the main expansion culture, is set as a control. With this sample, we monitor the proliferation of NK cells and the decrease of irradiated K562-mb15-411BBL using flow cytometry.
25. We also monitor glucose levels in culture supernatant from the G-Rex100 to determine the medium quality. When the glucose level is lower than 5.5 mM, half of the medium is exchanged. The half medium change is usually unnecessary in cultures prolonged for 10 days or less.
26. Sterility tests (bacteria, fungi, and mycoplasma) are performed using the cell culture supernatant 2–5 days prior to infusion.

27. We examine cell concentration, cell viability, and percentage of NK cells in the cell product. The presence of viable K562-mb15-41BBL cells is excluded using flow cytometry.
28. We examine the sterility (bacteria, fungi, mycoplasma, and endotoxin).
29. The cells' potency is assessed by an in vitro cytotoxicity assay. The expanded NK cells are tested against a cell line representative of the tumor in the protocol eligibility.
30. If the cell product is allogeneic to the patient, residual T lymphocytes are removed using the CliniMACS system to reduce the risk of graft-versus-host disease (Fig. 2b). In our experience, the percentage of T lymphocytes after this procedure reaches less than 0.05%. The number of T cells infused is consistently less than 5×10^4/kg [42].
31. After T lymphocytes depletion, the cells are washed and resuspended in saline containing human serum albumin for infusion as in Subheading 3.1.4. For genetic modification by electroporation of mRNA, proceed to Subheading 3.2.
32. Overview of electroporation processing is shown in Fig. 2b.
33. NK cells are seeded at $0.5–2 \times 10^9$ cells per G-Rex100.
34. Stimulation with a high dose of IL-2 (1000 IU/ml) increases the transfection efficacy by electroporation according to our experience.
35. Several GMP-grade electroporators are commercially available. Here, we describe a method of electroporation using MaxCyte GT system (MaxCyte).
36. Culture medium is warmed in preparation for cells after electroporation. The number of G-Rex100 and the volume of medium depend on the cell number for electroporation. We aim at a cell concentration after electroporation of 1×10^9 cells in 400 ml of SCGM medium per G-Rex100.
37. Resting NK cells or T cells can be electroporated using the same protocol. The transfection efficacy in expanded NK cells is higher than in resting NK cells [31].
38. To maximize cell viability, cells are centrifuged before electroporation at minimal speed and time.
39. An electroporation buffer compatible with the electroporator should be used. We use MaxCyte electroporation buffer for MaxCyte GT. For removal of culture medium, cells are washed in electroporation buffer at least once.
40. The recommended concentration of mRNA is 150–200 μg/ml. The mRNA for electroporation should have 5′ cap and 3′ polyadenylated tail.

41. Instead of mRNA, DNA such as a plasmid or dumbbell can be used for transfection. The transfection efficacy of DNA is lower than mRNA.
42. The voltage and capacitance of electroporation are critical parameters for high transfection efficacy. Electroporators typically have multiple program settings. The optimal setting should be selected.
43. Electroporation imposes stress on cells. For minimizing stress to the electroporated cells, the cells in CL-2 bag are placed directly in an incubator after electroporation.
44. We incubate electroporated NK cells for 4 h to allow for recovery before preparing cells for infusion. After incubation, cells are collected, washed, and resuspended in saline containing human serum albumin for infusion as per Subheading 3.1.4.

Acknowledgements

This work was supported by grant NMRC/STaR/0025/2015a from the National Medical Research Council of Singapore.

Conflict of Interest Statement: NS and DC are coinventors in patent applications describing some of the technologies used or related technologies. DC is scientific founder and stockholder of Nkarta Therapeutics, which holds the license for some of the technologies described.

References

1. Morvan MG, Lanier LL (2016) NK cells and cancer: you can teach innate cells new tricks. Nat Rev Cancer 16(1):7–19. https://doi.org/10.1038/nrc20155
2. Vivier E, Raulet DH, Moretta A, Caligiuri MA, Zitvogel L, Lanier LL, Yokoyama WM, Ugolini S (2011) Innate or adaptive immunity? The example of natural killer cells. Science 331(6013):44–49. https://doi.org/10.1126/science1198687
3. Guillerey C, Huntington ND, Smyth MJ (2016) Targeting natural killer cells in cancer immunotherapy. Nat Immunol 17(9):1025–1036. https://doi.org/10.1038/ni.3518
4. Martinet L, Smyth MJ (2015) Balancing natural killer cell activation through paired receptors. Nat Rev Immunol 15(4):243–254. https://doi.org/10.1038/nri3799
5. Shimasaki N, Coustan-Smith E, Kamiya T, Campana D (2016) Expanded and armed natural killer cells for cancer treatment. Cytotherapy 18(11):1422–1434. https://doi.org/10.1016/j.jcyt.2016.06.013
6. Weiner LM, Surana R, Wang S (2010) Monoclonal antibodies: versatile platforms for cancer immunotherapy. Nat Rev Immunol 10(5):317–327. https://doi.org/10.1038/nri2744
7. Gwalani LA, Orange JS (2018) Single Degranulations in NK cells can mediate target cell killing. J Immunol 200(9): 3231–3243. https://doi.org/10.4049/jimmunol.1701500
8. Wallin RP, Screpanti V, Michaelsson J, Grandien A, Ljunggren HG (2003) Regulation of perforin-independent NK cell-mediated cytotoxicity. Eur J Immunol 33(10):2727–2735. https://doi.org/10.1002/eji.200324070
9. Miller JS (2013) Therapeutic applications: natural killer cells in the clinic. Hematology Am Soc Hematol Educ Program 2013:247–253. https://doi.org/10.1182/asheducation-2013.1.247

10. Alici E, Sutlu T, Bjorkstrand B, Gilljam M, Stellan B, Nahi H, Quezada HC, Gahrton G, Ljunggren HG, Dilber MS (2008) Autologous antitumor activity by NK cells expanded from myeloma patients using GMP-compliant components. Blood 111(6):3155–3162. https://doi.org/10.1182/blood-2007-09-110312
11. Masuyama J, Murakami T, Iwamoto S, Fujita S (2016) Ex vivo expansion of natural killer cells from human peripheral blood mononuclear cells co-stimulated with anti-CD3 and anti-CD52 monoclonal antibodies. Cytotherapy 18(1):80–90. https://doi.org/10.1016/j.jcyt.2015.09.011
12. Romee R, Schneider SE, Leong JW, Chase JM, Keppel CR, Sullivan RP, Cooper MA, Fehniger TA (2012) Cytokine activation induces human memory-like NK cells. Blood 120(24):4751–4760. https://doi.org/10.1182/blood-2012-04-419283
13. Siegler U, Meyer-Monard S, Jorger S, Stern M, Tichelli A, Gratwohl A, Wodnar-Filipowicz A, Kalberer CP (2010) Good manufacturing practice-compliant cell sorting and large-scale expansion of single KIR-positive alloreactive human natural killer cells for multiple infusions to leukemia patients. Cytotherapy 12(6):750–763. https://doi.org/10.3109/14653241003786155
14. Berg M, Lundqvist A, McCoy P Jr, Samsel L, Fan Y, Tawab A, Childs R (2009) Clinical-grade ex vivo-expanded human natural killer cells up-regulate activating receptors and death receptor ligands and have enhanced cytolytic activity against tumor cells. Cytotherapy 11(3):341–355. https://doi.org/10.1080/14653240902807034
15. Ishikawa E, Tsuboi K, Saijo K, Harada H, Takano S, Nose T, Ohno T (2004) Autologous natural killer cell therapy for human recurrent malignant glioma. Anticancer Res 24(3b):1861–1871
16. Harada H, Watanabe S, Saijo K, Ishiwata I, Ohno T (2004) A Wilms tumor cell line, HFWT, can greatly stimulate proliferation of CD56+ human natural killer cells and their novel precursors in blood mononuclear cells. Exp Hematol 32(7):614–621. https://doi.org/10.1016/j.exphem.2004.03.011
17. Imai C, Iwamoto S, Campana D (2005) Genetic modification of primary natural killer cells overcomes inhibitory signals and induces specific killing of leukemic cells. Blood 106(1):376–383
18. Fujisaki H, Kakuda H, Shimasaki N, Imai C, Ma J, Lockey T, Eldridge P, Leung WH, Campana D (2009) Expansion of highly cytotoxic human natural killer cells for cancer cell therapy. Cancer Res 69(9):4010–4017
19. Lapteva N, Durett AG, Sun J, Rollins LA, Huye LL, Fang J, Dandekar V, Mei Z, Jackson K, Vera J, Ando J, Ngo MC, Coustan-Smith E, Campana D, Szmania S, Garg T, Moreno-Bost A, Vanrhee F, Gee AP, Rooney CM (2012) Large-scale ex vivo expansion and characterization of natural killer cells for clinical applications. Cytotherapy 14(9):1131–1143. https://doi.org/10.3109/14653249.2012.700767
20. Zhang H, Cui Y, Voong N, Sabatino M, Stroncek DF, Morisot S, Civin CI, Wayne AS, Levine BL, Mackall CL (2011) Activating signals dominate inhibitory signals in CD137L/IL-15 activated natural killer cells. J Immunother 34(2):187–195. https://doi.org/10.1097/CJI.0b013e31820d2a21
21. Denman CJ, Senyukov VV, Somanchi SS, Phatarpekar PV, Kopp LM, Johnson JL, Singh H, Hurton L, Maiti SN, Huls MH, Champlin RE, Cooper LJ, Lee DA (2012) Membrane-bound IL-21 promotes sustained ex vivo proliferation of human natural killer cells. PLoS One 7(1):e30264. https://doi.org/10.1371/journal.pone.0030264
22. Szmania S, Lapteva N, Garg T, Greenway A, Lingo J, Nair B, Stone K, Woods E, Khan J, Stivers J, Panozzo S, Campana D, Bellamy WT, Robbins M, Epstein J, Yaccoby S, Waheed S, Gee A, Cottler-Fox M, Rooney C, Barlogie B, van Rhee F (2015) Ex vivo-expanded natural killer cells demonstrate robust proliferation in vivo in high-risk relapsed multiple myeloma patients. J Immunother 38(1):24–36. https://doi.org/10.1097/cji.0000000000000059
23. Shah N, Martin-Antonio B, Yang H, Ku S, Lee DA, Cooper LJ, Decker WK, Li S, Robinson SN, Sekine T, Parmar S, Gribben J, Wang M, Rezvani K, Yvon E, Najjar A, Burks J, Kaur I, Champlin RE, Bollard CM, Shpall EJ (2013) Antigen presenting cell-mediated expansion of human umbilical cord blood yields log-scale expansion of natural killer cells with anti-myeloma activity. PLoS One 8(10):e76781. https://doi.org/10.1371/journal.pone.0076781
24. Knorr DA, Ni Z, Hermanson D, Hexum MK, Bendzick L, Cooper LJ, Lee DA, Kaufman DS (2013) Clinical-scale derivation of natural killer cells from human pluripotent stem cells for cancer therapy. Stem Cells Transl Med 2(4):274–283. https://doi.org/10.5966/sctm.2012-0084
25. Garg TK, Szmania SM, Khan JA, Hoering A, Malbrough PA, Moreno-Bost A, Greenway AD, Lingo JD, Li X, Yaccoby S, Suva LJ, Storrie B, Tricot G, Campana D, Shaughnessy JD Jr, Nair BP, Bellamy WT, Epstein J, Barlogie B, van Rhee F (2012) Highly activated and expanded natural killer cells for multiple myeloma immunotherapy. Haematologica

97(9):1348–1356. https://doi.org/10.3324/haematol.2011.056747

26. Cho FN, Chang TH, Shu CW, Ko MC, Liao SK, Wu KH, Yu MS, Lin SJ, Hong YC, Chen CH, Hung CH, Chang YH (2014) Enhanced cytotoxicity of natural killer cells following the acquisition of chimeric antigen receptors through trogocytosis. PLoS One 9(10):e109352. https://doi.org/10.1371/journal.pone.0109352
27. Kamiya T, Chang YH, Campana D (2016) Expanded and activated natural killer cells for immunotherapy of hepatocellular carcinoma. Cancer Immunol Res 4(7):574–581. https://doi.org/10.1158/2326-6066CIR-15-0229
28. Mimura K, Kamiya T, Shiraishi K, Kua LF, Shabbir A, So J, Yong WP, Suzuki Y, Yoshimoto Y, Nakano T, Fujii H, Campana D, Kono K (2014) Therapeutic potential of highly cytotoxic natural killer cells for gastric cancer. Int J Cancer 135(6):1390–1398. https://doi.org/10.1002/ijc.28780
29. Kruschinski A, Moosmann A, Poschke I, Norell H, Chmielewski M, Seliger B, Kiessling R, Blankenstein T, Abken H, Charo J (2008) Engineering antigen-specific primary human NK cells against HER-2 positive carcinomas. Proc Natl Acad Sci U S A 105(45):17481–17486. https://doi.org/10.1073/pnas.0804788105
30. Altvater B, Landmeier S, Pscherer S, Temme J, Schweer K, Kailayangiri S, Campana D, Juergens H, Pule M, Rossig C (2009) 2B4 (CD244) signaling by recombinant antigen-specific chimeric receptors costimulates natural killer cell activation to leukemia and neuroblastoma cells. Clin Cancer Res 15(15):4857–4866. https://doi.org/10.1158/1078-0432.ccr-08-2810
31. Shimasaki N, Fujisaki H, Cho D, Masselli M, Lockey T, Eldridge P, Leung W, Campana D (2012) A clinically adaptable method to enhance the cytotoxicity of natural killer cells against B-cell malignancies. Cytotherapy 14(7):830–840. https://doi.org/10.3109/14653249.2012.671519
32. Liu E, Tong Y, Dotti G, Shaim H, Savoldo B, Mukherjee M, Orange J, Wan X, Lu X, Reynolds A, Gagea M, Banerjee P, Cai R, Bdaiwi MH, Basar R, Muftuoglu M, Li L, Marin D, Wierda W, Keating M, Champlin R, Shpall E, Rezvani K (2018) Cord blood NK cells engineered to express IL-15 and a CD19-targeted CAR show long-term persistence and potent antitumor activity. Leukemia 32(2):520–531. https://doi.org/10.1038/leu.2017.226
33. Li Y, Hermanson DL, Moriarity BS, Kaufman DS (2018) Human iPSC-derived natural killer cells engineered with chimeric antigen receptors enhance anti-tumor activity. Cell Stem Cell 23(2):181–192. https://doi.org/10.1016/j.stem.2018.06.002
34. Chang YH, Connolly J, Shimasaki N, Mimura K, Kono K, Campana D (2013) A chimeric receptor with NKG2D specificity enhances natural killer cell activation and killing of tumor cells. Cancer Res 73(6):1777–1786. https://doi.org/10.1158/0008-5472CAN-12-3558
35. Imamura M, Shook D, Kamiya T, Shimasaki N, Chai SM, Coustan-Smith E, Imai C, Campana D (2014) Autonomous growth and increased cytotoxicity of natural killer cells expressing membrane-bound interleukin-15. Blood 124(7):1081–1088. https://doi.org/10.1182/blood-2014-02-556837
36. Shimasaki N, Campana D (2013) Natural killer cell reprogramming with chimeric immune receptors. Methods Mol Biol 969:203–220. https://doi.org/10.1007/978-1-62703-260-5_13
37. Li L, Liu LN, Feller S, Allen C, Shivakumar R, Fratantoni J, Wolfraim LA, Fujisaki H, Campana D, Chopas N, Dzekunov S, Peshwa M (2010) Expression of chimeric antigen receptors in natural killer cells with a regulatory-compliant non-viral method. Cancer Gene Ther 17(3):147–154. https://doi.org/10.1038/cgt200961
38. Regis S, Caliendo F, Dondero A, Casu B, Romano F, Loiacono F, Moretta A, Bottino C, Castriconi R (2017) TGF-beta1 Downregulates the expression of CX3CR1 by inducing miR-27a-5p in primary human NK cells. Front Immunol 8:868. https://doi.org/10.3389/fimmu.2017.00868
39. Somanchi SS, Somanchi A, Cooper LJ, Lee DA (2012) Engineering lymph node homing of ex vivo-expanded human natural killer cells via trogocytosis of the chemokine receptor CCR7. Blood 119(22):5164–5172. https://doi.org/10.1182/blood-2011-11-389924
40. Hank JA, Surfus J, Gan J, Albertini M, Lindstrom M, Schiller JH, Hotton KM, Khorsand M, Sondel PM (1999) Distinct clinical and laboratory activity of two recombinant interleukin-2 preparations. Clin Cancer Res 5(2):281–289
41. Hamelik RM, Krishan A (2009) Click-iT assay with improved DNA distribution histograms. Cytometry A 75(10):862–865. https://doi.org/10.1002/cytoa20780
42. Muller S, Schulz A, Reiss U, Schwarz K, Schreiner T, Wiesneth M, Debatin KM, Friedrich W (1999) Definition of a critical T cell threshold for prevention of GVHD after HLA non-identical PBPC transplantation in children. Bone Marrow Transplant 24(6):575–581. https://doi.org/10.1038/sj.bmt.1701970

Chapter 7

Dextran Enhances the Lentiviral Transduction Efficiency of Murine and Human Primary NK Cells

Arash Nanbakhsh and Subramaniam Malarkannan

Abstract

Recent advances in cancer immunotherapy emphasize the need for an efficient method to genetically modify effector lymphocytes to express exogenous "synthetic" genes. NK cells represent 10–20% of total lymphocytes in the peripheral blood of humans and play an essential role in clearing infections and malignant cells. A significant number of NK cells express and utilize non-clonotypic receptors that recognize cognate ligands expressed on a broad spectrum of target cells. Thus, NK cells can be utilized as potent immunotherapeutic tools with fewer limitations. Considerable amount of progress in improving effector functions through genetic manipulations has been centered around T cells. However, a similar technological and translational exploration on NK cells is lacking. One major constrain is the significantly low efficiency of lentiviral-mediated gene transductions into primary human or mouse NK cells. We found that dextran, a branched glucan polysaccharide, significantly improves the transduction efficiency of human and mouse primary NK cells. This highly reproducible methodology offers an approach that can help to improve gene delivery into NK cells and thereby cancer immunotherapy.

Key words Transduction, Primary NK cells, Dextran, Lentivirus, Genetically modified, Immunotherapy

1 Introduction

NK cells are a major lymphocyte subset of the innate immune system and form the first line of defense against bacterial and viral infections [1–3]. Importantly, NK cells do not require prior sensitization and produce multiple proinflammatory cytokines and chemokines [4]. By secreting these cytokines, NK cells are capable of activating and regulating the adoptive immune response primarily mediated by T and B cells [5]. Transient transduction is an essential step to genetically modify primary human NK cells for clinical utilization [6–9]. In addition, transient expression of genes in primary mouse or human NK cells is critical to define the development and effector functions through in vitro or in vivo murine experiments [10]. Transient transduction includes plasmid transfection EBV/retroviral

Samuel G. Katz and Peter M. Rabinovich (eds.), *Cell Reprogramming for Immunotherapy: Methods and Protocols*, Methods in Molecular Biology, vol. 2097, https://doi.org/10.1007/978-1-0716-0203-4_7,

hybrid vector, vaccinia vectors, and Ad5/F35 chimeric adenoviral vectors [11–13]. Although this technique has moderate success that could answer important questions in the NK cell biology field, the development of stable transduction methods with high ratio of efficiency is extremely crucial in chimeric antigen receptor-based adoptive immunotherapy applications. A difficulty in transduction of primary NK cells is that both lentiviruses and retroviruses use the host machinery to uptake viral particles and translocate the viral pre-integration complex into the nucleus, and this process is often limited in primary cells, including lymphocytes.

The uptake of the viral particles is the first major barrier in genetically modifying primary immune cells [14]. Specifically, the negative charge between host receptors and viral envelope proteins limits the interaction between the virus and host cell [15]. Cationic polymers such as polybrene (Pb), protamine sulfate (PS), or dextran could switch this negative charge to positive and increase the infusion between cell membrane and envelope proteins and consequently improve the fusion efficiency of the viral proteins to the cell membrane. Previous studies showed the efficiency of cationic polymers (Pb and PS) in T-cell transformation [16], but unfortunately, their efficacy to transduce NK cells is negligible. Here, we propose methods to use different cationic polymers and emphasize that dextran significantly enhances efficient gene transfer into mouse and human primary NK cells.

2 Materials

2.1 General Supplies/Reagents

1. 1× phosphate-buffered saline (1× PBS).
2. Fetal bovine serum (FBS).
3. Dimethyl sulfoxide (DMSO).
4. Isoflurane (or alternate institutional murine anesthetic).
5. Dextran (Sigma-Aldrich 90-64-91-9).
6. Polybrene (Sigma-Aldrich TR-1003).
7. Protamine sulfate (Sigma-Aldrich p3369).
8. NK cell negative selection kit (Stem Cell 19,855).
9. Sodium butyrate.
10. Ficoll (GE Life Science 17-1440-03).
11. Hank's Balanced Salt Solution (HBSS).
12. Annexin-V (PE)/7-AAD kit (BD).
13. Lentiviral preparations with known titers. Lentiviruses encoding genes of interest along with a GFP or other similar reporter genes.
14. Radiolabeled ^{51}chromium.

2.2 Human and Mouse Peripheral Blood Mononuclear Cells (PBMC)

1. C57BL/6 mice are obtained from commercial vendors. Maintain the mouse colonies in pathogen-free conditions, and use female and male mice between the ages of 6 and 12 weeks.
2. Obtain de-identified human peripheral blood mononuclear cell (PBMCs) from IRB (Institution Review Board)-approved sources.
3. To euthanize the animals, perform cervical dislocation after the induction of anesthesia as approved by IACUC protocols (*see* **Note 1**).
4. The complete medium for mouse NK cell culture includes RPMI1640 medium with 10% FBS, 100 U/mL penicillin, 100 μg/mL streptomycin, 1 mM sodium pyruvate, 5% of 7.5% sodium bicarbonate solution, and 0.001% β-mercaptoethanol.

2.3 Human and Mouse Antibodies

1. Anti-mouse CD3ε-eflour 450 (1/200).
2. Anti-mouse NK1.1-APC (1/200).
3. Anti-mouse IFN-γ-PECy770 (1/200).
4. Anti-human CD3ε-APC (1/200).
5. Anti-human CD56-FITC (1/200).
6. Anti-human IFN-γ-PECy770 (1/100).

2.4 Human and Mouse Target Cells

1. K562 (CCL-243) and YAC-1 (TIB-160) cells can be obtained from commercial vendors. Maintain these cell lines in RPMI 1640 medium containing 10% heat-inactivated FBS. Test these cell lines periodically to exclude the possibility of mycoplasma contamination.

3 Methods

3.1 Purification and Expansion of Murine Primary NK Cells

1. Murine primary NK cells can be purified using established methods [17]. Briefly, single-cell spleen suspensions are passed through nylon wool columns to deplete the populations adherent to the nylon wool, including B cells and macrophages.
2. Cell populations nonadherent to the nylon wool from the previous step are cultured with 1000 U/mL of recombinant interleukin (IL)-2 in complete medium.
3. On day 4, replace the complete medium in the flasks to remove the floating cells that contain most of the T and NKT cells. The majority of the cells in culture that are adherent to the tissue culture plate are NK cells. Add 20 mL of fresh complete medium and 1000 U/mL IL-2 to replenish the flasks.

4. Test the purity of the NK cell cultures on day 7 by flow cytometry using anti-CD3ε and anti-NK1.1 antibody. NK cells are characterized as CD3ε⁻ NK1.1⁺ population.
5. Perform experiments using preparations with more than 95% of CD3ε⁻NK1.1⁺ NK cell population. If purity of the NK cells are less than 95%, utilize either flow cytometry-based sorting or antibody-coated bead-based isolation kits.

3.2 Purification and Expansion of Human Primary NK Cells

1. Isolate lymphocytes from PBMCs using density gradient from the buffy coats of healthy human donors [17] (*see* **Note 2**).
2. Commercial antibody-based negative selection kits can be used to purify human primary NK cells following the manufacturer's protocol.
3. Isolated human NK cells are grown in 40 mL of complete medium (Subheading 2.2; **item 4**) with 200 U/mL of recombinant IL-2.
4. The purity of the isolated NK cell population should be analyzed by flow cytometry. Stain NK cells with 2 μg/mL anti-CD3ε and 2 μg/mL anti-CD56 antibodies at 4 °C for 20 min. After staining cells, wash twice with PBS and measure the purity of cell preparations by the absence of CD3ε and the presence of CD56 on cell surface by flow cytometry.

3.3 Transduction of Murine and Human Primary NK Cells with Lentivirus

1. Plate mouse or human primary NK cells in 24-well plates at 0.5×10^5 cells/mL in the presence of lentiviral vectors of interest and a control vector (with a GFP or mCherry reporter gene) preparations at 5, 10, and, 20 multiplicity of infection (MOI) in the presence of polybrene, (Pb, 8 μg/mL), protamine sulfate, (PS, 8 μg/mL), or dextran (8 μg/mL).
2. Spin the plates at 1000 × *g* for transduction (spinoculation) for 10 min (*see* **Note 3**).
3. Without decanting the supernatant, culture the cells overnight in a 37 °C incubator infused with 5.2% CO_2.
4. Spin the transduced cells the following day at 1000 × *g* for 60 min, and resuspend in 2 mL of complete medium in the presence of IL-2 (200 U/mL).
5. Determine the level of GFP (FITC channel) or mCherry (PE channel) in control samples to quantify the efficacy of viral transduction under different conditions (Pb, PS, or dextran).
6. Annexin-V (PE)/7-AAD kit (BD) can be used to determine the viability of transduced NK cells.
7. Functional status of transduced NK cells can be tested using radioactive ^{51}chromium (^{51}Cr)-release assays (*see* **Note 4**). Antitumor cytotoxicity of human or mouse primary NK cells can be quantified against K562 or YAC-1, respectively. Varied

effector-to-target-cell ratios should be employed to eliminate the suboptimal or above-maximal activation of NK cells.

8. Calculate specific tumor cell lysis by the amount of maximum, spontaneous, and experimental release of ^{51}Cr from the target cells (*see* **Note 5**).
9. To evaluate the functional status of the transduced mouse or human NK cells, use plate-bound mitogenic antibodies against activation receptors such as NKG2D, NCR1, or CD244. Coat 96-well Nunc ELISA plates with 2.5 μg/mL of mitogenic antibodies for overnight [17]. Wash the plates twice with PBS, and add 1×10^5 purified human or mouse NK cells and incubate for 18 h.
10. Collect culture supernatants using a multichannel pipette after 18 h post-activation to quantify cytokines including IFN-γ, GM-CSF, TNF-α, and chemokines CCL3, CCL4, CCL5, and CCL9 (*see* **Note 6**). Production of these cytokines and chemokines indicate the activation status of the genetically modified NK cells.

4 Notes

1. To produce the dislocation, push the hand restraining the head of the animal forward and push down while pulling backward with the hand holding the tail base. Verify the effectiveness of dislocation by feeling for a separation of cervical tissue.
2. Isolation of PBMCs by density gradient is realized by carefully layering 35 mL of half-diluted (with HBSS) buffy preparation over 15 mL of Ficoll-Paque density gradient in a 50 mL canonical tube. Centrifuge at 400 × *g* for 30 min at 20 °C. More importantly, do not use the brake in the centrifuge at the end of the run.
3. For spinoculation step, set the centrifuge at 1000 × *g* for 60 min at 37 °C.
4. One million target cells are incubated with 50 μCi of radiolabeled sodium chromate ^{51}Cr [17]. During the 1-h incubation time, cells uptake the ^{51}Cr which associates with cellular proteins, which helps to contain it intracellularly. At the end of the incubation, wash the cells to remove any unincorporated radioactivity.
5. The calculation of the percentage of specific lysis from pentaplicate experiments is done using the following equation: % specific lysis = (mean experimental release of ^{51}Cr – mean spontaneous release of ^{51}Cr)/(mean maximal release of ^{51}Cr – mean spontaneous release of ^{51}Cr) × 100, where "mean spontaneous

release of ^{51}Cr" is the ^{51}Cr released from target cells in the absence of NK cells and "mean maximal release of ^{51}Cr" is the ^{51}Cr released from target cells upon lysis by 2 N hydrochloric acid (HCl).

6. Cytokines and chemokines can be quantified by Enzyme-linked Immunosorbent Assay (ELISA) or using multiplex-assays.

Acknowledgements

Funding Support: We thank Lucia Sammarco and her Lulu's Lemonade Stand for inspiration, motivation, and support. This work was supported in part by Ann's Hope Melanoma Foundation (S.M. and M.S.T.); NIH R01 AI102893 (S.M.) and NCI R01 CA179363 (S.M. and M.S.T.); HRHM Program of MACC Fund/Children's Hospital of Wisconsin (S.M.), Nicholas Family Foundation (S.M.); Gardetto Family (S.M.); MCW-Cancer Center-Large Seed Grant (S.M. & M.S.T.); and MACC Fund/Children's Hospital of Wisconsin (M.S.T. and S.M.).

References

1. Vivier E, Tomasello E, Baratin M, Walzer T, Ugolini S (2008) Functions of natural killer cells. Nat Immunol 9(5):503–510
2. Zitvogel L, Terme M, Borg C, Trinchieri G (2006) Dendritic cell-NK cell cross-talk: regulation and physiopathology. Curr Top Microbiol Immunol 298:157–174
3. Abel AM, Yang C, Thakar MS, Malarkannan S (2018) Natural killer cells: development, maturation, and clinical utilization. Front Immunol 9:1869. https://doi.org/10.3389/fimmu.2018.01869
4. Raulet DH, Vance RE (2006) Self-tolerance of natural killer cells. Nat Rev Immunol 6(7):520–531
5. Schuster IS, Coudert JD, Andoniou CE, Degli-Esposti MA (2016) Natural regulators: NK cells as modulators of T cell immunity. Front Immunol 7:235. https://doi.org/10.3389/fimmu.2016.00235
6. Landoni E, Savoldo B (2018) Treating hematological malignancies with cell therapy: where are we now? Expert Opin Biol Ther 18(1):65–75. https://doi.org/10.1080/14712598.2018.1384810
7. Liu E, Tong Y, Dotti G, Shaim H, Savoldo B, Mukherjee M, Orange J, Wan X, Lu X, Reynolds A, Gagea M, Banerjee P, Cai R, Bdaiwi MH, Basar R, Muftuoglu M, Li L, Marin D, Wierda W, Keating M, Champlin R, Shpall E, Rezvani K (2018) Cord blood NK cells engineered to express IL-15 and a CD19-targeted CAR show long-term persistence and potent antitumor activity. Leukemia 32(2):520–531. https://doi.org/10.1038/leu.2017.226
8. Rezvani K, Rouce R, Liu E, Shpall E (2017) Engineering natural killer cells for cancer immunotherapy. Mol Ther 25(8):1769–1781. https://doi.org/10.1016/j.ymthe.2017.06.012
9. Zhang C, Oberoi P, Oelsner S, Waldmann A, Lindner A, Tonn T, Wels WS (2017) Chimeric antigen receptor-engineered NK-92 cells: an off-the-shelf cellular therapeutic for targeted elimination of cancer cells and induction of protective antitumor immunity. Front Immunol 8:533. https://doi.org/10.3389/fimmu.2017.00533
10. Regunathan J, Chen Y, Kutlesa S, Dai X, Bai L, Wen R, Wang D, Malarkannan S (2006) Differential and nonredundant roles of phospholipase Cgamma2 and phospholipase Cgamma1 in the terminal maturation of NK cells. J Immunol 177(8):5365–5376
11. Jiang K, Zhong B, Gilvary DL, Corliss BC, Vivier E, Hong-Geller E, Wei S, Djeu JY (2002) Syk regulation of phosphoinositide 3-kinase-dependent NK cell function. J Immunol 168(7):3155–3164

12. Maasho K, Marusina A, Reynolds NM, Coligan JE, Borrego F (2004) Efficient gene transfer into the human natural killer cell line, NKL, using the Amaxa nucleofection system. J Immunol Methods 284(1–2):133–140
13. Becknell B, Trotta R, Yu J, Ding W, Mao HC, Hughes T, Marburger T, Caligiuri MA (2005) Efficient infection of human natural killer cells with an EBV/retroviral hybrid vector. J Immunol Methods 296(1–2): 115–123. https://doi.org/10.1016/j.jim.2004.11.012
14. Simmons A, Alberola-Ila J (2016) Retroviral transduction of T cells and T cell precursors. Methods Mol Biol 1323:99–108. https://doi.org/10.1007/978-1-4939-2809-5_8
15. Froelich S, Tai A, Wang P (2010) Lentiviral vectors for immune cells targeting. Immunopharmacol Immunotoxicol 32(2):208–218. https://doi.org/10.3109/08923970903420582
16. Denning W, Das S, Guo S, Xu J, Kappes JC, Hel Z (2013) Optimization of the transductional efficiency of lentiviral vectors: effect of sera and polycations. Mol Biotechnol 53(3):308–314. https://doi.org/10.1007/s12033-012-9528-5
17. Rajasekaran K, Kumar P, Schuldt KM, Peterson EJ, Vanhaesebroeck B, Dixit V, Thakar MS, Malarkannan S (2013) Signaling by Fyn-ADAP via the Carma1-Bcl-10-MAP3K7 signalosome exclusively regulates inflammatory cytokine production in NK cells. Nat Immunol 14(11):1127–1136

Chapter 8

In Vivo Assessment of NK Cell-Mediated Cytotoxicity by Adoptively Transferred Splenocyte Rejection

Nathan J. Schloemer, Alex M. Abel, Monica S. Thakar, and Subramaniam Malarkannan

Abstract

NK cells are innate lymphocytes that are vital to clearance of virally infected or malignantly transformed cells. Assessment of the cytotoxic response is an important component of NK cell research and investigation of human disease. Standard assays of NK cell-mediated cytotoxicity of CD107a degranulation or ^{51}Cr release assay utilize cultured or freshly isolated NK cell populations in vitro. In addition to requirements to maintain multiple target cell lines and radioactivity precautions in the case of ^{51}Cr, these are in vitro evaluations of a complex in vivo function. Here, we describe the in vivo assessment of NK cell-mediated cytotoxicity through the adoptive transfer of splenocytes and their subsequent rejection. This protocol offers rapid, quantitative, and concurrent assessment of NK cell-mediated cytotoxicity against the prototypic NK stimulations of "*missing-self*" and "*nonself*."

Key words NK cell, Cytotoxicity, Transplant, Missing-self, Nonself, In vivo

1 Introduction

NK cells are innate immune system lymphocytes with germline-encoded activating and inhibitory receptors that allow a non-clonotypic response to cognate antigens and elicit effector functions including cell-mediated cytotoxicity and production of inflammatory cytokines [1]. NK cells are crucial to clearance of virally infected and malignant cells [2]. The cytolytic potential of NK cells is standardly assessed through a conventional in vitro cytotoxicity assay (^{51}chromium-release assay) or degranulation assay (CD107a/LAMP1) [3]. Degranulation assays provide in vitro response percentages to stimulations with cellular targets or matrix-bound mitogenic antibodies; however, they do not reflect actual cytotoxic effect of that response [4]. Cytotoxic assays such as ^{51}chromium-release assay (CRA) have long been utilized to describe the

Samuel G. Katz and Peter M. Rabinovich (eds.), *Cell Reprogramming for Immunotherapy: Methods and Protocols*, Methods in Molecular Biology, vol. 2097, https://doi.org/10.1007/978-1-0716-0203-4_8,

cytotoxic effect of immune cells [5]. The ^{51}chromium assay is considered the "gold standard" method of quantifying NK cells cytotoxicity [6]. The exposure to radioactivity and the considerations for radioactivity disposal have led to the development of multiple methods including lactate dehydrogenase (LDH) assay [7].

Direct cytolytic methods such as ^{51}chromium also rely on the maintenance of multiple tumor cell lines with potential antigenic expression alterations and are limited to in vitro evaluations. The evaluation of a complex cytotoxic process in the in vitro setting allows for narrow variable control and assessment of different components. Classic NK stimulations are "*self*" which represents the inhibitory interaction, "*missing-self*" or absence of "*self*" MHC presentation common as an immune evasion strategy in viral infections or malignant transformations, "*nonself*" or MHC mismatch, and "*induced-self*" where a stress response ligand such as H60 or Rae-1 can overcome inhibitory signals and induce NK cell response. The in vitro assessment of CRA can allow for assessments of different canonical NK cell activation methods through different cell lines but necessarily limits the complexity of the in vivo cytotoxic response.

Using an in vivo NK cell cytotoxic assessment with multiple concurrent adoptive splenocyte challenges allows quantification of multiple canonical NK cell activation methods in an immunocompetent setting. We describe a method with built-in controls to provide an in vivo, controlled, quantitative, and rapid NK cell-specific cytotoxicity. Control mice undergo NK cell depletion [8] or SHAM NK depletion to provide NK cell-specific experimental cytotoxicity ranges and allow quantification. The following day donor C57/BL6J (H-2^b, H2-K^b, and H2-D^b) "*self*," B6.β2^{mtm1Unc}/J (H-2^b but negative surface H2-K^b and K2-D^b) "*missing-self*," and BALB/c (H-2^d, H2-K^d and H2-D^d) "*nonself*" mice are procured and their donor splenocytes are labeled with CellTrace vital dyes. The uniquely stained donor cell populations are then combined into a 1:1:1 ratio (C57BL/6-CFSE, B6.β2^{mtm1Unc}/J-CFSE and CellTrace Violet (CTV)), BALB/cJ-CFSE, and CellTrace Red (CTR)) and injected into recipient control and experimental mice. The recipient mice spleens are collected 18 hours following injections, and NK cell-mediated rejection of the donor splenocytes is assessed by fluorescence-activated cell sorting (FACS). The CellTrace vital dyes applied to the donor splenocytes prior to adoptive transfer/transplant into the recipient mice allow for minimal "bench" time of samples following organ removal to FACS analysis. The resultant ratios of either "*self*":"*missing-self*" or "*self*":"*nonself*" in the control depleted mouse and the non-depleted experimental mice are entered into the equation: (1 − (Ratio Experimental /Ratio Depleted)) × 100%. This provides the in vivo, quantitative, rapid, and concurrent assessment of NK cell-mediated cytotoxicity against the prototypic NK cell stimulations of "*missing-self*" and "*nonself*."

2 Materials

2.1 General Supplies/Reagents

1. 1× phosphate-buffered saline (1× PBS).
2. 1% paraformaldehyde (PFA).
3. Fluorescence-activated cell sorting (FACS) buffer (1× PBS, 0.5–1% bovine serum albumin, 0.1% NaN_3 sodium azide).
4. 1 mL insulin syringes.
5. Fetal bovine serum (FBS).
6. Dimethyl sulfoxide (DMSO).
7. 70 μM cellular filters.
8. Lancets for facial vein bleed.
9. Isoflurane (or alternate institutional murine anesthetic).

2.2 Mice

1. All mice were maintained in pathogen-free conditions. Described includes mice C57BL/6, B6.β2$^{mtml Unc}$/J, and BALB/c all of which can be obtained from Jackson Laboratory (Bar Harbor, ME).
2. Target mutation/treated/tested mice: B6 background essential for inclusion of the B6 background B6.β2$^{mtml Unc}$/J but this can be adapted to include "nonself" cytotoxic rejection with any mix of backgrounds.
3. C57BL6J: Control or "*self*" mice for adoptive transfer (H-2^b, H2-K^b, and H2-D^b).
4. B6.β2$^{mtml Unc}$/J: "*Missing-self*" for adoptive transfer (H-2^b but negative surface H2-K^b and K2-D^b).
5. BALB/c: "*Non-self*" for adoptive transfer (H-2^d, H2-K^d, and H2-D^d).

2.3 NK Cell Depletion Antibodies

1. NK cell depletion antibody: anti-nk1.1 clone: PK136—200 μg of antibody per control mouse. Working concentration of 1 mg/mL in 200 μL aliquot per mouse to be administered intraperitoneal (I.P.) route in phosphate-buffered saline (PBS) (*see* **Note 1** for example calculations).
2. NK cell depletion isotype control: IgG2a clone (*see* **Note 2**). Working concentration of 1 mg/mL in 200 μL aliquot per mouse to be administered intraperitoneal (I.P.) route in phosphate-buffered saline (PBS).

2.4 Cell Labeling Dyes (*See* Note 3)

1. Stock solution: 5 mM CFSE in DMSO.
2. Stock solution: 5 mM CellTrace Violet (CTV) in DMSO.
3. Stock solution: 1 mM CellTrace Red (CTR) in DMSO.
4. Working solution A: 2× CFSE. Add 3 μL of CFSE stock to 6 mL pre-warmed PBS. Make this first and then add the CTV

and CTR to aliquots from solution A to ensure equal CFSE (*see* **Note 4**).

5. Working solution B: 2× CFSE + CTV. Add 1 μL CTV to 2 mL solution A.
6. Working solution C: 2× CFSE + CTR. Add 1 μL CTR to 2 mL solution A.
7. Working solution D: 2× CTV. If needed for flow cytometry compensation controls, add 1 μL CTV to 2 mL pre-warmed PBS.
8. Working solution E: 2× CTR. If needed for flow cytometry compensation controls, add 1 μL CTR to 2 mL pre-warmed PBS.

2.5 Antibodies (Optional, *See* Note 5)

1. CD3ε-eflour 450 (1/200).
2. NK1.1-APC (1/200).
3. CD69-PE (1/200).
4. NCR1-PerCPe710 (1/200).
5. IFN-γ-PECy770 (1/200).

3 Methods

3.1 Antibody-Mediated NK Cell Depletion Procedure for Controls (Day 1)

1. 24 h before injecting the splenocytes (day 1), inject experimental control mice (normally C57BL/6 or WT) with NK cell depletion OR isotype control antibodies.
2. Each mouse will be injected via intraperitoneal (I.P.) route with 200 μg of antibody diluted in 200 μL PBS.
3. Prepare antibodies per materials (*see* **Note 1**). Scale up for multiple mice/group.
4. Inject mice as soon as possible after diluting the antibody, and proceed with the experiment after 24 h.

3.2 Splenocyte Labeling and Injection Procedure (Day 0)

1. Warm 50 mL 1× PBS and 100% heat-inactivated FBS (5–10 mL depending on number of mice) in a 37 °C water bath.
2. Warm the dyes (CFSE, CTV, CTR) by incubating them at 37 °C for 10 min on a shaker.
3. Collect spleens from each donor mouse (C57/BL/6, B6.β2^{mtm1Unc}/J, and BALB/c) and homogenize into single-cell suspension in PBS and count (*see* **Note 6**).
4. Once spleens from each donor are collected, spin at 420 × *g* for 2 min, decant supernatant, and resuspend in PBS. Keep them in PBS until after the loading procedure.
5. After the PBS wash, immediately before loading, aspirate the supernatant with a vacuum line.

3.3 Splenocyte Loading Procedure

1. Resuspend splenocytes in pre-warmed PBS at a concentration of 4×10^7/mL with a minimum volume of 500 μL (*see* **Note 7**).
2. Mix the 2× dye solutions with the cells in a 1:1 ratio (500 μL cells +500 μL dye) and vortex briefly to mix. Add equal volume of dye solution A to C57B6/J splenocytes, dye solution B to BALB/cJ splenocytes, and dye solution C to B6.β2^{mtm1Unc}/J splenocytes. Use leftover C57/B6J or BALB/cJ or B6.β2^{mtm1Unc}/J splenocytes remaining for single dye controls solutions D and E, respectively.
3. Vortex briefly to mix and incubate at 37 °C for 10 min. Briefly vortex cells every 2 min during the 10 min incubation.
4. To inactivate the unbound dye, after the 10 min incubation, add at least an equal volume of pre-warmed 100% FBS to each tube, briefly vortex to mix, and let sit for 1 min at room temperature.
5. Rinse: Centrifuge at 420 × *g* for 2 min, aspirate supernatant, and resuspend in plain RPMI media.
6. Centrifuge at 420 × *g* for 2 min, aspirate supernatant, and resuspend in plain RPMI media, volume dependent on method of counting.
7. Count cells and make a 1:1:1 C57BL/6/BALB/c/B6.β2^{mtm1Unc}/J cell mixture from cells dyed in solution A, solution B, and solution C. Cells dyed in solutions D and E remain alone. Keep an aliquot of cells dyed in solution A alone and do not mix with other dyes.
8. Centrifuge at 420 × *g* for 2 min, aspirate supernatant, and resuspend 1:1:1 cell mix at ~37.5×10^7/mL in plain RPMI media.
9. Load 200 μL/syringe for intravenous retro-orbital injection of a total of 7.5×10^6 cells (2.5×10^6 C57BL/6 + 2.5×10^6 BALB/c + 2.5×10^6 B6.β2^{mtm1Unc}/J).
10. Fix a small aliquot of the 1:1:1 cell mixture in 1% paraformaldehyde for subsequent validation of ratio by FACS. May perform FACS analysis on this ratio now if desired without fixation.
11. Fix a small aliquot of the single stains of each CFSE, CTV, and CTR for subsequent compensation controls by FACS.
12. Anesthetize single mouse per institutional protocol for intravenous injection (*see* **Note 8**).
13. Inject 200 μL intravenously per institutional AUA-approved protocol.
14. Ensure recovery of mouse prior to anesthetic administration for next mouse.
15. Repeat for all experimental mice and controls of NK depleted and SHAM antibody injection.

3.4 Mouse Euthanasia, Splenocyte Collection, and Donor Splenocyte Evaluation (Day 1)

1. Procurement of mice: 18 h after splenocyte injections.
2. OPTIONAL: Perform cheek stick of facial vein to collect blood for cytokine measurement (*see* **Note 9**).
3. Euthanize the mice by the institutional AUA-approved method (i.e., CO_2 asphyxiation followed by cervical dislocation).
4. Collect spleens of each mouse and homogenize into a single-cell suspension in PBS, filter (70 μM), and count the number of white blood cells.
5. Cytotoxic assessment: FACS Assessment: Take 5×10^6 cells/spleen sample, filter into 5 mL tube, and resuspend in a final volume of 2 mL with FACS buffer (*see* **Note 10**). Use fixed single dye cells in solutions A, D, and E as compensation controls if required. If not completed prior, confirm preinjection 1:1:1 ratio with fixed sample from day 0. Collect 1×10^6 events per experimental sample on the cytometer (*see* **Note 11**).
6. OPTIONAL: Cytotoxic assessment of adoptively transferred splenocytes in peripheral blood by FACS (*see* **Note 12**).

3.5 Recipient NK Cell Depletion Confirmation (Optional, *See* Note 5)

1. Surface stain splenocytes from each spleen to determine % of NK cells to validate the IgG (SHAM) and PK136 injections and presence of splenic NK cells in experimental mice.
2. Since PK136 depletes by binding NKR-P1C (NK1.1), identify NK cells in the mice using an independent marker such as NCR1 or DX5. For example, use CD3-efluor 450 (1/200) + NCR1-PerCPe710 (1/200) and negative gate FITC (CFSE) to exclude all NK cells from donor spleens. Make single stain compensation controls as necessary for utilized fluorophores.

3.6 Recipient Splenic NK Cell Activation Assessment (Optional, *See* Note 5)

1. CD69: Surface stain splenocytes to determine recipient NK cell activation with CD69. Follow standard extracellular staining protocol. Example fluorophores: CD3-eflour 450 (1/200), NK1.1-APC (1/200), and CD69-PE (1/200) for surface stain. Will need to negative gate FITC (CFSE) to exclude all NK cells from donor spleens.
2. IFN-γ: Intracellular stain splenocytes to determine recipient NK cells inflammatory IFN-γ. Follow intracellular staining protocol. Example fluorophores: Surface: CD3-eflour 450 (1/200) and NK1.1-APC (1/200). Intracellular: IFN-gamma-PECy770 (1/200). Will need to negative gate FITC (CFSE) to exclude all NK cells from donor spleens (*see* **Note 13**).

3.7 Data Analysis

1. Cytotoxicity calculation: Calculate the ratio of C57BL/6: BALB/c: B6.$\beta 2^{\mathrm{mtml Unc}}$/J for each mouse. The percent cytotoxicity of the experimental mice can be calculated based on the ratio of the depleted mouse with the following calculation: $(1 - (\mathrm{Ratio}_{\mathrm{Experimental}}/\mathrm{Ratio}_{\mathrm{PK136\ Depleted}})) \times 100\%$.

2. Optional calculations include the NK depletion efficiency (*see* **Note 14**), the NK activation with CD69 and IFN-gamma, and the serum systemic cytokine response.

4 Notes

1. Example calculations.
   - Stock concentration: *X* mg/mL.
   - Goal working concentration: 1 mg/mL in a final volume of 200 μL (0.2 mL) in PBS.
   - Stock concentration volume: *Y* mL.
   - Dilution volume PBS: *Z* mL.
   - Calculation: *X* mg/mL × *Y* = 1 mg/mL × 0.2 mL.
     - *Y* = (0.2/*X*) = Volume of stock concentration.
     - *Z* = 0.2 mL − *Y* = Volume of PBS to add to stock Ab concentration.
     - Scale up to provide adequate antibody for each mouse.
2. We have used IgG2a clone: C1.18.4.
3. Prepare 3 color viability dyes per manufacture recommendations that your current flow cytometer can discriminate. We have utilized CFSE, CTV, and CTR which have like preparation and staining and are easily discernable by flow cytometry.
4. Generate the 2× CFSE first, and then remove aliquots and add CTV and CTR rather than separately dilute CFSE for each of solution B and solution C. Due to large dilution factors, the error within pipetting may cause the CFSE staining to be of different intensity within cell groups. If desire to test additional cytotoxic challenges beyond the 3 described in this experiment, may utilize this effect to differentiate additional populations adoptively transferred.
5. These antibodies are optional if the investigator chooses to confirm depletion or activation of the recipient NK cells.
6. Assume 30–60 × 10^6 splenocytes per donor mouse. With 2.5 × 10^6 to 5 × 10^6 transferred per recipient mouse and 10% loss with preparation and dye loading, anticipate ~5–10 recipient mouse per donor. Donor splenocytes to include are (a) C57BL/6 (H-2^b, H2-K^b, and H2-D^b) as "*self*" dyed with CFSE, (b) B6.β$2^{mtm1Unc}$/J (H-2^b but negative surface H2-K^b and K2-D^b) as "*missing-self*" dyed with CFSE and CTR, and (c) BALB/c (H-2^d, H2-K^d, and H2-D^d) as "*nonself*" dyed with CFSE and CTV.

7. Manufacturer recommendation to have minimum volume 500 μL but we have utilized lower volumes without complications.
8. We have utilized isoflurane for anesthesia and a retro-orbital injection.
9. Expect a minimal alteration of cytokines with a low cell load. If interest in systemic cytokine alterations is primary focus, may require increasing the number of adoptively transferred cells. Additionally, if a larger volume of blood is required, may euthanize mouse first and then perform alternate blood collection.
10. We have utilized this dilution for an appropriate run time for our cytometer. Alternate dilutions may be required for appropriate event rate on different cytometers. Goal of minimization "bench" time for samples.
11. Goal of minimization of bench time, if event rate is 5–10,000 events/second, would anticipate a run time of approximately 2–4 min/sample.
12. We have previously utilized RBC lysis of facial vein blood draw and subsequent FACS analysis of donor splenocyte ratios. These ratios have mirrored the spleen data. However, the "bench time" is slightly increased with an RBC lysis step added as opposed to the direct FACS analysis following splenectomy, homogenization, counting, and resuspension. We therefore have elected to utilize spleen assessment primarily.
13. Systemic cytokine response can also be performed. We utilize IFN-gamma ELISA on the serum from the pre-euthanasia bleed.
14. % NK in depleted mice is normally around 0.2.

Acknowledgements

Funding Support: We thank Lucia Sammarco and her Lulu's Lemonade Stand for inspiration, motivation, and support. This work was supported in part by Ann's Hope Melanoma Foundation (S.M. and M.S.T.); NIH R01 AI102893 (S.M.) and NCI R01 CA179363 (S.M. and M.S.T.); Alex Lemonade Stand Foundation (N.J.S); HRHM Program of MACC Fund/Children's Hospital of Wisconsin (S.M.), Nicholas Family Foundation (S.M.); Gardetto Family (S.M.); MCW-Cancer Center-Large Seed Grant (S.M. & M.S.T.); and MACC Fund/Children's Hospital of Wisconsin (M.S.T. and S.M.).

References

1. Abel AM, Yang C, Thakar MS, Malarkannan S (2018) Natural killer cells: development, maturation, and clinical utilization. Front Immunol 9:1869. https://doi.org/10.3389/fimmu.2018.01869
2. Vivier E, Tomasello E, Baratin M, Walzer T, Ugolini S (2008) Functions of natural killer cells. Nat Immunol 9(5):503–510. https://doi.org/10.1038/ni1582
3. Rajasekaran K, Kumar P, Schuldt KM, Peterson EJ, Vanhaesebroeck B, Dixit V, Thakar MS, Malarkannan S (2013) Signaling by Fyn-ADAP via the Carma1-Bcl-10-MAP3K7 signalosome exclusively regulates inflammatory cytokine production in NK cells. Nat Immunol 14(11):1127–1136. https://doi.org/10.1038/ni.2708
4. Alter G, Malenfant JM, Altfeld M (2004) CD107a as a functional marker for the identification of natural killer cell activity. J Immunol Methods 294(1-2):15–22. https://doi.org/10.1016/j.jim.2004.08.008
5. Brunner KT, Mauel J, Cerottini JC, Chapuis B (1968) Quantitative assay of the lytic action of immune lymphoid cells on ^{51}-Cr-labelled allogeneic target cells in vitro; inhibition by isoantibody and by drugs. Immunology 14(2):181–196
6. Hsu HT, Mace EM, Carisey AF, Viswanath DI, Christakou AE, Wiklund M, Onfelt B, Orange JS (2016) NK cells converge lytic granules to promote cytotoxicity and prevent bystander killing. J Cell Biol 215(6):875–889. https://doi.org/10.1083/jcb.201604136
7. Korzeniewski C, Callewaert DM (1983) An enzyme-release assay for natural cytotoxicity. J Immunol Methods 64(3):313–320
8. Wang M, Ellison CA, Gartner JG, HayGlass KT (1998) Natural killer cell depletion fails to influence initial CD4 T cell commitment in vivo in exogenous antigen-stimulated cytokine and antibody responses. J Immunol 160(3):1098–1105

Chapter 9

Immunomodulation of NK Cell Activity

Carolina I. Domaica, Jessica M. Sierra, Norberto W. Zwirner, and Mercedes B. Fuertes

Abstract

Natural killer (NK) cells can kill virus-infected cells and tumor cells without prior sensitization and secrete numerous cytokines and chemokines that modulate the activity of different cells of the immune system. The recognition of target cells is mediated by germ line-encoded receptors, and the activity of NK cells can be further regulated by soluble factors such as cytokines and Toll-like receptor ligands. Thus, NK cells display an exciting potential as a powerful immunotherapeutic tool against malignant diseases, and different strategies are being tested aiming to overcome tumor-induced NK cell suppression and restore NK-cell mediated antitumor activity. This section describes different flow cytometry-based protocols to study NK cell effector functions, which can be used to evaluate the immunomodulatory ability of different therapeutic compounds.

Key words Natural killer cells, Flow cytometry, Cytotoxicity, CD107a, Degranulation, Cytokines, IFN-γ, ADCC (antibody-dependent cellular cytotoxicity)

1 Introduction

Natural killer (NK) cells represent the cytotoxic lineage of innate lymphoid cells (ILCs) [1] and play an important role in the immune response against intracellular pathogens and tumors [2]. NK cells have the ability to kill susceptible target cells without prior sensitization and to secrete numerous cytokines and chemokines that modulate the activity of different cells of the immune system [2–4]. The activity of NK cells is regulated by the balance of signals transduced by germ line-encoded activating receptors (recognizing ligands induced on tumor cells or virus-infected cells) and inhibitory receptors (which predominantly recognize MHC class I molecules) and can be further modulated by soluble factors such as cytokines and Toll-like receptor ligands [5–8]. Human NK cells are usually defined as $CD3^{-}CD56^{+}$ cells and can be broadly divided into two phenotypically and functionally different subsets based

Samuel G. Katz and Peter M. Rabinovich (eds.), *Cell Reprogramming for Immunotherapy: Methods and Protocols*, Methods in Molecular Biology, vol. 2097, https://doi.org/10.1007/978-1-0716-0203-4_9,

upon the expression of CD56 and CD16 [9–11]. The majority of peripheral blood NK cells are highly cytotoxic and express low levels of CD56 (CD56^{dim}) and high levels of CD16 (the low-affinity receptor for IgG, Fcγ receptor IIIA), allowing them to mediate strong antibody-dependent cellular cytotoxicity (ADCC) against IgG-coated target cells. The other subset expresses high levels of CD56 (CD56^{bright}) and low levels of CD16, is less cytotoxic, accounts for 10% of circulating NK cells (but most NK cells in lymphoid tissues), and secretes high levels of cytokines in response to activating soluble factors such as cytokines (IL-1β, IL-2, IL-12, IL-15, IL-18, and/or IL-23) [7, 9] and Toll-like receptor ligands (ligands for TLR2, TLR3, TLR5, TLR7/8, and/or TLR9) [8]. In addition, recent studies have expanded the knowledge about the phenotype and function of NK cells, and a higher diversity in NK cell populations than traditionally appreciated is becoming evident [12, 13]. The stochastic and variegated expression pattern of the different receptors in each individual NK cell allows for a great phenotypic and functional heterogeneity within the human NK cell population. Accordingly, recent studies estimate that in any healthy adult, there may be between 6000 and 30,000 distinct NK cell subsets [14]. This combinatorial expression of NK cell receptors is influenced not only by genetic but also by environmental factors. It has been shown that an NK cell population with memory-like properties is expanded after HCMV infection. This population of self-renewing, long-lived "adaptive" NK cells express high levels of NKG2C and CD57 and shows enhanced IFN-γ production and cytotoxic capacity upon a secondary challenge [15]. Additionally, different tissue resident NK cell populations have been described in uterus, kidney, gut, bone marrow, spleen, lung, and liver, with diverse functions in reproduction and tissue remodeling [13, 16].

Although NK cells play a key role in immunosurveillance against tumor formation, the tolerogenic tumor microenvironment found in advanced cancers usually induces NK cell dysfunction. In this context, rational manipulation of NK cell activity aiming to overcome tumor-induced NK cell suppression and to harness NK cell-mediated tumor elimination becomes an exciting research field [17, 18]. Accordingly, different compounds, including activating cytokines, blocking antibodies targeting inhibitory receptors, activating antibodies targeting activating receptors, and small molecules, are being tested for their ability to modulate NK cell activity with the goal of developing an effective tool in cancer immunotherapy [19]. Moreover, therapeutic efficacy of many humanized monoclonal antibodies (mAbs) against tumor cells relies partially on ADCC [20, 21].

In this chapter, we present different flow cytometry-based assays to monitor the modulation of NK cell activity mediated by any of such therapeutic compounds. The study of NK cell activity is focused on the evaluation of their cytotoxic capacity; their ability to secrete its signature cytokine, IFN-γ (among others like TNF, GM-CSF, and chemokines like CCL3-5); and the expression of activating receptors (such as CD25 and CD69). Flow cytometry shows several advantages over other detection methods, and importantly, given the diversity of NK cells [12, 13], it allows for discrimination of different NK cell subsets.

2 Materials

2.1 General Reagents

1. Complete medium: RPMI-1640 medium, 10% heat-inactivated fetal bovine serum (FBS), 2 mM L-glutamine, 0.5 mM pyruvate, 40 μg/mL gentamicin.
2. 96 U-well and 96 V-well culture plates.
3. 1× phosphate-buffered saline (PBS).
4. 1× PBS with 2% FBS.
5. Staining buffer: 1× PBS, 1% FBS, 0.1% sodium azide.
6. Fixation buffer: 1× PBS, 1% (w/v) paraformaldehyde.
7. Fixation solution: 4× FOXP3 Fix/Perm buffer (BioLegend), diluted to 1× in PBS.
8. Permeabilization solution: 10× FOXP3 Perm buffer (BioLegend), diluted to 1× in PBS.
9. Blocking solution: Staining buffer, 10% normal mouse serum (*see* **Note 1**).
10. Monensin Solution (2 mM, 1000×) or GolgiStop (BD): Final concentration 2 μM diluted in complete medium (*see* **Note 2**).
11. Brefeldin A Solution (5 mg/mL, 1000×) or GolgiPlug (BD): Final concentration 5 μg/mL, diluted in complete medium (*see* **Note 3**).
12. Recombinant human IL-12 p70 (PeproTech), final concentration 10 ng/mL diluted in complete medium.
13. Recombinant human IL-15 (PeproTech), final concentration 1 ng/mL diluted in complete medium.
14. Recombinant human IL-18 (MBL), final concentration 10 ng/mL diluted in complete medium.

2.2 NK Cell Isolation and Cell Lines

1. Whole blood collected in sodium heparin-treated syringes (20–40 mL is enough for multiple functional experiments).
2. NK cell-negative isolation kit (RosetteSep™ Human NK Cell Enrichment Cocktail, STEMCELL Technologies).
3. Ficoll-Paque™ Plus (GE Healthcare).
4. Target cells: The human erythroleukemia cell line K562 [American Tissue Type Collection (ATCC), No. CCL-243], maintained in complete medium.

2.3 Antibodies and Fluorescent Reagents

1. CFSE (5-(and 6)-carboxyfluorescein diacetate succinimidyl ester or CFDA SE, Molecular Probes).
2. Cell Proliferation Dye eFluor™ 670 (eBioscience).
3. Fixable Viability Zombie Aqua™ Dye (BioLegend).
4. Fluorochrome-conjugated anti-CD3 mAb (Clone UCHT1).
5. Fluorochrome-conjugated anti-CD56 mAb (Clone N901).
6. Fluorochrome-conjugated anti-IFN-γ mAb (Clone 4S.B3).
7. Fluorochrome-conjugated anti-CD107a mAb (Clone H4A3).
8. Fluorochrome-conjugated anti-CD16 mAb (Clone 3G8).
9. Fluorochrome-conjugated anti-CD25 mAb (Clone BC96).
10. Fluorochrome-conjugated anti-CD69 mAb (Clone FN50).
11. Fluorochrome-conjugated mouse IgG1, κ (isotype control) (Clone MOPC-21).

3 Methods

In this section, we describe methods for in vitro flow cytometry-based phenotypic and functional analysis of NK cell responses after stimulation with any immunotherapeutic compound under study. First, we propose to isolate NK cells to ensure direct activity of the compound on NK cells. We describe the steps for *intracellular staining of IFN-γ*, a cytokine that mediates many of NK cell functions. Nevertheless, this general protocol can be used to measure other cytokines or chemokines of interest.

For years, the chromium-release assay used to be the gold standard method for *cytotoxicity*, but recently, other nonradioactive cytometric assays were developed that have proved to be as sensitive [22]. We describe one of such assays where target cells are labeled with a fluorescent dye, like CFSE, and following incubation with NK cells, dead cells are stained with a viability dye, allowing to asses cytotoxicity at the single-cell level.

During NK cell-mediated cytotoxicity, cytoplasmatic lytic granules containing perforin and granzyme are released in a polarized fashion toward the target cell in a process termed degranulation. These granules fuse to the plasma membrane of the effector cell, resulting in the expression of CD107a (lysosomal-associated membrane protein 1, LAMP-1) at the surface of NK cells. While the cytotoxicity assay provides information about the lysis of target cells, *CD107a staining* offers information about the activity of the effector population. Labeling responding cells to evaluate expression of CD107a by flow cytometry directly identifies degranulating NK cells [23] and can be used with other markers of NK cell function and/or markers that allow discrimination of different populations of NK cells.

3.1 NK Cell Isolation

Work in a laminar flow hood to prevent contamination.

1. Use a 20 mL syringe with a drop of heparin or other anticoagulant to obtain 10 mL of fresh blood. Transfer the blood into a 50 mL tube (*see* **Note 4**).
2. Add 50 μL of RosetteSep™ Human NK Cell Enrichment Cocktail for each mL of blood. Mix gently and incubate 20 min (min) at room temperature.
3. Dilute blood with one volume of PBS supplemented with 2% FBS and mix gently.
4. Add 10 mL of Ficoll-Paque Plus to a new 50 mL tube, and carefully layer the diluted blood on top of it. Centrifuge at 800 × *g* with the brake off for 20 min.
5. NK cells will be localized between the Ficoll layer (bottom) and the plasma layer (top) as a white ring. Using a pipette, aspirate and discard most of the upper plasma layer. Then collect the NK cell ring and transfer it into a 15 mL tube.
6. Wash the NK cells twice with PBS, centrifuge at 600 × *g* for 5 min, and resuspend the cells in 1 mL pre-warmed complete medium.
7. Count the number of NK cells and adjust to the desired concentration in complete medium.
8. Evaluate NK cell purity as percentage of $CD56^{+}CD3^{-}$ by flow cytometry following Subheading 3.3. The purity of NK cells achieved using this method is usually higher than 90%.

3.2 NK Cell Pre-Stimulation

1. Resuspend isolated NK cells at 2 × 10^6 cells/mL in complete medium, and plate 100 μL/well into 96 U-well plates (200,000 NK cells/well).
2. Prepare dilutions of the different stimuli (the compound/s being tested, and a positive control; *see* **Note 5**). Add 100 μL to the wells previously seeded with NK cells to reach a final volume of 200 μL/well. As a negative control, leave NK cells unstimulated by adding 100 μL of complete medium without stimulus.
3. Place the cells in an incubator at 37 °C for 16 h (*see* **Note 6**).

3.3 NK Cell Phenotype: Cell Surface Staining

1. Pre-stimulate NK cells following Subheading 3.2.
2. Transfer the cells from the 96 U-well plates to 96 V-well plates for staining.
3. Centrifuge at 600 × *g* for 5 min and discard the supernatant.
4. Wash once with 150 μL of staining buffer.
5. Resuspend the pellet in 10 μL of blocking solution, and incubate for 10–20 min at room temperature.

6. Prepare an antibody mix with the appropriate amount of anti-CD56 and anti-CD3 mAbs in staining buffer.
7. Add 10 μL of antibody mix to the resuspended cells, and incubate in the dark for 30 min at 4 °C.
8. *Optional*: Other markers can be included in this analysis. Usually, CD25 and CD69 are used as activation markers, and CD16 for better identification of $CD56^{dim}$ (which are also $CD16^{+}$) and $CD56^{bright}$ (which are also $CD16^{-}$). For this, prepare another antibody mix with the desired mAbs, and add 10 μL of mix to the resuspended cells. An FMO (Fluorescence Minus One) control should be included where NK cells are stained only for CD3 and CD56 (*see* **Note 7**).
9. Wash with 150 μL of staining buffer and centrifuge at 600 × *g* for 5 min.
10. *Optional*: Fix the cells with 100 μL of fixation buffer, and store at 4 °C until acquisition in a flow cytometer.
11. Resuspend the cells in 150 μL of PBS and acquire in a flow cytometer.

3.4 NK Cell Effector Functions: Intracellular Staining of Cytokines

1. Pre-stimulate NK cells following Subheading 3.2 (*see* **Note 8**).
2. During the last 5 h of culture, prepare a 1:50 dilution of Monensin and Brefeldin A in culture medium, and add 10 μL/well (final concentration of Monensin and Brefeldin A equal to 2 μM and 5 μg/mL, respectively) and return the plate to the incubator at 37 °C.
3. Perform cell surface staining for CD56 and CD3 (and other markers if desired) following Subheading 3.3, **steps 2** through **9**.
4. Add 100 μL of BioLegend fixation solution and incubate in the dark for 20 min.
5. Add 100 μL of staining buffer and centrifuge at 300 × *g* for 5 min.
6. Discard the supernatant, resuspend the cells in 100 μL of BioLegend permeabilization solution, and incubate in the dark for 15 min.
7. Centrifuge at 300 × *g* for 5 min.
8. Prepare a dilution of the anti-IFN-γ mAb (and the corresponding isotype control mAb) in BioLegend permeabilization solution, and add 20 μL to each well. Incubate in the dark for 30 min at room temperature.

   The following wells should be included as controls:

   Spontaneous IFN-γ production: Unstimulated NK cells labeled with anti-IFN-γ.

   Positive control: Overnight-stimulated NK cells (*see* **Note 5**) labeled with anti-IFN-γ

9. Wash with 150 μL of staining buffer and centrifuge at 300 × *g* for 5 min.
10. *Optional*: Fix the cells with 100 μL of fixation buffer, and store at 4 °C until acquisition in a flow cytometer.
11. Resuspend the cells in 150 μL of PBS and acquire in a flow cytometer.

3.5 NK Cell Effector Functions: Staining of Dead Target Cells to Evaluate Cytotoxicity

1. Harvest the target cells (K562 cell line or other cell line) from an exponentially growing culture, and transfer them into a new 50 mL tube. Centrifuge at 600 × *g* for 5 min, and wash the cells two more times with PBS to remove any serum. Before the last centrifugation, count the cells and transfer the appropriate number to a new tube. Centrifuge at 600 × *g* for 5 min.
2. During **step 1**, prechill PBS and FBS on ice.
3. Resuspend target cells at 20 × 10^6 cells/mL in cold PBS and keep on ice (*see* **Note 9**).
4. Prepare 1 mL of a 10 μM solution of CFSE in ice cold PBS (*see* **Note 10**).
5. While vortexing cells, add an equal volume of the 10 μM CFSE solution.
6. Incubate for 6 min in the dark at room temperature.
7. Stop labeling by adding at least 5 mL of ice cold FBS, and centrifuge at 600 × *g* for 5 min.
8. Wash cells four times with 40 mL ice cold PBS supplemented with 10% FBS.
9. Resuspend CFSE-labeled target cells in pre-warmed complete medium at a concentration of 6 × 10^5 cells/mL.
10. Add 50 μL of the CFSE-labeled target cell suspension (30,000 cells) per well into 96 U-well plates.
11. Optional: To evaluate ADCC, prepare a dilution of the mAbs under study and an isotype control (although the optimal concentration should be titrated, a starting point of 10 μg/mL is recommended), and add 50 μL to the wells seeded with target cells. Incubate for 60–120 min (*see* **Note 11**).
12. If **step 11** is skipped, add 50 μL of complete medium so that target cells end up in a volume of 100 μL.
13. (A) If NK cells were pre-stimulated (obtained following Subheading 3.2), resuspend pre-stimulated NK cells in pre-warmed complete medium at a concentration of 1.5 × 10^6 cells/mL and plate 100 μL of the effector cell suspension (150,000 NK cells) per well.

    (B) If NK cells are going to be stimulated only during the coculture, resuspend freshly isolated NK cells (obtained following Subheading 3.1) in pre-warmed complete medium at a concentration of 3×10^6 cells/mL, and plate 50 μL of the effector cell suspension (150,000 NK cells) per well. Prepare dilutions of the different stimuli (the compound/s being tested, and a positive control; *see* **Note 5**) in complete medium, and add 50 μL to the wells. As a negative control, leave NK cells unstimulated by adding 50 μL of complete medium without stimulus.

14. Mix gently each well with a 200 μL micropipette. The effector:target (E:T) ratio, proposed here is 5:1 in a final volume of 200 μL/well (*see* **Note 12**).
15. Centrifuge the cells at 200 × *g* for 30 s to favor cell contact.
16. Place the cells in an incubator at 37 °C for 4–6 h.
17. Transfer the cells from the 96 U-well plate to 96 V-well plate for staining.
18. Centrifuge at 600 × *g* for 5 min.
19. Wash once with 150 μL of PBS (use PBS without sodium azide as this would reduce cell viability and without serum or proteins as this could bind up some proportion of the dye), and centrifuge at 600 × *g* for 5 min.
20. Prepare a 1:200 dilution of Zombie Aqua Dye into PBS (*see* **Note 13**).
21. Resuspend the cells in 50 μL of the Zombie Aqua solution.
22. Incubate in the dark for 15 min at room temperature.
23. Wash once with 150 μL of staining buffer and centrifuge at 600 × *g* for 5 min.
24. *Optional*: Fix the cells with 100 μL of fixation buffer, and store at 4 °C until acquisition in a flow cytometer.
25. Resuspend the cells in 150 μL of PBS and acquire in a flow cytometer.

The following should be included as controls:

Spontaneous death: CFSE-labeled target cells cultured without NK cells to determine the percentage of target cells that die spontaneously ($CFSE^+ZA^+_{spontaneous}$).

Positive control: If using a target different from K562 cells, also include a well of NK cells incubated with K562 cells.

The percentage of cytotoxicity is calculated as:

$$\%\ \text{of cytotoxicity} = \frac{\%\,CFSE^+ZA^+_{(\text{treatment})} - \%\,CFSE^+ZA^+_{(\text{spontaneous})}}{100 - \%\,CFSE^+ZA^+_{(\text{spontaneous})}} \times 100$$

where % $CFSE^+ZA^+_{(\text{spontaneous})}$ is the percentage of target cells ($CFSE^+$ cells) dead in the absence of NK cells, and % $CFSE^+ZA^+_{(\text{treatment})}$ is the percentage of target cells ($CFSE^+$ cells) dead in the presence of NK cells and the stimulus under study.

3.6 NK Cell Effector Functions: Staining of CD107a to Identify Degranulating NK Cells

1. Harvest the target cells (K562 cell line or other cell line) from an exponentially growing culture, and transfer them into a new 15 mL tube. Centrifuge at 600 × *g* for 5 min.
2. Resuspend the cells in pre-warmed complete medium at a concentration of 6×10^6 cells/mL.
3. Plate 50 μL/well of the target cells suspension (300,000 cells) into 96 U-well plates.
4. Optional: To evaluate ADCC, prepare a dilution of the mAbs under study and an isotype control (although the optimal concentration should be titrated, a starting point of 10 μg/mL is recommended), and add 50 μL to the wells seeded with target cells. Incubate for 60–120 min (*see* **Note 11**). If this step is skipped, add 50 μL of complete medium so that target cells end up in a volume of 100 μL.
5. (A) If NK cells were pre-stimulated (obtained following Subheading 3.2), resuspend pre-stimulated NK cells in pre-warmed complete medium at a concentration of 1×10^6 cells/mL, and plate 100 μL/well of the effector cell suspension (1×10^5 NK cells).

   (B) If NK cells are going to be stimulated only during the coculture, resuspend freshly isolated NK cells (obtained following Subheading 3.1) in pre-warmed complete medium at a concentration of 2×10^6 cells/mL, and plate 50 μL/well of the effector cell suspension (100,000 NK cells). Prepare dilutions of the different stimuli (the compound/s being tested, and a positive control; *see* **Note 5**) in complete medium, and add 50 μL/well. As a negative control, leave NK cells unstimulated by adding 50 μL of complete medium without stimulus.
6. Mix gently each well with a 200 μL micropipette. The E:T ratio proposed here is 1:3 in a final volume of 200 μL/well (*see* **Note 14**).
7. Prepare a 1:50 dilution of Monensin and Brefeldin A in culture medium, and add 10 μL/well (final concentration of Monensin and Brefeldin A equal to 2 μM and 5 μg/mL, respectively). Finally, add 3 μL/well of anti-CD107a mAb or the corresponding isotype control to the different wells, and mix gently each well with a 200 μL micropipette.
8. Centrifuge the cells at 200 × *g* for 30 s to favor cell contact.
9. Place the cells in an incubator at 37 °C for 4–6 h.
10. Perform cell surface staining for CD56 and CD3 (and other markers if desired) following Subheading 3.3, **steps 2** through **11**.

The following should be included as controls:

Spontaneous degranulation: Unstimulated NK cells (NK cells that did not receive any stimulus and were not in contact with target cells) labeled with the anti-CD107a mAb.

Positive control: If using a target different from K562 cells, also include a well of NK cells incubated with K562 cells.

4 Notes

1. Normal mouse serum is used to block nonspecific Fc-mediated interactions.
2. It is recommended that cells are cultured with Monensin for no more than 6 h as this can become toxic for cell viability [24].
3. If required, cells can be cultured with Brefeldin A for at least 12 h, without a visible reduction of cell viability [24].
4. The yield of NK cells obtained will depend on the percentage of NK cells present in peripheral blood of each particular donor but usually is around 1×10^6 NK cells from 10 mL of blood.
5. For standard NK cell activation, stimulate with a cytokine combo containing a final concentration of 10 ng/mL of IL-12 and IL-18 and 1 ng/mL of IL-15 in complete medium.
6. The incubation period depends on the immunomodulator being evaluated. It is recommended to perform a dose-response and a time-response curve to establish the optimal conditions of the assay.
7. To further characterize/identify NK cell populations, cell surface receptors for activating (NKG2D, NKG2C, DNAM-1, NCRs) and inhibitory (NKG2A, ILT-2, TIGIT) NK cell receptors can be evaluated, among others. The number of markers analyzed in the same sample will depend on the number of lasers and detectors available in the flow cytometer.
8. IFN-γ production by NK cells is not immediate and requires pre-stimulation. Thus, usually when IFN-γ production is evaluated, NK cells are stimulated for a longer period of time than when cytotoxicity or degranulation are measured. Also, the stimulation of NK cells using soluble compounds usually induces the secretion of cytokines, preferentially by $CD56^{bright}$ NK cells, so it could be useful to analyze that population.
9. It is recommended to start with at least twice the number of cells required for the assay as many cells are lost during the extensive washing. And if labeling less than $4x10^6$ total cells, do not use less than 0.5 mL PBS.

10. It is recommended to use 5 μM as a starting point for labeling cells; however, it is recommended to determine the optimal concentration for the assay of interest. Other dyes can be used, like Cell Proliferation Dye eFluor 670, etc. Also, to avoid the step of CFSE labeling, a GFP-expressing K562 cell line (or other GFP-expressing target cell) can be used to discriminate effector from target cells. CFSE has an excitation maximum of 490 nm and an emission maximum of 520 nm.
11. To perform the ADCC assay, the target cell must express the antigen specifically recognized by the mAb. As a control, rituximab (anti-CD20 mAb)-coated Raji cells or cetuximab (anti-EGFR mAb)-coated HCT116 cells can be used.
12. For the cytotoxicity assay, the evaluation of more than one E:T ratio is recommended. Usually, the effector cells must be in excess to ensure that every target cell will be in contact with an effector cell.
13. Different cell types can show variability in staining degree; thus, optimal concentration for the assay of interest should be titrated. Zombie Aqua is an amine-reactive fluorescent dye that is non-permeant to live cells but permeant to cells with compromised membranes, which allows for dead/live discrimination of cells. The advantage of using Zombie dyes is that, as opposed to 7AAD or propidium iodide, they are fixable, allowing for later analysis. Zombie Aqua Dye is excited by the violet laser and has a maximum emission of 516 nm.
14. In the degranulation assay, NK cells (and not the target cells) are labeled, so it is recommended that the target cells are in excess to be sure that every NK cell in the coculture is able to interact with a target cell. Differently from IFN-γ production, cytotoxicity or degranulation by NK cells is immediate and does not require pre-stimulation. Thus, in these assays, NK cells can be directly stimulated with the compound of interest during the coculture. The responding NK cells after direct target recognition are preferentially $CD56^{dim}$ NK cells, so it could be useful to analyze specifically CD107a expression on that population.

References

1. Diefenbach A, Colonna M, Romagnani C (2017) The ILC world revisited. Immunity 46:327–332
2. Vivier E, Tomasello E, Baratin M, Walzer T, Ugolini S (2008) Functions of natural killer cells. Nat Immunol 9:503–510
3. Lanier LL (2008) Up on the tightrope: natural killer cell activation and inhibition. Nat Immunol 9:495–502
4. Fauriat C, Long EO, Ljunggren HG, Bryceson YT (2010) Regulation of human NK-cell cytokine and chemokine production by target cell recognition. Blood 115:2167–2176
5. Long EO, Kim HS, Liu D, Peterson ME, Rajagopalan S (2013) Controlling natural killer cell responses: integration of signals for activation and inhibition. Annu Rev Immunol 31:227–258

6. Lanier LL (2005) NK cell recognition. Annu Rev Immunol 23:225–274
7. Zwirner NW, Domaica CI (2010) Cytokine regulation of natural killer cell effector functions. Biofactors 36:274–288
8. Sivori S, Carlomagno S, Pesce S, Moretta A, Vitale M, Marcenaro E (2014) TLR/NCR/KIR: which one to use and when? Front Immunol 5:105
9. Caligiuri MA (2008) Human natural killer cells. Blood 112:461–469
10. Cooper MA, Fehniger TA, Caligiuri MA (2001) The biology of human natural killer-cell subsets. Trends Immunol 22:633–640
11. Montaldo E, Del Zotto G, Della Chiesa M, Mingari MC, Moretta A, De Maria A, Moretta L (2013) Human NK cell receptors/markers: a tool to analyze NK cell development, subsets and function. Cytometry A 83:702–713
12. Wilk AJ, Blish CA (2018) Diversification of human NK cells: lessons from deep profiling. J Leukoc Biol 103:629–641
13. Freud AG, Mundy-Bosse BL, Yu J, Caligiuri MA (2017) The broad Spectrum of human natural killer cell diversity. Immunity 47:820–833
14. Horowitz A, Strauss-Albee DM, Leipold M, Kubo J, Nemat-Gorgani N, Dogan OC, Dekker CL, Mackey S, Maecker H, Swan GE, Davis MM, Norman PJ, Guethlein LA, Desai M, Parham P, Blish CA (2013) Genetic and environmental determinants of human NK cell diversity revealed by mass cytometry. Sci Transl Med 5:208ra145
15. Lopez-Verges S, Milush JM, Schwartz BS, Pando MJ, Jarjoura J, York VA, Houchins JP, Miller S, Kang SM, Norris PJ, Nixon DF, Lanier LL (2011) Expansion of a unique CD57(+)NKG2Chi natural killer cell subset during acute human cytomegalovirus infection. Proc Natl Acad Sci U S A 108:14725–14732
16. Peng H, Tian Z (2017) Diversity of tissue-resident NK cells. Semin Immunol 31:3–10
17. Morvan MG, Lanier LL (2016) NK cells and cancer: you can teach innate cells new tricks. Nat Rev Cancer 16:7–19
18. Guillerey C, Huntington ND, Smyth MJ (2016) Targeting natural killer cells in cancer immunotherapy. Nat Immunol 17:1025–1036
19. Childs RW, Carlsten M (2015) Therapeutic approaches to enhance natural killer cell cytotoxicity against cancer: the force awakens. Nat Rev Drug Discov 14:487–498
20. Dall'Ozzo S, Tartas S, Paintaud G, Cartron G, Colombat P, Bardos P, Watier H, Thibault G (2004) Rituximab-dependent cytotoxicity by natural killer cells: influence of FCGR3A polymorphism on the concentration-effect relationship. Cancer Res 64:4664–4669
21. Kawaguchi Y, Kono K, Mimura K, Sugai H, Akaike H, Fujii H (2007) Cetuximab induce antibody-dependent cellular cytotoxicity against EGFR-expressing esophageal squamous cell carcinoma. Int J Cancer 120:781–787
22. Kim GG, Donnenberg VS, Donnenberg AD, Gooding W, Whiteside TL (2007) A novel multiparametric flow cytometry-based cytotoxicity assay simultaneously immunophenotypes effector cells: comparisons to a 4 h ^{51}Cr-release assay. J Immunol Methods 325:51–66
23. Alter G, Malenfant JM, Altfeld M (2004) CD107a as a functional marker for the identification of natural killer cell activity. J Immunol Methods 294:15–22
24. Schuerwegh AJ, Stevens WJ, Bridts CH, De Clerck LS (2001) Evaluation of monensin and brefeldin a for flow cytometric determination of interleukin-1 beta, interleukin-6, and tumor necrosis factor-alpha in monocytes. Cytometry 46:172–176

Part II

Reprogramming Diverse Immunocytes

Chapter 10

An Overview of Advances in Cell-Based Cancer Immunotherapies Based on the Multiple Immune-Cancer Cell Interactions

Jialing Zhang, Stephan S. Späth, Sherman M. Weissman, and Samuel G. Katz

Abstract

Tumors have a complex ecosystem in which behavior and fate are determined by the interaction of diverse cancerous and noncancerous cells at local and systemic levels. A number of studies indicate that various immune cells participate in tumor development (Fig. 1). In this review, we will discuss interactions among T lymphocytes (T cells), B cells, natural killer (NK) cells, dendritic cells (DCs), tumor-associated macrophages (TAMs), neutrophils, and myeloid-derived suppressor cells (MDSCs). In addition, we will touch upon attempts to either use or block subsets of immune cells to target cancer.

Key words Immunotherapy, Adoptive cell therapy, Immune cells

1 Introduction

Cancer is a complex ecosystem where the interaction of diverse cancer and noncancerous cells collectively determine the overall behavior and subsequent treatment response [1–3]. Many studies indicate that various immune cells (e.g., T cells, B cells, natural killer cells (NK cells), tumor-associated macrophages (TAMs), neutrophils, and dendritic cells (DCs)) have been found in and influence the natural course of tumors. The majority of immune cells reveal subsets that have both anti- and pro-tumorigenic roles in regulating tumor cell proliferation, angiogenesis, and metastasis [4, 5]. The idea of maximizing the antitumor activity and minimizing the tumor-promoting activity of immune cells to target and eliminate tumors has been investigated since the beginning of the twentieth century. An increased understanding of the processes that take place in the tumor immune microenvironment is rapidly advancing the translation toward novel therapeutic strategies in

Samuel G. Katz and Peter M. Rabinovich (eds.), *Cell Reprogramming for Immunotherapy: Methods and Protocols*, Methods in Molecular Biology, vol. 2097, https://doi.org/10.1007/978-1-0716-0203-4_10,

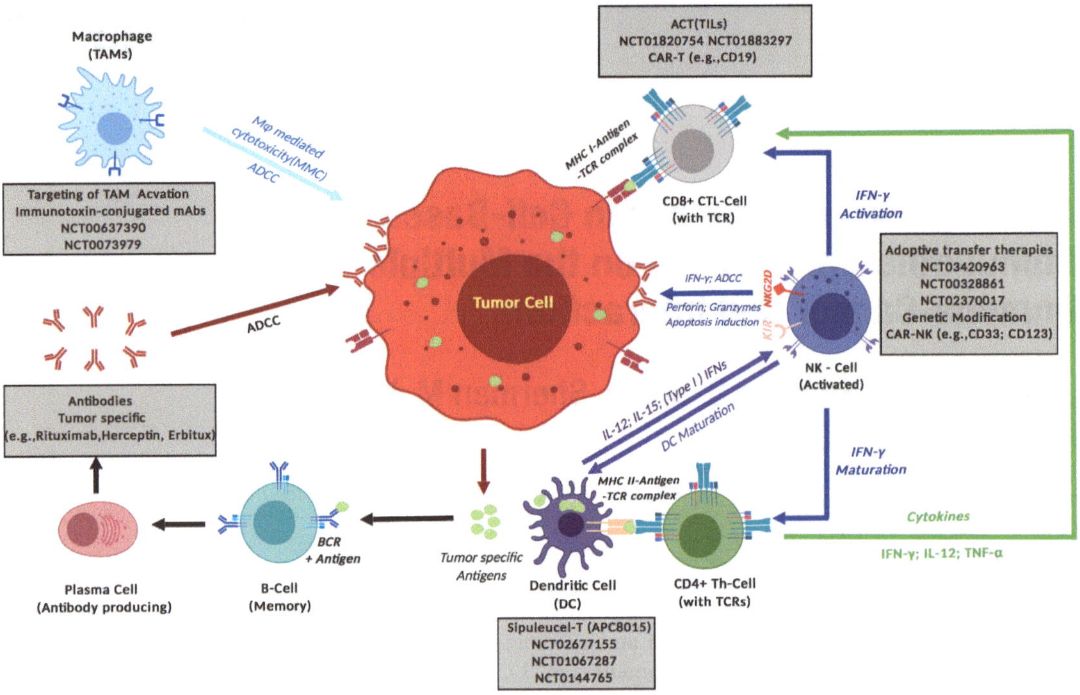

Fig. 1 Immune response in antitumor activity. Key cellular components of the innate and adaptive immune system as well as broad defense mechanisms against tumors are shown. Current targeting approaches and/or ongoing clinical trials are shown in gray boxes. *TAM* (tumor-associated macrophage), *ADCC* (antibody-dependent cellular cytotoxicity), *mAbs* (monoclonal antibodies), *BCR* (B-cell receptor), *TCR* (T-cell receptor), *MHC* (major histocompatibility), *NK* (natural killer), *CTL* (cytotoxic T lymphocyte), *ACT* (adoptive cell transfer), *CAR-T* (chimeric antigen receptor T cell)

cancer treatment. The progress of antitumor immune responses are coordinated through a set of defined, as well as yet unknown innate and adaptive immune components, comprised of immune cells [6], cell surface molecules [7], co-stimulatory receptors [8], ligands [6], cytokines, and chemokine signaling pathways (Fig. 1) [6, 9].

The collaboration of innate and adaptive immune cells has a significant impact on tumorigenesis and treatment response via cell-cell physical contacts, the production of cytokines (e.g., IL-6, interferon-α (IFN-α), tumor necrosis factor-α (TNF-α)), and the use of chemokines (e.g., CCL2, CXCL12, CX3C) [5, 6, 10]. Numerous studies on the effects of cytokines and chemokines in the development of primary tumors and metastases have been reported and even translated to adjuvant immunotherapeutic agents (e.g., IL-2, CCL21) [11]. These studies indicate that cytokines and chemokines impart a synergistic effect by attracting relevant subsets of the host immune cells to the site of antigen presentation. Enhancement of exposure of tumor-associated alterations to the host immune system during early tumorigenesis poses a vulnerability to the tumor's continued growth. Shedding malignant cell antigens into the blood system; the lymphatic circulation, including lymph nodes; and the extracellular matrix (ECM) ultimately activates the adaptive immune system [6, 12]. For cancers lacking major histocompatibility complexes-I (MHC-I) on the surface of their malignant cells, the innate immune system recognizes their absence and attacks them in a manner similar to the recognition of nonself pathogens [13]. Thus, it is not surprising to see the growing number of studies revealing that cellular components of both innate and adaptive immunity significantly contribute to disease outcomes [6]. Here, we review the recent findings of cellular components in the human immune system, which indicate promising potential in immunotherapy and perspectives for the continued translation from experimental evidence into clinical practice.

2 T Cells

T cells are involved in many different types of immune responses and are regulated by signals of T cell receptors and co-activating, inhibitory, and cytokine receptors [14]. Different subsets of T cells play crucial roles in mediating tumor growth control. Some T cells, like cytotoxic $CD8^+$ T cells and helper $CD4^+$ T cells, inhibit tumor growth, while others like regulatory T cells (Treg) support tumor progression, and still others, like T helper 17 (Th17) cells, do both. Helper $CD4^+$ T cells can recognize cancer-associated antigens presented on MHC class II molecules on antigen-presenting cells, and mediate antitumor immunity by secretion of IFN and inhibition of angiogenesis [15]. Cytotoxic $CD8^+$ T cells express T cell receptors (TCRs) that can bind to MHC class I molecules on cancer cells, subsequently targeting the cells for elimination (Fig. 1) [14]. Tumor-infiltrating $CD8^+$ T cells have been shown to slow tumor cell growth and significantly reduce metastasis [5, 16]. Increased infiltration of certain T cell subsets within several types of tumors has a favorable prognosis, particularly when those T cells express memory and activation markers [5, 16, 17]. This finding led to the development of an adoptive cell therapy (ACT) approach, in which tumor-infiltrating lymphocytes (TILs) harvested from the patient are expanded ex vivo and then injected back into the patient (Fig. 1) [18]. Clinical outcomes of these studies are variable and likely impacted by the TIL composition and prior chemotherapy and radiotherapy. Treg cells, through the production of immunosuppressive cytokines (e.g., TGFβ and IL-10) and the immune-inhibitory receptor cytotoxic T lymphocyte antigen 4 (CTLA4), negatively modulate T-cell activation, survival, and expansion [19–22]. Depletion of $CD25^+$ Treg cells in B16/BL6 syngeneic mouse melanoma models results in improved anti-melanoma $CD4^+$ and

CD8$^+$ T-cell responses and, if completed before tumor engraftment, also results in increased T-cell infiltration and tumor rejection [21, 23]. High numbers of tumor-infiltrating Th17 cells correlate with poor prognosis in colon and pancreatic cancer [24, 25], but improved survival in ovarian cancer [26–28]. These opposing regulatory and inflammatory properties of Th17 cells are thought to be influenced by the stimuli they encounter in different tumor microenvironments [29]. Taken together, the therapeutic value of TILs either alone or in combination with other treatments, such as Treg cell depletion, needs to be further elucidated.

A second form of ACT uses circulating T cells from patients' blood. These peripheral blood T cells can be reprogrammed ex vivo with specific antitumor T-cell receptors (TCRs) or chimeric antigen receptors (CARs) (*see* Chapter 1). The usefulness of T cells expressing ectopic TCRs and CARs against B-cell leukemias has been demonstrated in both preclinical and clinical trials (Fig. 1) [30]. In 2017, an autologous CAR T-cell therapy developed for children and young adults with relapsed and/or refractory CD19$^+$ acute lymphoblastic leukemia (ALL) was the first CAR-T cell therapy approved by the FDA in the USA. Extending this to solid tumors is still difficult. Some of the challenges that must be overcome include getting enough of the engineered T cells to infiltrate the site of solid tumors, enabling the CAR-T cells to survive in the inhospitable tumor microenvironment, and identifying homogenously expressed, unique target antigens. One approach that has already emerged in clinical testing for solid tumors is the combination of CAR-T cells with a checkpoint inhibitor antibody (e.g., PD-1, CTLA-4).

3 B Cells

After CD8$^+$ T cells, B cells are the second most abundant TIL population in lung cancer and melanoma [31, 32]. Whereas some studies have associated the presence of B cells within solid tumors with poor survival [33, 34], others have associated their presence with improved survival [35–38], suggesting that, like other immune cell types, B cells may have both tumor-inhibitory and tumor-promoting roles. One study showed that the presence of both B cells and T cells in ovarian cancer correlates with a better survival than if only B or T cells are present alone, suggesting important interactive functions [39]. B cells are chiefly known for producing antibodies through which they can influence all immune cells that express Fc receptors, including dendritic cells, granulocytes, NK cells, and myeloid-derived suppressor cells. B cells also interact with other immune cells as potent antigen-presenting cells and through the secretion of cytokines and chemokines [40].

B cells are able to inhibit tumor growth through several mechanisms. Autoantibodies can recognize tumor-associated antigens and discriminate between cancer and control cells [41]. Some autoantibodies are anti-tumorigenic by reducing invasiveness and increasing apoptosis [42]. In ovarian cancer, production of IFNγ, IL12, GM-CSF, and CXCL10 by B cells supports an antitumor response [37]. Cell communication between T and B cells is tightly linked through CD40L-CD40 and CD80-CD28 signaling. The cell surface protein CD40L serves as a crucial co-stimulatory factor for B cell activation by binding CD40, which promotes B-cell proliferation, germinal center formation, immunoglobulin class switching, somatic hypermutation, plasma cell and memory B-cell formation, and antigen presentation [43–49]. CD40-activated B-cell-based cancer immunotherapy induces effective antitumor immunity in mice and dogs [50].

B cells also perform multiple functions that can promote tumor growth. For example, some autoantibodies have been identified, which are pro-tumorigenic and can help form a pre-metastatic niche [51]. In addition, by production of TNFα and IL-21, tumor cells can induce the conversion of TIL B cells into Breg cells, a poorly defined subset of B cells [52, 53]. Breg cells promote tumor growth through the secretion of IL10 and TGFβ [54–56]. Through checkpoint receptors like PD-1, Breg cells inhibit T-cell functions in hepatocellular carcinoma and thyroid cancer [57, 58]. However, at least in melanomas, PD-1 inhibitors maintain activity even in the absence of B cells [59]. Additional pro-tumorigenic roles of B cells include reducing $CD8^+$ T-cell and NK cell infiltration [60], the polarization of immunosuppressive macrophages [61, 62], and the induction of cancer cells with stem cell-like properties in melanoma [63]. Future studies will be needed to identify the immunologic conditions that specifically enhance the effects of B cells on antitumor immunity in solid tumors, while avoiding these pro-tumorigenic aspects of their function as a form of cancer immunotherapy.

4 NK Cells

NK cells comprise 5–15% of circulating lymphocytes and are part of the first line of defense against cancer (Fig. 1) [6]. The infiltration of NK cells in the solid tumor microenvironment is a well-documented favorable prognostic sign in cancer patients [64, 65]. NK cells discriminate between cancerous and healthy cells based on a tightly regulated balance of the signaling produced by their activating (e.g., NKG2D) and inhibitory (e.g., KIRs) receptors (Fig. 1) [66]. Activated NK cells can kill tumor cells through various mechanisms, including the release of perforin and granzymes, expression of Fas antigen ligand (FasL) or TNF-related, apoptosis-inducing ligand (TRAIL), secretion of IFN-γ, and by antibody-dependent cellular cytotoxicity (ADCC) (Fig. 1) [67]. Through the production of IFN-γ, NK cells contribute to the activation of cytotoxic $CD8^+$ T cells and promote differentiation of $CD4^+$ T cells [68]. NK cell-mediated tumor cell lysis may also produce tumor antigens for phagocytic immune cells, which subsequently induce DC activation and maturation, with the ultimate goal of cross priming specific T cells [69].

More than a decade has passed since initial reports established the ability of NK cells to mediate tumor regression in patients with acute myeloid leukemia (AML) [70, 71]. Many discoveries on the pathways that activate and suppress NK cell function, and sensitize tumors to NK cell cytotoxicity, have led to the development of numerous pharmacological and genetic methods to enhance NK cell antitumor immunity (Fig. 1) [72–74]. Genetically modified NK cells and NK cell adoptive transfer therapies have shown promising results and are currently being tested in multiple ongoing phase I/II and III trials (e.g., NCT 03420963, NCT 00328861, NCT 02370017), involving several cancer types (Fig. 1) [66]. So far, the therapeutic exploitation of NK cell-based immunotherapy has achieved only limited clinical success in cancer patients. Further research toward a better understanding of NK cell biology and their function in cancer is needed to foster the development of novel NK cell-based therapeutic approaches.

5 Dendritic Cells (DCs)

DCs represent the most important subset of antigen-presenting cells that activate innate immune pathways and induce potent adaptive immune responses, and in doing so act as messenger cells between the innate and adaptive immune systems (Fig. 1) [75, 76]. There are many different types of DCs, including myeloid (mDCs), plasmacytoid DCs (pDCs), monocyte-related $CD14^+$ DCs, Langerhans cells, and microglia [76]. DCs detect invading pathogens through their expression of a variety of sensors for pathogen-derived components including Toll-like receptors (TLRs) [77], NOD-like receptors (NLRs) [78], RIG-I like helicases [79], and C-type lectin receptors [75, 80, 81]. The DCs are stimulated to mature and migrate to the T-cell zones of lymph nodes, where the resulting peptide-MHC complexes are transported to the plasma membrane [82, 83] and presented to naïve T cells [84–86].

Depending on multiple factors, DCs may either be immune stimulatory or suppressive. In general, mature DCs and mDCs are more immunostimulatory, whereas immature DCs and pDCs are tolerogenic [87]. Mature DCs, expressing co-stimulatory receptors (e.g., CD28), produce IL-12 for optimal T-cell stimulation and can even stimulate B cells [75, 88]. By the production of IL-12, IL-15, and type I IFNs, mature DCs trigger NK cell activation, while NK cells can induce further DC maturation through the secretion of TNFα (Fig. 1) [89]. Given these immune stimulatory functions, it is not surprising that the presence of mature antigen-presenting DCs leading to an enhanced immune response has been correlated with improved survival in many cancer patients [87, 90], including in melanoma [91, 92], breast cancer [93], and gastric cancer [94].

Other studies have shown that increased tumor-infiltrating DCs correlates with worse survival and have linked this to their immunosuppressive functions. For example, worse patient survival has been seen in colorectal cancer, where the DCs were more immature [95] and in breast cancer, where pDCs produced low levels of IFNα and sustained Treg expansion [96]. TGFβ, IL-10, and IL2 also induce DCs to stimulate Treg formation [97]. Tumors inhibit DC function and maturation through production of chemokines (e.g., CCL2, CXCL1, CXCL5) [98], by induction of PD-1 expression [99], and/or by induction of TIM-3 expression [100]. In addition, tumor cells induce transcription factors in DCs, like STAT3 [101–103] and FOXO3 [104], that make them immunosuppressive. DCs can suppress T cells both by increasing ROS [105] and by upregulation of arginase I, which depletes arginine, an essential amino acid that T cells cannot produce themselves [106].

DC-based immunotherapy attempts to exploit the ability of DCs to stimulate a T-cell-based antitumor response (Fig. 1). Several studies utilizing established DC-based vaccines for cancer treatment have confirmed their strong potential for clinical benefit [107, 108]. Sipuleucel-T (APC8015), the first and only approved DC vaccine, targets prostate-specific acid phosphatase in human prostate cancer. While effective, it only prolonged the median overall survival of castration-resistant prostate cancer to 25.8 months, compared with 21.7 months in the placebo group [109]. The majority of past melanoma vaccines have been ineffective in clinical trials [110]. Additional DC vaccine combination studies, specifically targeting checkpoint inhibitors, are underway (NCT02677155, NCT01067287, and NCT0144765) with notable preliminary efficacy in preclinical mouse models (Fig. 1) [107]. This is an interesting approach, given the ability of PD-1 expression on DCs to induce tolerogenic effects [99]. A variety of other methods including gene transfer, antigene pulsing, and DC/tumor cell fusion have also yielded some clinical evidence of antitumor responses [107]. More trials are needed to further evaluate the efficacy of DC vaccines and draw meaningful conclusions on their ability to prevent cancer recurrence.

6 Macrophages

Macrophages are highly heterogeneous with multiple functions, origins, and locations of action within the tumor. Historically, macrophages were predominantly separated into two major groups: a classically activated M1 type that is pro-inflammatory and tumoricidal, and an alternatively activated M2 type that is anti-inflammatory and pro-tumorigenic [111, 112]. M1-polarized macrophages are activated by lipopolysaccharide (LPS) and Th1 cytokines IFNγ and TNFα [113, 114]. M1 macrophages secrete pro-inflammatory cytokines IL-12 and IL-23, they present antigens through MHC class II, and they help $CD4^+$ T cells differentiate into Th1 and Th17 cells [112]. M2 macrophages are activated by numerous stimuli, including M-CSF/CSF1 produced by the tumor cells [115], immunoglobulin produced by B cells [61], IL-10 produced by Treg cells [116], IL-4 and IL-13 produced by $CD4^+$ Th2 T cells [117–120], and MIF and CXCL12 produced by the monocytes/macrophages themselves [121, 122]. Activated M2 macrophages are anti-inflammatory, have relatively decreased phagocytic activity, stimulate angiogenesis, and help Treg differentiation [123, 124]. However, as the various stimuli for M1 and M2 macrophages do not exist in isolation in vivo, macrophages exist along a spectrum with various expression levels of markers associated with both of these phenotypes [125, 126]. In addition, there is plasticity in the macrophage activation state, with COX2, TLR/Myd88 signaling, and Notch, all favoring an M2 to M1 switch [127–129]. Most tumor-associated macrophages (TAM) lean more toward the so-called M2-like, pro-tumorigenic phenotype [130], and concordantly, high numbers of TAMs are typically considered a poor prognostic factor in breast [131], lung [132, 133], liver [134], gastric [135], and bladder cancer [136].

Differences in the macrophage's developmental origin also impacts on their function. Both macrophages recruited into the tumor that are derived from adult peripheral blood monocytes and tissue resident macrophages that are derived from embryonic macrophages and proliferate in situ, as the tissue develops have been identified in glioblastomas, pancreatic cancers, and lung cancers [137–140]. For pancreatic cancers, the embryonically derived, tissue-resident macrophages were more important in establishing the fibrotic tumor microenvironment, whereas the peripheral blood monocyte-derived macrophages were more important in antigen presentation [140].

TAM function is also dependent on the macrophage's specific location within the tumor, with differences found in invasive, intra-tumoral, stromal, hypoxic, and perivascular regions [126, 141–143]. At the invasive front, macrophages can help invasive potential by the expression of cathepsins B and S, matrix metalloproteinase 9, and urokinase-type plasminogen activator [117, 144]. Conversely, TAMs at the invasive front of human colon carcinomas express the T-cell co-stimulatory signals CD80 and CD86 [145], which may explain why high macrophage numbers at this location correlate with decreased metastasis and improved survival [142, 146].

TAMs within tumor nests are in close contact with cancer cells and have both tumoricidal and immunosuppressive properties. In prostate cancer, TAMs express NOS2, which has been linked to their cytotoxic potential due to the production of nitric oxide [147]. In line with this, high numbers of TAMs in tumor nests of endometrial or gastric cancer correlate with improved prognosis and reduced recurrence, respectively [148, 149]. However, several cancers, including bladder cancer, breast cancer, non-Hodgkin lymphomas, and acute myeloid leukemia, express high levels of the "don't eat me" receptor CD47, which associates with poor prognosis [150, 151]. Also, in hepatocellular carcinoma, TAMs localized to the tumor nests preferentially recruit Treg cells through the expression of IL-10 [152]. Although this was not correlated with prognosis, high numbers of nest-located TAMs have been correlated with reduced overall survival in melanoma and breast and esophageal tumors [153–155].

Stromal macrophages have also been shown to both promote and inhibit tumor growth. Macrophage phenotypes can be regulated by extracellular matrix (ECM) components, like fibronectin and laminin-10 [156]. Decellularized human colorectal cancer ECM stimulates macrophages to increase anti-inflammatory cytokines like IL-10 and TGFβ and decrease pro-inflammatory cytokines like TNFα and IL-6 [157]. However, the biophysical properties of the ECM also regulates TAM phenotype. Increased substrate stiffness activates TLR4 pathways and pro-inflammatory gene expression [158]. Given these opposing stimuli and phenotypes, it is not surprising that high numbers of stromal TAMs have been correlated with a poor overall survival in breast, esophageal, gastric, pancreatic, oral, and skin tumors [153, 155, 159–162], an improved survival in bladder cancer [163], and no correlation in endometrial, cervical, and lung cancer [147, 164, 165].

Under hypoxia, TAMs increase expression of IL-10, PD-L1, and indoleamine 2,3-dioxygenase (IDO), which suppress T-cell activation [166, 167]. Likewise, high numbers of TAMs in hypoxic regions are associated with increased angiogenesis and metastasis and reduced overall survival in endometrial, breast, and cervical cancer [147, 168, 169]. Perivascular TAMs are also tumor promoting. These TIE2-expressing macrophages help vascularize tumors [170–173], increase lymphovascular invasion [174], and increase metastasis [175, 176].

Given the generally pro-tumorigenic role of TAMs, one therapeutic goal is to eliminate them by targeting their cell surface markers (e.g., CSF1, CSF1R, scavenger receptor A, and CD52), which are being investigated in ongoing clinical trials (i.e., NCT00637390, NCT0073979) (Fig. 1) [114, 177]. Furthermore, trabectedin was shown to decrease the number of TAMs in tumor tissues by inducing the apoptosis of macrophages [178]. Trabectedin was approved by the European commission for use in the treatment of ovarian cancer and soft tissue sarcomas, and by the FDA for use in unremovable or metastatic liposarcoma or leiomyosarcoma [179].

An alternative approach has been to generate reprogrammed macrophages. One reprogramming scheme is to convert M2 TAMs to M1 TAMs since macrophages exhibit high plasticity [126]. Reprogramming macrophages to an M1 phenotype has been accomplished by *Pseudomonas aeruginosa* mannose-sensitive hemagglutinin [180], various nanoparticles [181–183], CD40 agonists [184], and yeast-derived β-glucan [185–187]. Genetic reprogramming of macrophages and using them as a cell-based delivery system is also being explored in several studies [188–190]. Some approaches that have led to mild to no improvement, include expressing IL-21 to activate T and NK cells, a soluble TGFβ receptor to act as a decoy and reduce TGFβ signaling, and cytochrome P450 under control of a hypoxia-regulated promoter to locally convert the prodrug cyclophosphamide [189, 190]. Recently, Morrissey et al. have reprogrammed macrophages with a chimeric antigen receptor that triggers phagocytosis using intracellular domains from the engulfment receptors Megf10 and FcRγ [191]. Alternatively, blocking CD47 is another promising avenue being explored to increase macrophage phagocytosis. In combination trials with rituximab in diffuse large B-cell lymphoma, CD47 blockade achieved a 71% objective response rate [192, 193]. Recently, another pathway shown to block TAM phagocytosis is tumor cell expression of CD24, which binds Siglec-10 on TAMs [194]. Genetic ablation of CD24 or Siglec-10, or antibody blockade reduced tumor growth and increased survival in a mouse model. In summary, macrophages perform numerous functions in the development of tumors, and multiple approaches to manipulate them are only beginning to be explored.

7 Neutrophils

Neutrophils are the most abundant white blood cells, representing 50–70% of all leukocytes in humans. They are one of the first responders of inflammatory cells and can quickly migrate toward the site of damaged tissue through the blood vessels following chemotaxis signals, such as IL-8, C5a, and fMLP [195]. Some oncogenes have been shown to produce the CXC chemokine, CXCL8, which recruits neutrophils [196, 197]. Activated neutrophils can also attract and activate other cells of the immune system, including DCs, mast cells, T cells, and macrophages [195]. Patients with various cancer types often exhibit increased numbers of circulating neutrophils, and the ratio of neutrophils to lymphocytes in the peripheral blood has been shown to correlate with better patient outcome in breast cancer [198–200]. Conversely, a meta-analysis of 100 studies comprising over 40,000 patients showed that the neutrophil to lymphocyte ratio correlated with an adverse overall survival in many solid tumors [199], and another meta-analysis of nearly 4000 patients revealed that high numbers of intra-tumoral neutrophils were significantly associated with an unfavorable survival [201].

Early observations from Anton Chekhov, William Coley, and others suggested that the neutrophilic infiltration associated with inflammation to an infection had tumoricidal properties [202]. Neutrophils are an important component of ADCC, particularly immature neutrophils mediated through IgA Fc receptors [203, 204]. Antitumor neutrophils can be recruited and activated by TGF-β-blockade or can be isolated and used from healthy donors [205, 206]. Similarly, studies of the antitumor response of *Mycobacterium bovis bacillus* Calmette-Guérin (BCG) therapy for bladder cancer revealed that BCG stimulates a massive infiltration of neutrophils into the bladder wall, which express the apoptosis-inducing molecule TNF-related ligand (TRAIL) [207].

Nonetheless, several observations have supported a pro-tumorigenic role for neutrophils [208]. Neutrophils can induce the apoptosis of $CD8^+$ T cells in a TNF-α- and nitric oxide (NO)-dependent manner [209]. These neutrophils are expanded and polarized to suppress $CD8^+$ T cells by IL-17-producing γδ T cells, leading to increased breast cancer metastasis in a mouse model [210]. Recently, pro-metastatic neutrophils were found to be stimulated by loss of p53 in cancer cells in 16 distinct mouse models of breast cancer [211]. Loss of p53 stimulated the cancer cells to secrete WNT ligands that stimulated TAMs to secrete IL-1β and attract neutrophils. In a mouse lung cancer model, elastase released from neutrophil granules was taken up by the tumor cells and stimulated the PDGFR-PI3K pathway [212]. In another mouse lung cancer study, neutrophils were shown to reduce T-cell homing, prevent successful anti-PD1 immunotherapy, and induce the expression of Snail, which accelerated tumor growth [213]. Similarly, patients with lower neutrophil to lymphocyte ratios responded better to PD-1 and PD-L1 inhibitors [214].

Human neutrophils have been shown to induce angiogenesis in multiple ways. CXCL1 is a neutrophil-attracting chemokine that increases angiogenesis via neutrophil production of VEGF-A [215, 216]. A second mechanism by which neutrophils increase angiogenesis is through the release of the matrix metalloproteinase MMP9 [217]. Expression of MMP9 from neutrophils was shown to increase tumor production of VEGF in a mouse pancreatic cancer model and to correlate with neutrophil number and angiogenesis in human hepatocellular carcinoma [218, 219]. A third mechanism neutrophils use to increase angiogenesis is through the production of BV8/prokineticin 2 [220].

Given the multiple pro-tumorigenic roles of neutrophils, eliminating them may offer therapeutic benefit. The CXCR1/CXCR2 antagonist reparixin, IL-23 antagonists, and IL-17 antagonists are all antineutrophil therapies approved for autoimmune diseases that might be of interest to test in the cancer setting [208]. However, there are mixed findings on the effects of neutrophil elimination in cancer. Some studies show that the presence of neutrophils are beneficial to chemotherapy responses, whereas others demonstrate the reverse [221]. Similarly, the use of G-CSF to increase neutrophil numbers and prevent infections during chemotherapy also polarizes neutrophils toward a pro-tumorigenic phenotype. Overall, the various neutrophil subpopulations and behavior of peripheral vs. tumor-infiltrating neutrophils need to be better characterized to design better neutrophil-targeted therapies.

8 Myeloid-Derived Suppressor Cells (MDSCs)

MDSCs are distinct from macrophages and neutrophils insofar as they maintain an immature differentiation state due to the chronic inflammatory conditions common in cancers and constitutive activation of STAT-3 [222–224]. Both monocytic and granulocytic lineages contribute to MDSCs and can be identified in humans as $HLADR^-$, $CD33^+$, $CD11b^+$, and either $CD14^+$ or $CD33^+$, respectively [225]. Although these surface markers alone cannot distinguish MDSCs from mature monocytes and granulocytes, the expression of other enzymes, cytokines, chemo-attractants, and transcription factors that mediate their immunosuppressive properties can discriminate them [226]. In addition, granulocytic MDSCs are distinguishable by their mononuclear variants, as opposed to the more mature polymorphonuclear forms [227].

The tumor microenvironment helps to recruit and expand MDSCs. Tumor cell production of IDO, CCL26, and CCL2 helps to attract MDSCs [228–231]. CCL26 is induced in hypoxic tumor cells by HIF-1 and binds the MDSC chemokine receptor CX3CR1 [229]. MDSC recruitment is impaired by deficiency of CCL2 in glioblastoma multiforme (GBM) and colorectal cancer mouse models, whereas CCL2 accumulation is a poor prognostic marker in human GBM [228, 230]. Tumor cells continue to drive the formation and expansion of MDSCs through the release of numerous factors, including granulocyte/monocyte-colony-stimulating factor (GM-CSF), G-CSF, M-CSF, PGE2, IL-6, IL-10, IL-1β, TGF-β, SCF, and VEGF, both as soluble molecules and within extracellular vesicles [232–237]. These stimuli then activate several transcriptional programs in MDSCs that stimulate their immunoregulatory function [222]. Most notably, high dose IL-6 increases STAT3 activation, correlating with MDSC suppressive function and number, and when blocked reduces tumor progression in mice [238–240]. Similarly, blockade of GM-CSF prevents the immunosuppressive features of human MDSCs in vitro [241]. STAT3 is an important MDSC regulatory factor, activating cell survival proteins, cell cycle proteins, NOX2 that suppresses T cells, and calcium binding proteins that suppress DC differentiation [101, 103, 242, 243]. Other STAT proteins that contribute to MDSC activity include STAT1, which increases ROS activity [244, 245], and STAT6, which increases TGFβ and ARG1 [246]. A second pathway important in MDSC activity is CK2 phosphorylation and inhibition of Notch [247–249]. Restoration of Notch activity with a CK2 inhibitor decreases MDSCs and reduces tumor growth in a mouse model [248]. Finally, there is some evidence that MDSCs can be generated in tumor-bearing mice, by direct conversion of adoptively transferred NK cells [250].

One of the ways MDSCs support tumor growth is by inhibition of cytotoxic T and NK cells [251]. MDSCs produce cytokines, growth factors, and MMPs that convert naïve $CD4^+$ T cells and Th17 cells into Tregs [252–254], as well as shift macrophages to a more M2-like phenotype [255]. Reactive oxygen species (ROS) produced by NOX1–4 in MDSCs trigger NK and T-cell apoptosis [242, 256–258]. In contrast, the ROS do not harm the MDSCs because of buffering by increased Nrf2 and increased phosphoenolpyruvate, due to their relatively high glycolytic metabolism [259, 260]. Rather, ROS prevents MDSC differentiation and increases their recruitment through VEGF receptors [242, 261]. Inhibition of ROS by NOX2 knockout or treatment with catalase limits MDSC immunosuppressive abilities [257]. Reactive nitrogen species (RNS) produced by MDSCs also cause T-cell apoptosis and block T-cell activation through peroxynitrites and nitration of the T-cell receptor [262]. In addition, RNS prevent FcR-mediated NK cell function [263], decrease lymphocyte recruitment by nitration of CCL2 [258], and increase immunosuppressive molecules IDO, IL-10, and ARG1, through induction of COX-2 and HIF-1 [264–266]. By expression of the metalloprotease ADAM17, MDSCs also decrease L-selectin and reduce T- and B-cell homing and activation in lymph nodes [264, 266]. Consumption of cysteine and arginine by MDSCs reduces T-cell proliferation and TCRζ expression, as T cells lack the ability to make cysteine themselves [266–268]. Priming of naïve T cells is also inhibited by extracellular adenosine, which is converted from ATP by CD39 and CD73 expressed on the cell surface of MDSCs [269, 270]. Downregulation of CD39 and CD73 by metformin in ovarian cancer can decrease MDSCs, enhance antitumor T cells, and prolong survival [271]. MDSCs also inhibit T cells by overexpression of HIF-1-induced PD-L1 [272] and NK cells by direct cell-cell contact with NKp30 [273].

These direct immunosuppressive properties, as well as a few other functions, help MDSCs promote metastases and establish the pre-metastatic niche. In a breast cancer mouse model, MDSCs are abundant in pre-metastatic lung tissue, and their numbers correlate with the subsequent metastatic cancer burden [274]. MDSCs accumulate in the pre-metastatic niche due to CCL12 [275], lysyl oxidase (LOX) produced by the primary tumor [276], STAT3 [277, 278], prokineticin (Bv8) [279], carbonic anhydrase IX [280, 281], cancer exosomes [237, 282], C5a receptor [283], and calprotectin (S100A8/A9) [284, 285]. Once there, MDSCs promote an immune-permissive environment, induce vascular leakage, remodel the extracellular matrix, produce CCL9 to attract cancer cells, and induce endothelium to upregulate E-selectin to help cancer cells adhere [275, 286–290]. MDSCs further support metastasis by STAT3 induction of VEGF and FGF-2 to increase angiogenesis [291–293], by increasing the number of cancer stem cells in pancreatic and ovarian cancer [294, 295], and by inducing epithelial-mesenchymal transition (EMT) through Cox-2 and NOS2 in a nasopharyngeal carcinoma model [296, 297]. Given these many forms of promoting tumor growth, it is not surprising that a meta-analysis of 16 studies and 1864 patients found that elevated MDSC frequency correlated with shorter overall survival [298].

Several different approaches have been taken to decrease MDSC function. Although perhaps only part of its mechanism of action, gemcitabine was noted to decrease both MDSCs and Tregs in pancreatic cancer [299]. To more directly block MDSC expansion and mobilization, several groups are exploring the therapeutic effects of blocking M-CSF, G-CSF, GM-CSF, SCF, VEGF-A, and IL-6 signaling [253, 300–303]. Unfortunately, STAT3 and MMP inhibitors have had limited efficacy in clinical trials, with severe adverse effects [304, 305]. Combinations being explored include IL-2 and an anti-CD40 antibody, which increases Fas-mediated apoptosis of MDSCs [306] and inhibition of PI3K, which reduces MDSC accumulation in tumor-bearing mice, and prolongs survival when combined with a PD-L1 antibody [307, 308]. In addition, depletion of MDSCs in small cell lung cancer patients was observed when combining a DC vaccine with all-trans retinoic acid (ATRA), an agent known to promote myeloid differentiation and used as differentiation therapy in acute promyelocytic leukemia [309, 310]. In summary, MDSCs play an important role in tumor progression, and several promising approaches to specifically target them, particularly as a combination therapy, are in the early phase of investigation.

9 Summary

A hallmark of solid tumors is their infiltration by immune cells that both inhibit and promote tumor cell growth. Despite a large number of promising immune cell-based therapies, only a very few products have been approved for clinical use, with a limited number of patients showing substantial benefit from cancer immunotherapies [311]. The major successes have been either by directly programming T cells to attack known tumor constituents or by blocking effectors that might inhibit T cells. However, the majority of patients are either unresponsive or relapse after a variable period of response to the immunotherapy [66, 190, 311, 312]. Often, it is the highly immunogenic tumors that respond, and there may be a role for tumor therapy that spreads immunity after increasing immunogenicity of some of the target cells. This may depend on manipulating the broad variety of immune cell components present in the tumor to enrich for the more tumoricidal subsets. Achieving this goal will require a better understanding of the biological mechanisms that determine the safety, efficacy, and potential of immune cellular components in cancer treatment.

References

1. Brady SW, McQuerry JA, Qiao Y, Piccolo SR, Shrestha G, Jenkins DF, Layer RM, Pedersen BS, Miller RH, Esch A, Selitsky SR, Parker JS, Anderson LA, Dalley BK, Factor RE, Reddy CB, Boltax JP, Li DY, Moos PJ, Gray JW, Heiser LM, Buys SS, Cohen AL, Johnson WE, Quinlan AR, Marth G, Werner TL, Bild AH (2017) Combating subclonal evolution of resistant cancer phenotypes. Nat Commun 8(1):1231. https://doi.org/10.1038/s41467-017-01174-3
2. Kuipers J, Jahn K, Raphael BJ, Beerenwinkel N (2017) Single-cell sequencing data reveal widespread recurrence and loss of mutational hits in the life histories of tumors. Genome Res 27(11):1885–1894. https://doi.org/10.1101/gr.220707.117
3. Valkenburg KC, de Groot AE, Pienta KJ (2018) Targeting the tumour stroma to improve cancer therapy. Nat Rev Clin Oncol 15(6):366–381. https://doi.org/10.1038/s41571-018-0007-1
4. Nosho K, Baba Y, Tanaka N, Shima K, Hayashi M, Meyerhardt JA, Giovannucci E, Dranoff G, Fuchs CS, Ogino S (2010) Tumour-infiltrating T-cell subsets, molecular changes in colorectal cancer, and prognosis: cohort study and literature review. J Pathol 222(4):350–366. https://doi.org/10.1002/path.2774
5. Pages F, Berger A, Camus M, Sanchez-Cabo F, Costes A, Molidor R, Mlecnik B, Kirilovsky A, Nilsson M, Damotte D, Meatchi T, Bruneval P, Cugnenc PH, Trajanoski Z, Fridman WH, Galon J (2005) Effector memory T cells, early metastasis, and survival in colorectal cancer. N Engl J Med 353(25):2654–2666. https://doi.org/10.1056/NEJMoa051424
6. Gajewski TF, Schreiber H, Fu YX (2013) Innate and adaptive immune cells in the tumor microenvironment. Nat Immunol 14(10):1014–1022. https://doi.org/10.1038/ni.2703
7. Scott AM, Wolchok JD, Old LJ (2012) Antibody therapy of cancer. Nat Rev Cancer 12(4):278–287. https://doi.org/10.1038/nrc3236
8. Nimmerjahn F, Ravetch JV (2007) Antibodies, Fc receptors and cancer. Curr Opin Immunol 19(2):239–245. https://doi.org/10.1016/j.coi.2007.01.005
9. Kessenbrock K, Plaks V, Werb Z (2010) Matrix metalloproteinases: regulators of the tumor microenvironment. Cell 141(1):52–67. https://doi.org/10.1016/j.cell.2010.03.015
10. Liu RB, Engels B, Arina A, Schreiber K, Hyjek E, Schietinger A, Binder DC, Butz E, Krausz T, Rowley DA, Jabri B, Schreiber H (2012) Densely granulated murine NK cells eradicate

large solid tumors. Cancer Res 72(8):1964–1974. https://doi.org/10.1158/0008-5472.CAN-11-3208

11. Homey B, Muller A, Zlotnik A (2002) Chemokines: agents for the immunotherapy of cancer? Nat Rev Immunol 2(3):175–184. https://doi.org/10.1038/nri748
12. Kitamura T, Qian BZ, Pollard JW (2015) Immune cell promotion of metastasis. Nat Rev Immunol 15(2):73–86. https://doi.org/10.1038/nri3789
13. Garrido F (2019) MHC/HLA class I loss in cancer cells. Adv Exp Med Biol 1151:15–78. https://doi.org/10.1007/978-3-030-17864-2_2
14. Golstein P, Griffiths GM (2018) An early history of T cell-mediated cytotoxicity. Nat Rev Immunol 18(8):527–535. https://doi.org/10.1038/s41577-018-0009-3
15. Whitmire JK, Eam B, Benning N, Whitton JL (2007) Direct interferon-gamma signaling dramatically enhances CD4+ and CD8+ T cell memory. J Immunol 179(2):1190–1197. https://doi.org/10.4049/jimmunol.179.2.1190
16. Reading JL, Galvez-Cancino F, Swanton C, Lladser A, Peggs KS, Quezada SA (2018) The function and dysfunction of memory CD8(+) T cells in tumor immunity. Immunol Rev 283(1):194–212. https://doi.org/10.1111/imr.12657
17. Han S, Zhang C, Li Q, Dong J, Liu Y, Huang Y, Jiang T, Wu A (2014) Tumour-infiltrating CD4(+) and CD8(+) lymphocytes as predictors of clinical outcome in glioma. Br J Cancer 110(10):2560–2568. https://doi.org/10.1038/bjc.2014.162
18. Yang JC, Rosenberg SA (2016) Adoptive T-cell therapy for cancer. Adv Immunol 130:279–294. https://doi.org/10.1016/bs.ai.2015.12.006
19. Curiel TJ, Coukos G, Zou L, Alvarez X, Cheng P, Mottram P, Evdemon-Hogan M, Conejo-Garcia JR, Zhang L, Burow M, Zhu Y, Wei S, Kryczek I, Daniel B, Gordon A, Myers L, Lackner A, Disis ML, Knutson KL, Chen L, Zou W (2004) Specific recruitment of regulatory T cells in ovarian carcinoma fosters immune privilege and predicts reduced survival. Nat Med 10(9):942–949. https://doi.org/10.1038/nm1093
20. Peggs KS, Quezada SA, Chambers CA, Korman AJ, Allison JP (2009) Blockade of CTLA-4 on both effector and regulatory T cell compartments contributes to the antitumor activity of anti-CTLA-4 antibodies. J Exp Med 206(8):1717–1725. https://doi.org/10.1084/jem.20082492
21. Quezada SA, Peggs KS, Simpson TR, Shen Y, Littman DR, Allison JP (2008) Limited tumor infiltration by activated T effector cells restricts the therapeutic activity of regulatory T cell depletion against established melanoma. J Exp Med 205(9):2125–2138. https://doi.org/10.1084/jem.20080099
22. Speiser DE, Ho PC, Verdeil G (2016) Regulatory circuits of T cell function in cancer. Nat Rev Immunol 16(10):599–611. https://doi.org/10.1038/nri.2016.80
23. Shimizu J, Yamazaki S, Sakaguchi S (1999) Induction of tumor immunity by removing CD25+CD4+ T cells: a common basis between tumor immunity and autoimmunity. J Immunol 163(10):5211–5218
24. Dai Y, Jiao H, Teng G, Wang W, Zhang R, Wang Y, Hebbard L, George J, Qiao L (2014) Embelin reduces colitis-associated tumorigenesis through limiting IL-6/STAT3 signaling. Mol Cancer Ther 13(5):1206–1216. https://doi.org/10.1158/1535-7163.MCT-13-0378
25. Lanca T, Silva-Santos B (2012) The split nature of tumor-infiltrating leukocytes: implications for cancer surveillance and immunotherapy. Oncoimmunology 1(5):717–725. https://doi.org/10.4161/onci.20068
26. Bremnes RM, Al-Shibli K, Donnem T, Sirera R, Al-Saad S, Andersen S, Stenvold H, Camps C, Busund LT (2011) The role of tumor-infiltrating immune cells and chronic inflammation at the tumor site on cancer development, progression, and prognosis: emphasis on non-small cell lung cancer. J Thorac Oncol 6(4):824–833. https://doi.org/10.1097/JTO.0b013e3182037b76
27. Fialova A, Partlova S, Sojka L, Hromadkova H, Brtnicky T, Fucikova J, Kocian P, Rob L, Bartunkova J, Spisek R (2013) Dynamics of T-cell infiltration during the course of ovarian cancer: the gradual shift from a Th17 effector cell response to a predominant infiltration by regulatory T-cells. Int J Cancer 132(5):1070–1079. https://doi.org/10.1002/ijc.27759
28. Jochems C, Schlom J (2011) Tumor-infiltrating immune cells and prognosis: the potential link between conventional cancer therapy and immunity. Exp Biol Med (Maywood) 236(5):567–579. https://doi.org/10.1258/ebm.2011.011007
29. Bailey SR, Nelson MH, Himes RA, Li Z, Mehrotra S, Paulos CM (2014) Th17 cells in cancer: the ultimate identity crisis. Front Immunol 5:276. https://doi.org/10.3389/fimmu.2014.00276
30. June CH, O'Connor RS, Kawalekar OU, Ghassemi S, Milone MC (2018) CAR T cell immunotherapy for human cancer. Science 359(6382):1361–1365. https://doi.org/10.1126/science.aar6711

31. Bruno TC, Ebner PJ, Moore BL, Squalls OG, Waugh KA, Eruslanov EB, Singhal S, Mitchell JD, Franklin WA, Merrick DT, McCarter MD, Palmer BE, Kern JA, Slansky JE (2017) Antigen-presenting intratumoral B cells affect CD4(+) TIL phenotypes in non-small cell lung cancer patients. Cancer Immunol Res 5(10):898–907. https://doi.org/10.1158/2326-6066.CIR-17-0075
32. Chiaruttini G, Mele S, Opzoomer J, Crescioli S, Ilieva KM, Lacy KE, Karagiannis SN (2017) B cells and the humoral response in melanoma: the overlooked players of the tumor microenvironment. Oncoimmunology 6(4):e1294296. https://doi.org/10.1080/2162402X.2017.1294296
33. Dong HP, Elstrand MB, Holth A, Silins I, Berner A, Trope CG, Davidson B, Risberg B (2006) NK- and B-cell infiltration correlates with worse outcome in metastatic ovarian carcinoma. Am J Clin Pathol 125(3):451–458
34. Yang C, Lee H, Pal S, Jove V, Deng J, Zhang W, Hoon DS, Wakabayashi M, Forman S, Yu H (2013) B cells promote tumor progression via STAT3 regulated-angiogenesis. PLoS One 8(5):e64159. https://doi.org/10.1371/journal.pone.0064159
35. Santoiemma PP, Reyes C, Wang LP, McLane MW, Feldman MD, Tanyi JL, Powell DJ Jr (2016) Systematic evaluation of multiple immune markers reveals prognostic factors in ovarian cancer. Gynecol Oncol 143(1):120–127. https://doi.org/10.1016/j.ygyno.2016.07.105
36. Nielsen JS, Sahota RA, Milne K, Kost SE, Nesslinger NJ, Watson PH, Nelson BH (2012) CD20+ tumor-infiltrating lymphocytes have an atypical CD27− memory phenotype and together with CD8+ T cells promote favorable prognosis in ovarian cancer. Clin Cancer Res 18(12):3281–3292. https://doi.org/10.1158/1078-0432.CCR-12-0234
37. Montfort A, Pearce O, Maniati E, Vincent BG, Bixby L, Bohm S, Dowe T, Wilkes EH, Chakravarty P, Thompson R, Topping J, Cutillas PR, Lockley M, Serody JS, Capasso M, Balkwill FR (2017) A strong B-cell response is part of the immune landscape in human high-grade serous ovarian metastases. Clin Cancer Res 23(1):250–262. https://doi.org/10.1158/1078-0432.CCR-16-0081
38. Milne K, Kobel M, Kalloger SE, Barnes RO, Gao D, Gilks CB, Watson PH, Nelson BH (2009) Systematic analysis of immune infiltrates in high-grade serous ovarian cancer reveals CD20, FoxP3 and TIA-1 as positive prognostic factors. PLoS One 4(7):e6412. https://doi.org/10.1371/journal.pone.0006412
39. Kroeger DR, Milne K, Nelson BH (2016) Tumor-infiltrating plasma cells are associated with tertiary lymphoid structures, cytolytic T-cell responses, and superior prognosis in ovarian cancer. Clin Cancer Res 22(12):3005–3015. https://doi.org/10.1158/1078-0432.CCR-15-2762
40. Sarvaria A, Madrigal JA, Saudemont A (2017) B cell regulation in cancer and anti-tumor immunity. Cell Mol Immunol 14(8):662–674. https://doi.org/10.1038/cmi.2017.35
41. Fortner RT, Damms-Machado A, Kaaks R (2017) Systematic review: tumor-associated antigen autoantibodies and ovarian cancer early detection. Gynecol Oncol 147(2):465–480. https://doi.org/10.1016/j.ygyno.2017.07.138
42. Cohen M, Petignat P (2011) Purified autoantibodies against glucose-regulated protein 78 (GRP78) promote apoptosis and decrease invasiveness of ovarian cancer cells. Cancer Lett 309(1):104–109. https://doi.org/10.1016/j.canlet.2011.05.022
43. Van Belle K, Herman J, Boon L, Waer M, Sprangers B, Louat T (2016) Comparative in vitro immune stimulation analysis of primary human B cells and B cell lines. J Immunol Res 5281823:2016. https://doi.org/10.1155/2016/5281823
44. Renshaw BR, Fanslow WC 3rd, Armitage RJ, Campbell KA, Liggitt D, Wright B, Davison BL, Maliszewski CR (1994) Humoral immune responses in CD40 ligand-deficient mice. J Exp Med 180(5):1889–1900. https://doi.org/10.1084/jem.180.5.1889
45. Kornbluth RS, Stempniak M, Stone GW (2012) Design of CD40 agonists and their use in growing B cells for cancer immunotherapy. Int Rev Immunol 31(4):279–288. https://doi.org/10.3109/08830185.2012.703272
46. Kawabe T, Naka T, Yoshida K, Tanaka T, Fujiwara H, Suematsu S, Yoshida N, Kishimoto T, Kikutani H (1994) The immune responses in CD40-deficient mice: impaired immunoglobulin class switching and germinal center formation. Immunity 1(3):167–178
47. Danese S, Sans M, Fiocchi C (2004) The CD40/CD40L costimulatory pathway in inflammatory bowel disease. Gut 53(7):1035–1043. https://doi.org/10.1136/gut.2003.026278
48. Clark EA (2014) A short history of the B-cell-associated surface molecule CD40. Front Immunol 5:472. https://doi.org/10.3389/fimmu.2014.00472
49. Castigli E, Alt FW, Davidson L, Bottaro A, Mizoguchi E, Bhan AK, Geha RS (1994) CD40-deficient mice generated by recombination-activating gene-2-deficient

blastocyst complementation. Proc Natl Acad Sci U S A 91(25):12135–12139. https://doi.org/10.1073/pnas.91.25.12135
50. Sorenmo KU, Krick E, Coughlin CM, Overley B, Gregor TP, Vonderheide RH, Mason NJ (2011) CD40-activated B cell cancer vaccine improves second clinical remission and survival in privately owned dogs with non-Hodgkin's lymphoma. PLoS One 6(8):e24167. https://doi.org/10.1371/journal.pone.0024167
51. Gu Y, Liu Y, Fu L, Zhai L, Zhu J, Han Y, Jiang Y, Zhang Y, Zhang P, Jiang Z, Zhang X, Cao X (2019) Tumor-educated B cells selectively promote breast cancer lymph node metastasis by HSPA4-targeting IgG. Nat Med 25(2):312–322. https://doi.org/10.1038/s41591-018-0309-y
52. Schioppa T, Moore R, Thompson RG, Rosser EC, Kulbe H, Nedospasov S, Mauri C, Coussens LM, Balkwill FR (2011) B regulatory cells and the tumor-promoting actions of TNF-alpha during squamous carcinogenesis. Proc Natl Acad Sci U S A 108(26):10662–10667. https://doi.org/10.1073/pnas.1100994108
53. Lindner S, Dahlke K, Sontheimer K, Hagn M, Kaltenmeier C, Barth TF, Beyer T, Reister F, Fabricius D, Lotfi R, Lunov O, Nienhaus GU, Simmet T, Kreienberg R, Moller P, Schrezenmeier H, Jahrsdorfer B (2013) Interleukin 21-induced granzyme B-expressing B cells infiltrate tumors and regulate T cells. Cancer Res 73(8):2468–2479. https://doi.org/10.1158/0008-5472.CAN-12-3450
54. Schwartz M, Zhang Y, Rosenblatt JD (2016) B cell regulation of the anti-tumor response and role in carcinogenesis. J Immunother Cancer 4:40. https://doi.org/10.1186/s40425-016-0145-x
55. Mion F, Tonon S, Valeri V, Pucillo CE (2017) Message in a bottle from the tumor microenvironment: tumor-educated DCs instruct B cells to participate in immunosuppression. Cell Mol Immunol 14(9):730–732. https://doi.org/10.1038/cmi.2017.63
56. He Y, Qian H, Liu Y, Duan L, Li Y, Shi G (2014) The roles of regulatory B cells in cancer. J Immunol Res 2014:215471. https://doi.org/10.1155/2014/215471
57. Xiao X, Lao XM, Chen MM, Liu RX, Wei Y, Ouyang FZ, Chen DP, Zhao XY, Zhao Q, Li XF, Liu CL, Zheng L, Kuang DM (2016) PD-1hi identifies a novel regulatory B-cell population in human hepatoma that promotes disease progression. Cancer Discov 6(5):546–559. https://doi.org/10.1158/2159-8290.CD-15-1408
58. Wang X, Wang G, Wang Z, Liu B, Han N, Li J, Lu C, Liu X, Zhang Q, Yang Q, Wang G (2019) PD-1-expressing B cells suppress CD4(+) and CD8(+) T cells via PD-1/PD-L1-dependent pathway. Mol Immunol 109:20–26. https://doi.org/10.1016/j.molimm.2019.02.009
59. Damsky W, Jilaveanu L, Turner N, Perry C, Zito C, Tomayko M, Leventhal J, Herold K, Meffre E, Bosenberg M, Kluger HM (2019) B cell depletion or absence does not impede anti-tumor activity of PD-1 inhibitors. J Immunother Cancer 7(1):153. https://doi.org/10.1186/s40425-019-0613-1
60. Zhang Y, Morgan R, Chen C, Cai Y, Clark E, Khan WN, Shin SU, Cho HM, Al Bayati A, Pimentel A, Rosenblatt JD (2016) Mammary-tumor-educated B cells acquire LAP/TGF-beta and PD-L1 expression and suppress anti-tumor immune responses. Int Immunol 28(9):423–433. https://doi.org/10.1093/intimm/dxw007
61. Andreu P, Johansson M, Affara NI, Pucci F, Tan T, Junankar S, Korets L, Lam J, Tawfik D, DeNardo DG, Naldini L, de Visser KE, De Palma M, Coussens LM (2010) FcRgamma activation regulates inflammation-associated squamous carcinogenesis. Cancer Cell 17(2):121–134. https://doi.org/10.1016/j.ccr.2009.12.019
62. Affara NI, Ruffell B, Medler TR, Gunderson AJ, Johansson M, Bornstein S, Bergsland E, Steinhoff M, Li Y, Gong Q, Ma Y, Wiesen JF, Wong MH, Kulesz-Martin M, Irving B, Coussens LM (2014) B cells regulate macrophage phenotype and response to chemotherapy in squamous carcinomas. Cancer Cell 25(6):809–821. https://doi.org/10.1016/j.ccr.2014.04.026
63. Somasundaram R, Zhang G, Fukunaga-Kalabis M, Perego M, Krepler C, Xu X, Wagner C, Hristova D, Zhang J, Tian T, Wei Z, Liu Q, Garg K, Griss J, Hards R, Maurer M, Hafner C, Mayerhofer M, Karanikas G, Jalili A, Bauer-Pohl V, Weihsengruber F, Rappersberger K, Koller J, Lang R, Hudgens C, Chen G, Tetzlaff M, Wu L, Frederick DT, Scolyer RA, Long GV, Damle M, Ellingsworth C, Grinman L, Choi H, Gavin BJ, Dunagin M, Raj A, Scholler N, Gross L, Beqiri M, Bennett K, Watson I, Schaider H, Davies MA, Wargo J, Czerniecki BJ, Schuchter L, Herlyn D, Flaherty K, Herlyn M, Wagner SN (2017) Tumor-associated B-cells induce tumor heterogeneity and therapy resistance. Nat Commun 8(1):607. https://doi.org/10.1038/s41467-017-00452-4
64. Woo SR, Corrales L, Gajewski TF (2015) Innate immune recognition of

cancer. Annu Rev Immunol 33:445–474. https://doi.org/10.1146/annurev-immunol-032414-112043
65. Villegas FR, Coca S, Villarrubia VG, Jimenez R, Chillon MJ, Jareno J, Zuil M, Callol L (2002) Prognostic significance of tumor infiltrating natural killer cells subset CD57 in patients with squamous cell lung cancer. Lung Cancer 35(1):23–28
66. Veluchamy JP, Kok N, van der Vliet HJ, Verheul HMW, de Gruijl TD, Spanholtz J (2017) The rise of allogeneic natural killer cells as a platform for cancer immunotherapy: recent innovations and future developments. Front Immunol 8:631. https://doi.org/10.3389/fimmu.2017.00631
67. Childs RW, Carlsten M (2015) Therapeutic approaches to enhance natural killer cell cytotoxicity against cancer: the force awakens. Nat Rev Drug Discov 14(7):487–498. https://doi.org/10.1038/nrd4506
68. Martin-Fontecha A, Thomsen LL, Brett S, Gerard C, Lipp M, Lanzavecchia A, Sallusto F (2004) Induced recruitment of NK cells to lymph nodes provides IFN-gamma for T(H)1 priming. Nat Immunol 5(12):1260–1265. https://doi.org/10.1038/ni1138
69. Kelly JM, Darcy PK, Markby JL, Godfrey DI, Takeda K, Yagita H, Smyth MJ (2002) Induction of tumor-specific T cell memory by NK cell-mediated tumor rejection. Nat Immunol 3(1):83–90. https://doi.org/10.1038/ni746
70. Miller JS, Soignier Y, Panoskaltsis-Mortari A, McNearney SA, Yun GH, Fautsch SK, McKenna D, Le C, Defor TE, Burns LJ, Orchard PJ, Blazar BR, Wagner JE, Slungaard A, Weisdorf DJ, Okazaki IJ, McGlave PB (2005) Successful adoptive transfer and in vivo expansion of human haploidentical NK cells in patients with cancer. Blood 105(8):3051–3057. https://doi.org/10.1182/blood-2004-07-2974
71. Ruggeri L, Capanni M, Urbani E, Perruccio K, Shlomchik WD, Tosti A, Posati S, Rogaia D, Frassoni F, Aversa F, Martelli MF, Velardi A (2002) Effectiveness of donor natural killer cell alloreactivity in mismatched hematopoietic transplants. Science 295(5562):2097–2100. https://doi.org/10.1126/science.1068440
72. Leong JW, Schneider SE, Sullivan RP, Parikh BA, Anthony BA, Singh A, Jewell BA, Schappe T, Wagner JA, Link DC, Yokoyama WM, Fehniger TA (2015) PTEN regulates natural killer cell trafficking in vivo. Proc Natl Acad Sci U S A 112(7):E700–E709. https://doi.org/10.1073/pnas.1413886112
73. Jiang K, Zhong B, Gilvary DL, Corliss BC, Hong-Geller E, Wei S, Djeu JY (2000) Pivotal role of phosphoinositide-3 kinase in regulation of cytotoxicity in natural killer cells. Nat Immunol 1(5):419–425. https://doi.org/10.1038/80859
74. Chen X, Trivedi PP, Ge B, Krzewski K, Strominger JL (2007) Many NK cell receptors activate ERK2 and JNK1 to trigger microtubule organizing center and granule polarization and cytotoxicity. Proc Natl Acad Sci U S A 104(15):6329–6334. https://doi.org/10.1073/pnas.0611655104
75. Mellman I (2013) Dendritic cells: master regulators of the immune response. Cancer Immunol Res 1(3):145–149. https://doi.org/10.1158/2326-6066.CIR-13-0102
76. Collin M, McGovern N, Haniffa M (2013) Human dendritic cell subsets. Immunology 140(1):22–30. https://doi.org/10.1111/imm.12117
77. Iwasaki A, Medzhitov R (2004) Toll-like receptor control of the adaptive immune responses. Nat Immunol 5(10):987–995. https://doi.org/10.1038/ni1112
78. Fritz JH, Ferrero RL, Philpott DJ, Girardin SE (2006) Nod-like proteins in immunity, inflammation and disease. Nat Immunol 7(12):1250–1257. https://doi.org/10.1038/ni1412
79. Meylan E, Tschopp J (2006) Toll-like receptors and RNA helicases: two parallel ways to trigger antiviral responses. Mol Cell 22(5):561–569. https://doi.org/10.1016/j.molcel.2006.05.012
80. Reis e Sousa C (2004) Activation of dendritic cells: translating innate into adaptive immunity. Curr Opin Immunol 16(1):21–25
81. Geijtenbeek TB, van Vliet SJ, Engering A, t Hart BA, van Kooyk Y (2004) Self- and nonself-recognition by C-type lectins on dendritic cells. Annu Rev Immunol 22:33–54. https://doi.org/10.1146/annurev.immunol.22.012703.104558
82. Pierre P, Turley SJ, Gatti E, Hull M, Meltzer J, Mirza A, Inaba K, Steinman RM, Mellman I (1997) Developmental regulation of MHC class II transport in mouse dendritic cells. Nature 388(6644):787–792. https://doi.org/10.1038/42039
83. Cella M, Engering A, Pinet V, Pieters J, Lanzavecchia A (1997) Inflammatory stimuli induce accumulation of MHC class II complexes on dendritic cells. Nature 388(6644):782–787. https://doi.org/10.1038/42030
84. Yanagihara S, Komura E, Nagafune J, Watarai H, Yamaguchi Y (1998) EBI1/CCR7 is a new member of dendritic cell chemokine receptor that is up-regulated upon maturation. J Immunol 161(6):3096–3102
85. Sallusto F, Schaerli P, Loetscher P, Schaniel C, Lenig D, Mackay CR, Qin

S, Lanzavecchia A (1998) Rapid and coordinated switch in chemokine receptor expression during dendritic cell maturation. Eur J Immunol 28(9):2760–2769. https://doi.org/10.1002/(SICI)1521-4141(199809)28:09<2760::AID-IMMU2760>3.0.CO;2-N
86. Dieu MC, Vanbervliet B, Vicari A, Bridon JM, Oldham E, Ait-Yahia S, Briere F, Zlotnik A, Lebecque S, Caux C (1998) Selective recruitment of immature and mature dendritic cells by distinct chemokines expressed in different anatomic sites. J Exp Med 188(2):373–386. https://doi.org/10.1084/jem.188.2.373
87. Tran Janco JM, Lamichhane P, Karyampudi L, Knutson KL (2015) Tumor-infiltrating dendritic cells in cancer pathogenesis. J Immunol 194(7):2985–2991. https://doi.org/10.4049/jimmunol.1403134
88. Batista FD, Harwood NE (2009) The who, how and where of antigen presentation to B cells. Nat Rev Immunol 9(1):15–27. https://doi.org/10.1038/nri2454
89. Chijioke O, Munz C (2013) Dendritic cell derived cytokines in human natural killer cell differentiation and activation. Front Immunol 4:365. https://doi.org/10.3389/fimmu.2013.00365
90. Lin A, Schildknecht A, Nguyen LT, Ohashi PS (2010) Dendritic cells integrate signals from the tumor microenvironment to modulate immunity and tumor growth. Immunol Lett 127(2):77–84. https://doi.org/10.1016/j.imlet.2009.09.003
91. Ladanyi A, Kiss J, Somlai B, Gilde K, Fejos Z, Mohos A, Gaudi I, Timar J (2007) Density of DC-LAMP(+) mature dendritic cells in combination with activated T lymphocytes infiltrating primary cutaneous melanoma is a strong independent prognostic factor. Cancer Immunol Immunother 56(9):1459–1469. https://doi.org/10.1007/s00262-007-0286-3
92. Ma Y, Shurin GV, Peiyuan Z, Shurin MR (2013) Dendritic cells in the cancer microenvironment. J Cancer 4(1):36–44. https://doi.org/10.7150/jca.5046
93. Iwamoto M, Shinohara H, Miyamoto A, Okuzawa M, Mabuchi H, Nohara T, Gon G, Toyoda M, Tanigawa N (2003) Prognostic value of tumor-infiltrating dendritic cells expressing CD83 in human breast carcinomas. Int J Cancer 104(1):92–97. https://doi.org/10.1002/ijc.10915
94. Tsukayama S, Omura K, Yoshida K, Tanaka Y, Watanabe G (2005) Prognostic value of CD83-positive mature dendritic cells and their relation to vascular endothelial growth factor in advanced human gastric cancer. Oncol Rep 14(2):369–375
95. Kocian P, Sedivcova M, Drgac J, Cerna K, Hoch J, Kodet R, Bartunkova J, Spisek R, Fialova A (2011) Tumor-infiltrating lymphocytes and dendritic cells in human colorectal cancer: their relationship to KRAS mutational status and disease recurrence. Hum Immunol 72(11):1022–1028. https://doi.org/10.1016/j.humimm.2011.07.312
96. Sisirak V, Faget J, Gobert M, Goutagny N, Vey N, Treilleux I, Renaudineau S, Poyet G, Labidi-Galy SI, Goddard-Leon S, Durand I, Le Mercier I, Bajard A, Bachelot T, Puisieux A, Puisieux I, Blay JY, Menetrier-Caux C, Caux C, Bendriss-Vermare N (2012) Impaired IFN-alpha production by plasmacytoid dendritic cells favors regulatory T-cell expansion that may contribute to breast cancer progression. Cancer Res 72(20):5188–5197. https://doi.org/10.1158/0008-5472.CAN-11-3468
97. Ramos RN, de Moraes CJ, Zelante B, Barbuto JA (2013) What are the molecules involved in regulatory T-cells induction by dendritic cells in cancer? Clin Dev Immunol 2013:806025. https://doi.org/10.1155/2013/806025
98. Michielsen AJ, Hogan AE, Marry J, Tosetto M, Cox F, Hyland JM, Sheahan KD, O'Donoghue DP, Mulcahy HE, Ryan EJ, O'Sullivan JN (2011) Tumour tissue microenvironment can inhibit dendritic cell maturation in colorectal cancer. PLoS One 6(11):e27944. https://doi.org/10.1371/journal.pone.0027944
99. Krempski J, Karyampudi L, Behrens MD, Erskine CL, Hartmann L, Dong H, Goode EL, Kalli KR, Knutson KL (2011) Tumor-infiltrating programmed death receptor-1+ dendritic cells mediate immune suppression in ovarian cancer. J Immunol 186(12):6905–6913. https://doi.org/10.4049/jimmunol.1100274
100. Chiba S, Baghdadi M, Akiba H, Yoshiyama H, Kinoshita I, Dosaka-Akita H, Fujioka Y, Ohba Y, Gorman JV, Colgan JD, Hirashima M, Uede T, Takaoka A, Yagita H, Jinushi M (2012) Tumor-infiltrating DCs suppress nucleic acid-mediated innate immune responses through interactions between the receptor TIM-3 and the alarmin HMGB1. Nat Immunol 13(9):832–842. https://doi.org/10.1038/ni.2376
101. Cheng P, Corzo CA, Luetteke N, Yu B, Nagaraj S, Bui MM, Ortiz M, Nacken W, Sorg C, Vogl T, Roth J, Gabrilovich DI (2008) Inhibition of dendritic cell differentiation and accumulation of myeloid-derived suppressor cells in cancer is regulated by S100A9 protein. J Exp Med 205(10):2235–2249. https://doi.org/10.1084/jem.20080132
102. Nefedova Y, Huang M, Kusmartsev S, Bhattacharya R, Cheng P, Salup R, Jove R,

Gabrilovich D (2004) Hyperactivation of STAT3 is involved in abnormal differentiation of dendritic cells in cancer. J Immunol 172(1):464–474. https://doi.org/10.4049/jimmunol.172.1.464

103. Yu H, Pardoll D, Jove R (2009) STATs in cancer inflammation and immunity: a leading role for STAT3. Nat Rev Cancer 9(11):798–809. https://doi.org/10.1038/nrc2734

104. Watkins SK, Zhu Z, Riboldi E, Shafer-Weaver KA, Stagliano KE, Sklavos MM, Ambs S, Yagita H, Hurwitz AA (2011) FOXO3 programs tumor-associated DCs to become tolerogenic in human and murine prostate cancer. J Clin Invest 121(4):1361–1372. https://doi.org/10.1172/JCI44325

105. Kusmartsev S, Nefedova Y, Yoder D, Gabrilovich DI (2004) Antigen-specific inhibition of CD8+ T cell response by immature myeloid cells in cancer is mediated by reactive oxygen species. J Immunol 172(2):989–999. https://doi.org/10.4049/jimmunol.172.2.989

106. Liu Q, Zhang C, Sun A, Zheng Y, Wang L, Cao X (2009) Tumor-educated CD11bhighIalow regulatory dendritic cells suppress T cell response through arginase I. J Immunol 182(10):6207–6216. https://doi.org/10.4049/jimmunol.0803926

107. Melero I, Gaudernack G, Gerritsen W, Huber C, Parmiani G, Scholl S, Thatcher N, Wagstaff J, Zielinski C, Faulkner I, Mellstedt H (2014) Therapeutic vaccines for cancer: an overview of clinical trials. Nat Rev Clin Oncol 11(9):509–524. https://doi.org/10.1038/nrclinonc.2014.111

108. Trumpfheller C, Longhi MP, Caskey M, Idoyaga J, Bozzacco L, Keler T, Schlesinger SJ, Steinman RM (2012) Dendritic cell-targeted protein vaccines: a novel approach to induce T-cell immunity. J Intern Med 271(2):183–192. https://doi.org/10.1111/j.1365-2796.2011.02496.x

109. Kantoff PW, Higano CS, Shore ND, Berger ER, Small EJ, Penson DF, Redfern CH, Ferrari AC, Dreicer R, Sims RB, Xu Y, Frohlich MW, Schellhammer PF, Investigators IS (2010) Sipuleucel-T immunotherapy for castration-resistant prostate cancer. N Engl J Med 363(5):411–422. https://doi.org/10.1056/NEJMoa1001294

110. Ozao-Choy J, Lee DJ, Faries MB (2014) Melanoma vaccines: mixed past, promising future. Surg Clin North Am 94(5):1017–1030, viii. https://doi.org/10.1016/j.suc.2014.07.005

111. Cassetta L, Pollard JW (2018) Targeting macrophages: therapeutic approaches in cancer. Nat Rev Drug Discov 17(12):887–904. https://doi.org/10.1038/nrd.2018.169

112. Mills CD, Kincaid K, Alt JM, Heilman MJ, Hill AM (2000) M-1/M-2 macrophages and the Th1/Th2 paradigm. J Immunol 164(12):6166–6173. https://doi.org/10.4049/jimmunol.164.12.6166

113. Salmaninejad A, Valilou SF, Soltani A, Ahmadi S, Abarghan YJ, Rosengren RJ, Sahebkar A (2019) Tumor-associated macrophages: role in cancer development and therapeutic implications. Cell Oncol (Dordr) 42:591–608. https://doi.org/10.1007/s13402-019-00453-z

114. Yang L, Zhang Y (2017) Tumor-associated macrophages: from basic research to clinical application. J Hematol Oncol 10(1):58. https://doi.org/10.1186/s13045-017-0430-2

115. Wang H, Shao Q, Sun J, Ma C, Gao W, Wang Q, Zhao L, Qu X (2016) Interactions between colon cancer cells and tumor-infiltrated macrophages depending on cancer cell-derived colony stimulating factor 1. Oncoimmunology 5(4):e1122157. https://doi.org/10.1080/2162402X.2015.1122157

116. DeNardo DG, Barreto JB, Andreu P, Vasquez L, Tawfik D, Kolhatkar N, Coussens LM (2009) CD4(+) T cells regulate pulmonary metastasis of mammary carcinomas by enhancing protumor properties of macrophages. Cancer Cell 16(2):91–102. https://doi.org/10.1016/j.ccr.2009.06.018

117. Gocheva V, Wang HW, Gadea BB, Shree T, Hunter KE, Garfall AL, Berman T, Joyce JA (2010) IL-4 induces cathepsin protease activity in tumor-associated macrophages to promote cancer growth and invasion. Genes Dev 24(3):241–255. https://doi.org/10.1101/gad.1874010

118. Martinez FO, Helming L, Milde R, Varin A, Melgert BN, Draijer C, Thomas B, Fabbri M, Crawshaw A, Ho LP, Ten Hacken NH, Cobos Jimenez V, Kootstra NA, Hamann J, Greaves DR, Locati M, Mantovani A, Gordon S (2013) Genetic programs expressed in resting and IL-4 alternatively activated mouse and human macrophages: similarities and differences. Blood 121(9):e57–e69. https://doi.org/10.1182/blood-2012-06-436212

119. Scotton CJ, Martinez FO, Smelt MJ, Sironi M, Locati M, Mantovani A, Sozzani S (2005) Transcriptional profiling reveals complex regulation of the monocyte IL-1 beta system by IL-13. J Immunol 174(2):834–845. https://doi.org/10.4049/jimmunol.174.2.834

120. Zhao P, Gao D, Wang Q, Song B, Shao Q, Sun J, Ji C, Li X, Li P, Qu X (2015) Response gene to complement 32 (RGC-32) expression on M2-polarized and tumor-associated macrophages is M-CSF-dependent and enhanced by tumor-derived IL-4. Cell Mol Immunol

12(6):692–699. https://doi.org/10.1038/cmi.2014.108
121. Sanchez-Martin L, Estecha A, Samaniego R, Sanchez-Ramon S, Vega MA, Sanchez-Mateos P (2011) The chemokine CXCL12 regulates monocyte-macrophage differentiation and RUNX3 expression. Blood 117(1):88–97. https://doi.org/10.1182/blood-2009-12-258186
122. Yaddanapudi K, Putty K, Rendon BE, Lamont GJ, Faughn JD, Satoskar A, Lasnik A, Eaton JW, Mitchell RA (2013) Control of tumor-associated macrophage alternative activation by macrophage migration inhibitory factor. J Immunol 190(6):2984–2993. https://doi.org/10.4049/jimmunol.1201650
123. Cua DJ, Stohlman SA (1997) In vivo effects of T helper cell type 2 cytokines on macrophage antigen-presenting cell induction of T helper subsets. J Immunol 159(12):5834–5840
124. Savage ND, de Boer T, Walburg KV, Joosten SA, van Meijgaarden K, Geluk A, Ottenhoff TH (2008) Human anti-inflammatory macrophages induce Foxp3+ GITR+ CD25+ regulatory T cells, which suppress via membrane-bound TGFbeta-1. J Immunol 181(3):2220–2226. https://doi.org/10.4049/jimmunol.181.3.2220
125. Martinez FO, Sica A, Mantovani A, Locati M (2008) Macrophage activation and polarization. Front Biosci 13:453–461
126. Yang M, McKay D, Pollard JW, Lewis CE (2018) Diverse functions of macrophages in different tumor microenvironments. Cancer Res 78(19):5492–5503. https://doi.org/10.1158/0008-5472.CAN-18-1367
127. Banerjee S, Halder K, Bose A, Bhattacharya P, Gupta G, Karmahapatra S, Das S, Chaudhuri S, Bhattacharyya Majumdar S, Majumdar S (2011) TLR signaling-mediated differential histone modification at IL-10 and IL-12 promoter region leads to functional impairments in tumor-associated macrophages. Carcinogenesis 32(12):1789–1797. https://doi.org/10.1093/carcin/bgr208
128. Nakanishi Y, Nakatsuji M, Seno H, Ishizu S, Akitake-Kawano R, Kanda K, Ueo T, Komekado H, Kawada M, Minami M, Chiba T (2011) COX-2 inhibition alters the phenotype of tumor-associated macrophages from M2 to M1 in ApcMin/+ mouse polyps. Carcinogenesis 32(9):1333–1339. https://doi.org/10.1093/carcin/bgr128
129. Wang YC, He F, Feng F, Liu XW, Dong GY, Qin HY, Hu XB, Zheng MH, Liang L, Feng L, Liang YM, Han H (2010) Notch signaling determines the M1 versus M2 polarization of macrophages in antitumor immune responses. Cancer Res 70(12):4840–4849. https://doi.org/10.1158/0008-5472.CAN-10-0269
130. Mantovani A, Sozzani S, Locati M, Allavena P, Sica A (2002) Macrophage polarization: tumor-associated macrophages as a paradigm for polarized M2 mononuclear phagocytes. Trends Immunol 23(11):549–555
131. Qian BZ, Li J, Zhang H, Kitamura T, Zhang J, Campion LR, Kaiser EA, Snyder LA, Pollard JW (2011) CCL2 recruits inflammatory monocytes to facilitate breast-tumour metastasis. Nature 475(7355):222–225. https://doi.org/10.1038/nature10138
132. Wang R, Zhang J, Chen S, Lu M, Luo X, Yao S, Liu S, Qin Y, Chen H (2011) Tumor-associated macrophages provide a suitable microenvironment for non-small lung cancer invasion and progression. Lung Cancer 74(2):188–196. https://doi.org/10.1016/j.lungcan.2011.04.009
133. Zhang S, Che D, Yang F, Chi C, Meng H, Shen J, Qi L, Liu F, Lv L, Li Y, Meng Q, Liu J, Shang L, Yu Y (2017) Tumor-associated macrophages promote tumor metastasis via the TGF-beta/SOX9 axis in non-small cell lung cancer. Oncotarget 8(59):99801–99815. https://doi.org/10.18632/oncotarget.21068
134. Capece D, Fischietti M, Verzella D, Gaggiano A, Cicciarelli G, Tessitore A, Zazzeroni F, Alesse E (2013) The inflammatory microenvironment in hepatocellular carcinoma: a pivotal role for tumor-associated macrophages. Biomed Res Int 2013:187204. https://doi.org/10.1155/2013/187204
135. Kim KJ, Wen XY, Yang HK, Kim WH, Kang GH (2015) Prognostic implication of M2 macrophages are determined by the proportional balance of tumor associated macrophages and tumor infiltrating lymphocytes in microsatellite-unstable gastric carcinoma. PLoS One 10(12):e0144192. https://doi.org/10.1371/journal.pone.0144192
136. Bostrom MM, Irjala H, Mirtti T, Taimen P, Kauko T, Algars A, Jalkanen S, Bostrom PJ (2015) Tumor-associated macrophages provide significant prognostic information in urothelial bladder cancer. PLoS One 10(7):e0133552. https://doi.org/10.1371/journal.pone.0133552
137. Bowman RL, Klemm F, Akkari L, Pyonteck SM, Sevenich L, Quail DF, Dhara S, Simpson K, Gardner EE, Iacobuzio-Donahue CA, Brennan CW, Tabar V, Gutin PH, Joyce JA (2016) Macrophage ontogeny underlies differences in tumor-specific education in brain malignancies. Cell Rep 17(9):2445–2459. https://doi.org/10.1016/j.celrep.2016.10.052
138. Chen Z, Feng X, Herting CJ, Garcia VA, Nie K, Pong WW, Rasmussen R, Dwivedi B, Seby S, Wolf SA, Gutmann DH, Hambardzumyan

D (2017) Cellular and molecular identity of tumor-associated macrophages in glioblastoma. Cancer Res 77(9):2266–2278. https://doi.org/10.1158/0008-5472.CAN-16-2310
139. Loyher PL, Hamon P, Laviron M, Meghraoui-Kheddar A, Goncalves E, Deng Z, Torstensson S, Bercovici N, Baudesson de Chanville C, Combadiere B, Geissmann F, Savina A, Combadiere C, Boissonnas A (2018) Macrophages of distinct origins contribute to tumor development in the lung. J Exp Med 215(10):2536–2553. https://doi.org/10.1084/jem.20180534
140. Zhu Y, Herndon JM, Sojka DK, Kim KW, Knolhoff BL, Zuo C, Cullinan DR, Luo J, Bearden AR, Lavine KJ, Yokoyama WM, Hawkins WG, Fields RC, Randolph GJ, DeNardo DG (2017) Tissue-resident macrophages in pancreatic ductal adenocarcinoma originate from embryonic hematopoiesis and promote tumor progression. Immunity 47(3):597. https://doi.org/10.1016/j.immuni.2017.08.018
141. Coffelt SB, Tal AO, Scholz A, De Palma M, Patel S, Urbich C, Biswas SK, Murdoch C, Plate KH, Reiss Y, Lewis CE (2010) Angiopoietin-2 regulates gene expression in TIE2-expressing monocytes and augments their inherent proangiogenic functions. Cancer Res 70(13):5270–5280. https://doi.org/10.1158/0008-5472.CAN-10-0012
142. Forssell J, Oberg A, Henriksson ML, Stenling R, Jung A, Palmqvist R (2007) High macrophage infiltration along the tumor front correlates with improved survival in colon cancer. Clin Cancer Res 13(5):1472–1479. https://doi.org/10.1158/1078-0432.CCR-06-2073
143. Laoui D, Van Overmeire E, Di Conza G, Aldeni C, Keirsse J, Morias Y, Movahedi K, Houbracken I, Schouppe E, Elkrim Y, Karroum O, Jordan B, Carmeliet P, Gysemans C, De Baetselier P, Mazzone M, Van Ginderachter JA (2014) Tumor hypoxia does not drive differentiation of tumor-associated macrophages but rather fine-tunes the M2-like macrophage population. Cancer Res 74(1):24–30. https://doi.org/10.1158/0008-5472.CAN-13-1196
144. Migita T, Sato E, Saito K, Mizoi T, Shiiba K, Matsuno S, Nagura H, Ohtani H (1999) Differing expression of MMPs-1 and -9 and urokinase receptor between diffuse- and intestinal-type gastric carcinoma. Int J Cancer 84(1):74–79. https://doi.org/10.1002/(sici)1097-0215(19990219)84:1<74::aid-ijc14>3.0.co;2-i
145. Ohtani H, Naito Y, Saito K, Nagura H (1997) Expression of costimulatory molecules B7-1 and B7-2 by macrophages along invasive margin of colon cancer: a possible antitumor immunity? Lab Investig 77(3):231–241
146. Zhou Q, Peng RQ, Wu XJ, Xia Q, Hou JH, Ding Y, Zhou QM, Zhang X, Pang ZZ, Wan DS, Zeng YX, Zhang XS (2010) The density of macrophages in the invasive front is inversely correlated to liver metastasis in colon cancer. J Transl Med 8:13. https://doi.org/10.1186/1479-5876-8-13
147. Shimura S, Yang G, Ebara S, Wheeler TM, Frolov A, Thompson TC (2000) Reduced infiltration of tumor-associated macrophages in human prostate cancer: association with cancer progression. Cancer Res 60(20):5857–5861
148. Ohno S, Inagawa H, Dhar DK, Fujii T, Ueda S, Tachibana M, Suzuki N, Inoue M, Soma G, Nagasue N (2003) The degree of macrophage infiltration into the cancer cell nest is a significant predictor of survival in gastric cancer patients. Anticancer Res 23(6D):5015–5022
149. Ohno S, Ohno Y, Suzuki N, Kamei T, Koike K, Inagawa H, Kohchi C, Soma G, Inoue M (2004) Correlation of histological localization of tumor-associated macrophages with clinicopathological features in endometrial cancer. Anticancer Res 24(5C):3335–3342
150. Chao MP, Weissman IL, Majeti R (2012) The CD47-SIRPalpha pathway in cancer immune evasion and potential therapeutic implications. Curr Opin Immunol 24(2):225–232. https://doi.org/10.1016/j.coi.2012.01.010
151. Willingham SB, Volkmer JP, Gentles AJ, Sahoo D, Dalerba P, Mitra SS, Wang J, Contreras-Trujillo H, Martin R, Cohen JD, Lovelace P, Scheeren FA, Chao MP, Weiskopf K, Tang C, Volkmer AK, Naik TJ, Storm TA, Mosley AR, Edris B, Schmid SM, Sun CK, Chua MS, Murillo O, Rajendran P, Cha AC, Chin RK, Kim D, Adorno M, Raveh T, Tseng D, Jaiswal S, Enger PO, Steinberg GK, Li G, So SK, Majeti R, Harsh GR, van de Rijn M, Teng NN, Sunwoo JB, Alizadeh AA, Clarke MF, Weissman IL (2012) The CD47-signal regulatory protein alpha (SIRPa) interaction is a therapeutic target for human solid tumors. Proc Natl Acad Sci U S A 109(17):6662–6667. https://doi.org/10.1073/pnas.1121623109
152. Kuang DM, Zhao Q, Peng C, Xu J, Zhang JP, Wu C, Zheng L (2009) Activated monocytes in peritumoral stroma of hepatocellular carcinoma foster immune privilege and disease progression through PD-L1. J Exp Med 206(6):1327–1337. https://doi.org/10.1084/jem.20082173
153. Gwak JM, Jang MH, Kim DI, Seo AN, Park SY (2015) Prognostic value of tumor-associated macrophages according to histologic locations

and hormone receptor status in breast cancer. PLoS One 10(4):e0125728. https://doi.org/10.1371/journal.pone.0125728
154. Hu JM, Liu K, Liu JH, Jiang XL, Wang XL, Yang L, Chen YZ, Liu CX, Li SG, Cui XB, Zou H, Pang LJ, Zhao J, Qi Y, Liang WH, Yuan XL, Li F (2017) The increased number of tumor-associated macrophage is associated with overexpression of VEGF-C, plays an important role in Kazakh ESCC invasion and metastasis. Exp Mol Pathol 102(1):15–21. https://doi.org/10.1016/j.yexmp.2016.12.001
155. Jensen TO, Schmidt H, Moller HJ, Hoyer M, Maniecki MB, Sjoegren P, Christensen IJ, Steiniche T (2009) Macrophage markers in serum and tumor have prognostic impact in American Joint Committee on Cancer stage I/II melanoma. J Clin Oncol 27(20):3330–3337. https://doi.org/10.1200/JCO.2008.19.9919
156. Varol C, Sagi I (2018) Phagocyte-extracellular matrix crosstalk empowers tumor development and dissemination. FEBS J 285(4):734–751. https://doi.org/10.1111/febs.14317
157. Pinto ML, Rios E, Silva AC, Neves SC, Caires HR, Pinto AT, Duraes C, Carvalho FA, Cardoso AP, Santos NC, Barrias CC, Nascimento DS, Pinto-do OP, Barbosa MA, Carneiro F, Oliveira MJ (2017) Decellularized human colorectal cancer matrices polarize macrophages towards an anti-inflammatory phenotype promoting cancer cell invasion via CCL18. Biomaterials 124:211–224. https://doi.org/10.1016/j.biomaterials.2017.02.004
158. Previtera ML, Sengupta A (2015) Substrate stiffness regulates proinflammatory mediator production through TLR4 activity in macrophages. PLoS One 10(12):e0145813. https://doi.org/10.1371/journal.pone.0145813
159. Hu H, Hang JJ, Han T, Zhuo M, Jiao F, Wang LW (2016) The M2 phenotype of tumor-associated macrophages in the stroma confers a poor prognosis in pancreatic cancer. Tumour Biol 37(7):8657–8664. https://doi.org/10.1007/s13277-015-4741-z
160. Medrek C, Ponten F, Jirstrom K, Leandersson K (2012) The presence of tumor associated macrophages in tumor stroma as a prognostic marker for breast cancer patients. BMC Cancer 12:306. https://doi.org/10.1186/1471-2407-12-306
161. Ni YH, Ding L, Huang XF, Dong YC, Hu QG, Hou YY (2015) Microlocalization of CD68+ tumor-associated macrophages in tumor stroma correlated with poor clinical outcomes in oral squamous cell carcinoma patients. Tumour Biol 36(7):5291–5298. https://doi.org/10.1007/s13277-015-3189-5
162. Park JY, Sung JY, Lee J, Park YK, Kim YW, Kim GY, Won KY, Lim SJ (2016) Polarized CD163+ tumor-associated macrophages are associated with increased angiogenesis and CXCL12 expression in gastric cancer. Clin Res Hepatol Gastroenterol 40(3):357–365. https://doi.org/10.1016/j.clinre.2015.09.005
163. Aljabery F, Olsson H, Gimm O, Jahnson S, Shabo I (2018) M2-macrophage infiltration and macrophage traits of tumor cells in urinary bladder cancer. Urol Oncol 36(4):159 e119–159 e126. https://doi.org/10.1016/j.urolonc.2017.11.020
164. Carus A, Ladekarl M, Hager H, Nedergaard BS, Donskov F (2013) Tumour-associated CD66b+ neutrophil count is an independent prognostic factor for recurrence in localised cervical cancer. Br J Cancer 108(10):2116–2122. https://doi.org/10.1038/bjc.2013.167
165. Carus A, Ladekarl M, Hager H, Pilegaard H, Nielsen PS, Donskov F (2013) Tumor-associated neutrophils and macrophages in non-small cell lung cancer: no immediate impact on patient outcome. Lung Cancer 81(1):130–137. https://doi.org/10.1016/j.lungcan.2013.03.003
166. Casazza A, Laoui D, Wenes M, Rizzolio S, Bassani N, Mambretti M, Deschoemaeker S, Van Ginderachter JA, Tamagnone L, Mazzone M (2013) Impeding macrophage entry into hypoxic tumor areas by Sema3A/Nrp1 signaling blockade inhibits angiogenesis and restores antitumor immunity. Cancer Cell 24(6):695–709. https://doi.org/10.1016/j.ccr.2013.11.007
167. Ye LY, Chen W, Bai XL, Xu XY, Zhang Q, Xia XF, Sun X, Li GG, Hu QD, Fu QH, Liang TB (2016) Hypoxia-induced epithelial-to-mesenchymal transition in hepatocellular carcinoma induces an immunosuppressive tumor microenvironment to promote metastasis. Cancer Res 76(4):818–830. https://doi.org/10.1158/0008-5472.CAN-15-0977
168. Kawanaka T, Kubo A, Ikushima H, Sano T, Takegawa Y, Nishitani H (2008) Prognostic significance of HIF-2alpha expression on tumor infiltrating macrophages in patients with uterine cervical cancer undergoing radiotherapy. J Med Investig 55(1–2):78–86
169. Leek RD, Lewis CE, Whitehouse R, Greenall M, Clarke J, Harris AL (1996) Association of macrophage infiltration with angiogenesis and prognosis in invasive breast carcinoma. Cancer Res 56(20):4625–4629

170. De Palma M, Venneri MA, Galli R, Sergi Sergi L, Politi LS, Sampaolesi M, Naldini L (2005) Tie2 identifies a hematopoietic lineage of proangiogenic monocytes required for tumor vessel formation and a mesenchymal population of pericyte progenitors. Cancer Cell 8(3):211–226. https://doi.org/10.1016/j.ccr.2005.08.002
171. Ji J, Zhang G, Sun B, Yuan H, Huang Y, Zhang J, Wei X, Zhang X, Hou J (2013) The frequency of tumor-infiltrating Tie-2-expressing monocytes in renal cell carcinoma: its relationship to angiogenesis and progression. Urology 82(4):974–e979-913. https://doi.org/10.1016/j.urology.2013.05.026
172. Matsubara T, Kanto T, Kuroda S, Yoshio S, Higashitani K, Kakita N, Miyazaki M, Sakakibara M, Hiramatsu N, Kasahara A, Tomimaru Y, Tomokuni A, Nagano H, Hayashi N, Takehara T (2013) TIE2-expressing monocytes as a diagnostic marker for hepatocellular carcinoma correlates with angiogenesis. Hepatology 57(4):1416–1425. https://doi.org/10.1002/hep.25965
173. Mazzieri R, Pucci F, Moi D, Zonari E, Ranghetti A, Berti A, Politi LS, Gentner B, Brown JL, Naldini L, De Palma M (2011) Targeting the ANG2/TIE2 axis inhibits tumor growth and metastasis by impairing angiogenesis and disabling rebounds of proangiogenic myeloid cells. Cancer Cell 19(4):512–526. https://doi.org/10.1016/j.ccr.2011.02.005
174. Harney AS, Arwert EN, Entenberg D, Wang Y, Guo P, Qian BZ, Oktay MH, Pollard JW, Jones JG, Condeelis JS (2015) Real-time imaging reveals local, transient vascular permeability, and tumor cell intravasation stimulated by TIE2hi macrophage-derived VEGFA. Cancer Discov 5(9):932–943. https://doi.org/10.1158/2159-8290.CD-15-0012
175. Linde N, Casanova-Acebes M, Sosa MS, Mortha A, Rahman A, Farias E, Harper K, Tardio E, Reyes Torres I, Jones J, Condeelis J, Merad M, Aguirre-Ghiso JA (2018) Macrophages orchestrate breast cancer early dissemination and metastasis. Nat Commun 9(1):21. https://doi.org/10.1038/s41467-017-02481-5
176. Rohan TE, Xue X, Lin HM, D'Alfonso TM, Ginter PS, Oktay MH, Robinson BD, Ginsberg M, Gertler FB, Glass AG, Sparano JA, Condeelis JS, Jones JG (2014) Tumor microenvironment of metastasis and risk of distant metastasis of breast cancer. J Natl Cancer Inst 106(8). https://doi.org/10.1093/jnci/dju136
177. Ahn GO, Tseng D, Liao CH, Dorie MJ, Czechowicz A, Brown JM (2010) Inhibition of Mac-1 (CD11b/CD18) enhances tumor response to radiation by reducing myeloid cell recruitment. Proc Natl Acad Sci U S A 107(18):8363–8368. https://doi.org/10.1073/pnas.0911378107
178. Germano G, Frapolli R, Belgiovine C, Anselmo A, Pesce S, Liguori M, Erba E, Uboldi S, Zucchetti M, Pasqualini F, Nebuloni M, van Rooijen N, Mortarini R, Beltrame L, Marchini S, Fuso Nerini I, Sanfilippo R, Casali PG, Pilotti S, Galmarini CM, Anichini A, Mantovani A, D'Incalci M, Allavena P (2013) Role of macrophage targeting in the antitumor activity of trabectedin. Cancer Cell 23(2):249–262. https://doi.org/10.1016/j.ccr.2013.01.008
179. Gordon EM, Sankhala KK, Chawla N, Chawla SP (2016) Trabectedin for soft tissue sarcoma: current status and future perspectives. Adv Ther 33(7):1055–1071. https://doi.org/10.1007/s12325-016-0344-3
180. Yang L, Wang F, Wang L, Huang L, Wang J, Zhang B, Zhang Y (2015) CD163+ tumor-associated macrophage is a prognostic biomarker and is associated with therapeutic effect on malignant pleural effusion of lung cancer patients. Oncotarget 6(12):10592–10603. https://doi.org/10.18632/oncotarget.3547
181. Song M, Liu T, Shi C, Zhang X, Chen X (2016) Bioconjugated manganese dioxide nanoparticles enhance chemotherapy response by priming tumor-associated macrophages toward M1-like phenotype and attenuating tumor hypoxia. ACS Nano 10(1):633–647. https://doi.org/10.1021/acsnano.5b06779
182. Wang Y, Lin YX, Qiao SL, An HW, Ma Y, Qiao ZY, Rajapaksha RP, Wang H (2017) Polymeric nanoparticles promote macrophage reversal from M2 to M1 phenotypes in the tumor microenvironment. Biomaterials 112:153–163. https://doi.org/10.1016/j.biomaterials.2016.09.034
183. Zanganeh S, Hutter G, Spitler R, Lenkov O, Mahmoudi M, Shaw A, Pajarinen JS, Nejadnik H, Goodman S, Moseley M, Coussens LM, Daldrup-Link HE (2016) Iron oxide nanoparticles inhibit tumour growth by inducing pro-inflammatory macrophage polarization in tumour tissues. Nat Nanotechnol 11(11):986–994. https://doi.org/10.1038/nnano.2016.168
184. Beatty GL, Chiorean EG, Fishman MP, Saboury B, Teitelbaum UR, Sun W, Huhn RD, Song W, Li D, Sharp LL, Torigian DA, O'Dwyer PJ, Vonderheide RH (2011) CD40 agonists alter tumor stroma and show efficacy against pancreatic carcinoma in mice and humans. Science 331(6024):1612–1616. https://doi.org/10.1126/science.1198443

185. Chan GC, Chan WK, Sze DM (2009) The effects of beta-glucan on human immune and cancer cells. J Hematol Oncol 2:25. https://doi.org/10.1186/1756-8722-2-25
186. Kushner BH, Cheung IY, Modak S, Kramer K, Ragupathi G, Cheung NK (2014) Phase I trial of a bivalent gangliosides vaccine in combination with beta-glucan for high-risk neuroblastoma in second or later remission. Clin Cancer Res 20(5):1375–1382. https://doi.org/10.1158/1078-0432.CCR-13-1012
187. Segal NH, Gada P, Senzer N, Gargano MA, Patchen ML, Saltz LB (2016) A phase II efficacy and safety, open-label, multicenter study of imprime PGG injection in combination with cetuximab in patients with stage IV KRAS-mutant colorectal cancer. Clin Colorectal Cancer 15(3):222–227. https://doi.org/10.1016/j.clcc.2016.02.013
188. Burke B, Sumner S, Maitland N, Lewis CE (2002) Macrophages in gene therapy: cellular delivery vehicles and in vivo targets. J Leukoc Biol 72(3):417–428
189. Griffiths L, Binley K, Iqball S, Kan O, Maxwell P, Ratcliffe P, Lewis C, Harris A, Kingsman S, Naylor S (2000) The macrophage – a novel system to deliver gene therapy to pathological hypoxia. Gene Ther 7(3):255–262. https://doi.org/10.1038/sj.gt.3301058
190. Moyes KW, Lieberman NA, Kreuser SA, Chinn H, Winter C, Deutsch G, Hoglund V, Watson R, Crane CA (2017) Genetically engineered macrophages: a potential platform for cancer immunotherapy. Hum Gene Ther 28(2):200–215. https://doi.org/10.1089/hum.2016.060
191. Morrissey MA, Williamson AP, Steinbach AM, Roberts EW, Kern N, Headley MB, Vale RD (2018) Chimeric antigen receptors that trigger phagocytosis. elife 7. https://doi.org/10.7554/eLife.36688
192. Advani R, Flinn I, Popplewell L, Forero A, Bartlett NL, Ghosh N, Kline J, Roschewski M, LaCasce A, Collins GP, Tran T, Lynn J, Chen JY, Volkmer JP, Agoram B, Huang J, Majeti R, Weissman IL, Takimoto CH, Chao MP, Smith SM (2018) CD47 blockade by Hu5F9-G4 and rituximab in non-Hodgkin's lymphoma. N Engl J Med 379(18):1711–1721. https://doi.org/10.1056/NEJMoa1807315
193. Yang H, Shao R, Huang H, Wang X, Rong Z, Lin Y (2019) Engineering macrophages to phagocytose cancer cells by blocking the CD47/SIRPa axis. Cancer Med 8:4245–4253. https://doi.org/10.1002/cam4.2332
194. Barkal AA, Brewer RE, Markovic M, Kowarsky M, Barkal SA, Zaro BW, Krishnan V, Hatakeyama J, Dorigo O, Barkal LJ, Weissman IL (2019) CD24 signalling through macrophage Siglec-10 is a target for cancer immunotherapy. Nature. https://doi.org/10.1038/s41586-019-1456-0
195. Scapini P, Lapinet-Vera JA, Gasperini S, Calzetti F, Bazzoni F, Cassatella MA (2000) The neutrophil as a cellular source of chemokines. Immunol Rev 177:195–203
196. Borrello MG, Alberti L, Fischer A, Degl'innocenti D, Ferrario C, Gariboldi M, Marchesi F, Allavena P, Greco A, Collini P, Pilotti S, Cassinelli G, Bressan P, Fugazzola L, Mantovani A, Pierotti MA (2005) Induction of a proinflammatory program in normal human thyrocytes by the RET/PTC1 oncogene. Proc Natl Acad Sci U S A 102(41):14825–14830. https://doi.org/10.1073/pnas.0503039102
197. Sparmann A, Bar-Sagi D (2004) Ras-induced interleukin-8 expression plays a critical role in tumor growth and angiogenesis. Cancer Cell 6(5):447–458. https://doi.org/10.1016/j.ccr.2004.09.028
198. Gentles AJ, Newman AM, Liu CL, Bratman SV, Feng W, Kim D, Nair VS, Xu Y, Khuong A, Hoang CD, Diehn M, West RB, Plevritis SK, Alizadeh AA (2015) The prognostic landscape of genes and infiltrating immune cells across human cancers. Nat Med 21(8):938–945. https://doi.org/10.1038/nm.3909
199. Templeton AJ, McNamara MG, Seruga B, Vera-Badillo FE, Aneja P, Ocana A, Leibowitz-Amit R, Sonpavde G, Knox JJ, Tran B, Tannock IF, Amir E (2014) Prognostic role of neutrophil-to-lymphocyte ratio in solid tumors: a systematic review and meta-analysis. J Natl Cancer Inst 106(6):dju124. https://doi.org/10.1093/jnci/dju124
200. Wei B, Yao M, Xing C, Wang W, Yao J, Hong Y, Liu Y, Fu P (2016) The neutrophil lymphocyte ratio is associated with breast cancer prognosis: an updated systematic review and meta-analysis. Onco Targets Ther 9:5567–5575. https://doi.org/10.2147/OTT.S108419
201. Shen M, Hu P, Donskov F, Wang G, Liu Q, Du J (2014) Tumor-associated neutrophils as a new prognostic factor in cancer: a systematic review and meta-analysis. PLoS One 9(6):e98259. https://doi.org/10.1371/journal.pone.0098259
202. McCarthy EF (2006) The toxins of William B. Coley and the treatment of bone and soft-tissue sarcomas. Iowa Orthop J 26:154–158
203. Otten MA, Rudolph E, Dechant M, Tuk CW, Reijmers RM, Beelen RH, van de Winkel JG, van Egmond M (2005) Immature neutrophils mediate tumor cell killing via IgA but not IgG Fc receptors. J Immunol 174(9):5472–5480. https://doi.org/10.4049/jimmunol.174.9.5472

204. Stockmeyer B, Dechant M, van Egmond M, Tutt AL, Sundarapandiyan K, Graziano RF, Repp R, Kalden JR, Gramatzki M, Glennie MJ, van de Winkel JG, Valerius T (2000) Triggering Fc alpha-receptor I (CD89) recruits neutrophils as effector cells for CD20-directed antibody therapy. J Immunol 165(10):5954–5961. https://doi.org/10.4049/jimmunol.165.10.5954
205. Fridlender ZG, Sun J, Kim S, Kapoor V, Cheng G, Ling L, Worthen GS, Albelda SM (2009) Polarization of tumor-associated neutrophil phenotype by TGF-beta: "N1" versus "N2" TAN. Cancer Cell 16(3):183–194. https://doi.org/10.1016/j.ccr.2009.06.017
206. Yan J, Kloecker G, Fleming C, Bousamra M II, Hansen R, Hu X, Ding C, Cai Y, Xiang D, Donninger H, Eaton JW, Clark GJ (2014) Human polymorphonuclear neutrophils specifically recognize and kill cancerous cells. Oncoimmunology 3(7):e950163. https://doi.org/10.4161/15384101.2014.950163
207. Simons MP, O'Donnell MA, Griffith TS (2008) Role of neutrophils in BCG immunotherapy for bladder cancer. Urol Oncol 26(4):341–345. https://doi.org/10.1016/j.urolonc.2007.11.031
208. Coffelt SB, Wellenstein MD, de Visser KE (2016) Neutrophils in cancer: neutral no more. Nat Rev Cancer 16(7):431–446. https://doi.org/10.1038/nrc.2016.52
209. Michaeli J, Shaul ME, Mishalian I, Hovav AH, Levy L, Zolotriov L, Granot Z, Fridlender ZG (2017) Tumor-associated neutrophils induce apoptosis of non-activated CD8 T-cells in a TNFalpha and NO-dependent mechanism, promoting a tumor-supportive environment. Oncoimmunology 6(11):e1356965. https://doi.org/10.1080/2162402X.2017.1356965
210. Coffelt SB, Kersten K, Doornebal CW, Weiden J, Vrijland K, Hau CS, Verstegen NJM, Ciampricotti M, Hawinkels L, Jonkers J, de Visser KE (2015) IL-17-producing gammadelta T cells and neutrophils conspire to promote breast cancer metastasis. Nature 522(7556):345–348. https://doi.org/10.1038/nature14282
211. Wellenstein MD, Coffelt SB, Duits DEM, van Miltenburg MH, Slagter M, de Rink I, Henneman L, Kas SM, Prekovic S, Hau CS, Vrijland K, Drenth AP, de Korte-Grimmerink R, Schut E, van der Heijden I, Zwart W, Wessels LFA, Schumacher TN, Jonkers J, de Visser KE (2019) Loss of p53 triggers WNT-dependent systemic inflammation to drive breast cancer metastasis. Nature 572:538–542. https://doi.org/10.1038/s41586-019-1450-6
212. Houghton AM, Rzymkiewicz DM, Ji H, Gregory AD, Egea EE, Metz HE, Stolz DB, Land SR, Marconcini LA, Kliment CR, Jenkins KM, Beaulieu KA, Mouded M, Frank SJ, Wong KK, Shapiro SD (2010) Neutrophil elastase-mediated degradation of IRS-1 accelerates lung tumor growth. Nat Med 16(2):219–223. https://doi.org/10.1038/nm.2084
213. Faget J, Groeneveld S, Boivin G, Sankar M, Zangger N, Garcia M, Guex N, Zlobec I, Steiner L, Piersigilli A, Xenarios I, Meylan E (2017) Neutrophils and snail orchestrate the establishment of a pro-tumor microenvironment in lung cancer. Cell Rep 21(11):3190–3204. https://doi.org/10.1016/j.celrep.2017.11.052
214. Moschetta M, Uccello M, Kasenda B, Mak G, McClelland A, Boussios S, Forster M, Arkenau HT (2017) Dynamics of neutrophils-to-lymphocyte ratio predict outcomes of PD-1/PD-L1 blockade. Biomed Res Int 2017:1506824. https://doi.org/10.1155/2017/1506824
215. Keeley EC, Mehrad B, Strieter RM (2010) CXC chemokines in cancer angiogenesis and metastases. Adv Cancer Res 106:91–111. https://doi.org/10.1016/S0065-230X(10)06003-3
216. Scapini P, Morini M, Tecchio C, Minghelli S, Di Carlo E, Tanghetti E, Albini A, Lowell C, Berton G, Noonan DM, Cassatella MA (2004) CXCL1/macrophage inflammatory protein-2-induced angiogenesis in vivo is mediated by neutrophil-derived vascular endothelial growth factor-A. J Immunol 172(8):5034–5040. https://doi.org/10.4049/jimmunol.172.8.5034
217. Ardi VC, Kupriyanova TA, Deryugina EI, Quigley JP (2007) Human neutrophils uniquely release TIMP-free MMP-9 to provide a potent catalytic stimulator of angiogenesis. Proc Natl Acad Sci U S A 104(51):20262–20267. https://doi.org/10.1073/pnas.0706438104
218. Kuang DM, Zhao Q, Wu Y, Peng C, Wang J, Xu Z, Yin XY, Zheng L (2011) Peritumoral neutrophils link inflammatory response to disease progression by fostering angiogenesis in hepatocellular carcinoma. J Hepatol 54(5):948–955. https://doi.org/10.1016/j.jhep.2010.08.041
219. Nozawa H, Chiu C, Hanahan D (2006) Infiltrating neutrophils mediate the initial angiogenic switch in a mouse model of multistage carcinogenesis. Proc Natl Acad Sci U S A 103(33):12493–12498. https://doi.org/10.1073/pnas.0601807103
220. Shojaei F, Singh M, Thompson JD, Ferrara N (2008) Role of Bv8 in neutrophil-dependent angiogenesis in a transgenic model of cancer progression. Proc Natl Acad

Sci U S A 105(7):2640–2645. https://doi.org/10.1073/pnas.0712185105
221. Coffelt SB, de Visser KE (2015) Immune-mediated mechanisms influencing the efficacy of anticancer therapies. Trends Immunol 36(4):198–216. https://doi.org/10.1016/j.it.2015.02.006
222. Condamine T, Mastio J, Gabrilovich DI (2015) Transcriptional regulation of myeloid-derived suppressor cells. J Leukoc Biol 98(6):913–922. https://doi.org/10.1189/jlb.4RI0515-204R
223. Jordan KR, Kapoor P, Spongberg E, Tobin RP, Gao D, Borges VF, McCarter MD (2017) Immunosuppressive myeloid-derived suppressor cells are increased in splenocytes from cancer patients. Cancer Immunol Immunother 66(4):503–513. https://doi.org/10.1007/s00262-016-1953-z
224. Meyer C, Sevko A, Ramacher M, Bazhin AV, Falk CS, Osen W, Borrello I, Kato M, Schadendorf D, Baniyash M, Umansky V (2011) Chronic inflammation promotes myeloid-derived suppressor cell activation blocking antitumor immunity in transgenic mouse melanoma model. Proc Natl Acad Sci U S A 108(41):17111–17116. https://doi.org/10.1073/pnas.1108121108
225. Ioannou M, Alissafi T, Lazaridis I, Deraos G, Matsoukas J, Gravanis A, Mastorodemos V, Plaitakis A, Sharpe A, Boumpas D, Verginis P (2012) Crucial role of granulocytic myeloid-derived suppressor cells in the regulation of central nervous system autoimmune disease. J Immunol 188(3):1136–1146. https://doi.org/10.4049/jimmunol.1101816
226. Safari E, Ghorghanlu S, Ahmadi-Khiavi H, Mehranfar S, Rezaei R, Motallebnezhad M (2019) Myeloid-derived suppressor cells and tumor: current knowledge and future perspectives. J Cell Physiol 234(7):9966–9981. https://doi.org/10.1002/jcp.27923
227. Dumitru CA, Moses K, Trellakis S, Lang S, Brandau S (2012) Neutrophils and granulocytic myeloid-derived suppressor cells: immunophenotyping, cell biology and clinical relevance in human oncology. Cancer Immunol Immunother 61(8):1155–1167. https://doi.org/10.1007/s00262-012-1294-5
228. Chang AL, Miska J, Wainwright DA, Dey M, Rivetta CV, Yu D, Kanojia D, Pituch KC, Qiao J, Pytel P, Han Y, Wu M, Zhang L, Horbinski CM, Ahmed AU, Lesniak MS (2016) CCL2 produced by the glioma microenvironment is essential for the recruitment of regulatory T cells and myeloid-derived suppressor cells. Cancer Res 76(19):5671–5682. https://doi.org/10.1158/0008-5472.CAN-16-0144
229. Chiu DK, Xu IM, Lai RK, Tse AP, Wei LL, Koh HY, Li LL, Lee D, Lo RC, Wong CM, Ng IO, Wong CC (2016) Hypoxia induces myeloid-derived suppressor cell recruitment to hepatocellular carcinoma through chemokine (C-C motif) ligand 26. Hepatology 64(3):797–813. https://doi.org/10.1002/hep.28655
230. Chun E, Lavoie S, Michaud M, Gallini CA, Kim J, Soucy G, Odze R, Glickman JN, Garrett WS (2015) CCL2 promotes colorectal carcinogenesis by enhancing polymorphonuclear myeloid-derived suppressor cell population and function. Cell Rep 12(2):244–257. https://doi.org/10.1016/j.celrep.2015.06.024
231. Holmgaard RB, Zamarin D, Li Y, Gasmi B, Munn DH, Allison JP, Merghoub T, Wolchok JD (2015) Tumor-expressed IDO recruits and activates MDSCs in a Treg-dependent manner. Cell Rep 13(2):412–424. https://doi.org/10.1016/j.celrep.2015.08.077
232. Barreda DR, Hanington PC, Belosevic M (2004) Regulation of myeloid development and function by colony stimulating factors. Dev Comp Immunol 28(5):509–554. https://doi.org/10.1016/j.dci.2003.09.010
233. Bronte V, Chappell DB, Apolloni E, Cabrelle A, Wang M, Hwu P, Restifo NP (1999) Unopposed production of granulocyte-macrophage colony-stimulating factor by tumors inhibits CD8+ T cell responses by dysregulating antigen-presenting cell maturation. J Immunol 162(10):5728–5737
234. Dolcetti L, Peranzoni E, Ugel S, Marigo I, Fernandez Gomez A, Mesa C, Geilich M, Winkels G, Traggiai E, Casati A, Grassi F, Bronte V (2010) Hierarchy of immunosuppressive strength among myeloid-derived suppressor cell subsets is determined by GM-CSF. Eur J Immunol 40(1):22–35. https://doi.org/10.1002/eji.200939903
235. Lechner MG, Liebertz DJ, Epstein AL (2010) Characterization of cytokine-induced myeloid-derived suppressor cells from normal human peripheral blood mononuclear cells. J Immunol 185(4):2273–2284. https://doi.org/10.4049/jimmunol.1000901
236. Morales JK, Kmieciak M, Knutson KL, Bear HD, Manjili MH (2010) GM-CSF is one of the main breast tumor-derived soluble factors involved in the differentiation of CD11b-Gr1-bone marrow progenitor cells into myeloid-derived suppressor cells. Breast Cancer Res Treat 123(1):39–49. https://doi.org/10.1007/s10549-009-0622-8
237. Xiang X, Poliakov A, Liu C, Liu Y, Deng ZB, Wang J, Cheng Z, Shah SV, Wang GJ, Zhang L, Grizzle WE, Mobley J, Zhang HG (2009) Induction of myeloid-derived suppressor cells by tumor exosomes. Int

J Cancer 124(11):2621–2633. https://doi.org/10.1002/ijc.24249
238. Jiang M, Chen J, Zhang W, Zhang R, Ye Y, Liu P, Yu W, Wei F, Ren X, Yu J (2017) Interleukin-6 trans-signaling pathway promotes immunosuppressive myeloid-derived suppressor cells via suppression of suppressor of cytokine signaling 3 in breast cancer. Front Immunol 8:1840. https://doi.org/10.3389/fimmu.2017.01840
239. Wang Y, Shen Y, Wang S, Shen Q, Zhou X (2018) The role of STAT3 in leading the crosstalk between human cancers and the immune system. Cancer Lett 415:117–128. https://doi.org/10.1016/j.canlet.2017.12.003
240. Wu CT, Hsieh CC, Lin CC, Chen WC, Hong JH, Chen MF (2012) Significance of IL-6 in the transition of hormone-resistant prostate cancer and the induction of myeloid-derived suppressor cells. J Mol Med (Berl) 90(11):1343–1355. https://doi.org/10.1007/s00109-012-0916-x
241. Gargett T, Christo SN, Hercus TR, Abbas N, Singhal N, Lopez AF, Brown MP (2016) GM-CSF signalling blockade and chemotherapeutic agents act in concert to inhibit the function of myeloid-derived suppressor cells in vitro. Clin Transl Immunol 5(12):e119. https://doi.org/10.1038/cti.2016.80
242. Corzo CA, Cotter MJ, Cheng P, Cheng F, Kusmartsev S, Sotomayor E, Padhya T, McCaffrey TV, McCaffrey JC, Gabrilovich DI (2009) Mechanism regulating reactive oxygen species in tumor-induced myeloid-derived suppressor cells. J Immunol 182(9):5693–5701. https://doi.org/10.4049/jimmunol.0900092
243. Poschke I, Mougiakakos D, Hansson J, Masucci GV, Kiessling R (2010) Immature immunosuppressive CD14+HLA–DR–/low cells in melanoma patients are Stat3hi and overexpress CD80, CD83, and DC-sign. Cancer Res 70(11):4335–4345. https://doi.org/10.1158/0008-5472.CAN-09-3767
244. Kusmartsev S, Gabrilovich DI (2005) STAT1 signaling regulates tumor-associated macrophage-mediated T cell deletion. J Immunol 174(8):4880–4891. https://doi.org/10.4049/jimmunol.174.8.4880
245. Movahedi K, Guilliams M, Van den Bossche J, Van den Bergh R, Gysemans C, Beschin A, De Baetselier P, Van Ginderachter JA (2008) Identification of discrete tumor-induced myeloid-derived suppressor cell subpopulations with distinct T cell-suppressive activity. Blood 111(8):4233–4244. https://doi.org/10.1182/blood-2007-07-099226
246. Dardalhon V, Awasthi A, Kwon H, Galileos G, Gao W, Sobel RA, Mitsdoerffer M, Strom TB, Elyaman W, Ho IC, Khoury S, Oukka M, Kuchroo VK (2008) IL-4 inhibits TGF-beta-induced Foxp3+ T cells and, together with TGF-beta, generates IL-9+ IL-10+ Foxp3(−) effector T cells. Nat Immunol 9(12):1347–1355. https://doi.org/10.1038/ni.1677
247. Ramain P, Khechumian K, Seugnet L, Arbogast N, Ackermann C, Heitzler P (2001) Novel notch alleles reveal a Deltex-dependent pathway repressing neural fate. Curr Biol 11(22):1729–1738. https://doi.org/10.1016/s0960-9822(01)00562-0
248. Cheng P, Kumar V, Liu H, Youn JI, Fishman M, Sherman S, Gabrilovich D (2014) Effects of notch signaling on regulation of myeloid cell differentiation in cancer. Cancer Res 74(1):141–152. https://doi.org/10.1158/0008-5472.CAN-13-1686
249. Gibb DR, Saleem SJ, Kang DJ, Subler MA, Conrad DH (2011) ADAM10 overexpression shifts lympho- and myelopoiesis by dysregulating site 2/site 3 cleavage products of Notch. J Immunol 186(7):4244–4252. https://doi.org/10.4049/jimmunol.1003318
250. Park YJ, Song B, Kim YS, Kim EK, Lee JM, Lee GE, Kim JO, Kim YJ, Chang WS, Kang CY (2013) Tumor microenvironmental conversion of natural killer cells into myeloid-derived suppressor cells. Cancer Res 73(18):5669–5681. https://doi.org/10.1158/0008-5472.CAN-13-0545
251. Gabrilovich DI (2017) Myeloid-derived suppressor cells. Cancer Immunol Res 5(1):3–8. https://doi.org/10.1158/2326-6066.CIR-16-0297
252. Hoechst B, Gamrekelashvili J, Manns MP, Greten TF, Korangy F (2011) Plasticity of human Th17 cells and iTregs is orchestrated by different subsets of myeloid cells. Blood 117(24):6532–6541. https://doi.org/10.1182/blood-2010-11-317321
253. Pan PY, Wang GX, Yin B, Ozao J, Ku T, Divino CM, Chen SH (2008) Reversion of immune tolerance in advanced malignancy: modulation of myeloid-derived suppressor cell development by blockade of stem-cell factor function. Blood 111(1):219–228. https://doi.org/10.1182/blood-2007-04-086835
254. Qu P, Yan C, Du H (2011) Matrix metalloproteinase 12 overexpression in myeloid lineage cells plays a key role in modulating myelopoiesis, immune suppression, and lung tumorigenesis. Blood 117(17):4476–4489. https://doi.org/10.1182/blood-2010-07-298380
255. Beury DW, Parker KH, Nyandjo M, Sinha P, Carter KA, Ostrand-Rosenberg S (2014) Cross-talk among myeloid-derived suppressor cells, macrophages, and tumor cells impacts the inflammatory milieu of solid tumors. J Leukoc Biol 96(6):1109–1118. https://doi.org/10.1189/jlb.3A0414-210R

256. Bogdan C (2001) Nitric oxide and the immune response. Nat Immunol 2(10):907–916. https://doi.org/10.1038/ni1001-907
257. Liu Y, Wei J, Guo G, Zhou J (2015) Norepinephrine-induced myeloid-derived suppressor cells block T-cell responses via generation of reactive oxygen species. Immunopharmacol Immunotoxicol 37(4):359–365. https://doi.org/10.3109/08923973.2015.1059442
258. Molon B, Ugel S, Del Pozzo F, Soldani C, Zilio S, Avella D, De Palma A, Mauri P, Monegal A, Rescigno M, Savino B, Colombo P, Jonjic N, Pecanic S, Lazzarato L, Fruttero R, Gasco A, Bronte V, Viola A (2011) Chemokine nitration prevents intratumoral infiltration of antigen-specific T cells. J Exp Med 208(10):1949–1962. https://doi.org/10.1084/jem.20101956
259. Beury DW, Carter KA, Nelson C, Sinha P, Hanson E, Nyandjo M, Fitzgerald PJ, Majeed A, Wali N, Ostrand-Rosenberg S (2016) Myeloid-derived suppressor cell survival and function are regulated by the transcription factor Nrf2. J Immunol 196(8):3470–3478. https://doi.org/10.4049/jimmunol.1501785
260. Jian SL, Chen WW, Su YC, Su YW, Chuang TH, Hsu SC, Huang LR (2017) Glycolysis regulates the expansion of myeloid-derived suppressor cells in tumor-bearing hosts through prevention of ROS-mediated apoptosis. Cell Death Dis 8(5):e2779. https://doi.org/10.1038/cddis.2017.192
261. Kusmartsev S, Eruslanov E, Kubler H, Tseng T, Sakai Y, Su Z, Kaliberov S, Heiser A, Rosser C, Dahm P, Siemann D, Vieweg J (2008) Oxidative stress regulates expression of VEGFR1 in myeloid cells: link to tumor-induced immune suppression in renal cell carcinoma. J Immunol 181(1):346–353. https://doi.org/10.4049/jimmunol.181.1.346
262. Nagaraj S, Gupta K, Pisarev V, Kinarsky L, Sherman S, Kang L, Herber DL, Schneck J, Gabrilovich DI (2007) Altered recognition of antigen is a mechanism of CD8+ T cell tolerance in cancer. Nat Med 13(7):828–835. https://doi.org/10.1038/nm1609
263. Stiff A, Trikha P, Mundy-Bosse B, McMichael E, Mace TA, Benner B, Kendra K, Campbell A, Gautam S, Abood D, Landi I, Hsu V, Duggan M, Wesolowski R, Old M, Howard JH, Yu L, Stasik N, Olencki T, Muthusamy N, Tridandapani S, Byrd JC, Caligiuri M, Carson WE (2018) Nitric oxide production by myeloid-derived suppressor cells plays a role in impairing Fc receptor-mediated natural killer cell function. Clin Cancer Res 24(8):1891–1904. https://doi.org/10.1158/1078-0432.CCR-17-0691
264. Ku AW, Muhitch JB, Powers CA, Diehl M, Kim M, Fisher DT, Sharda AP, Clements VK, O'Loughlin K, Minderman H, Messmer MN, Ma J, Skitzki JJ, Steeber DA, Walcheck B, Ostrand-Rosenberg S, Abrams SI, Evans SS (2016) Tumor-induced MDSC act via remote control to inhibit L-selectin-dependent adaptive immunity in lymph nodes. elife 5. https://doi.org/10.7554/eLife.17375
265. Obermajer N, Kalinski P (2012) Generation of myeloid-derived suppressor cells using prostaglandin E2. Transplant Res 1(1):15. https://doi.org/10.1186/2047-1440-1-15
266. Ostrand-Rosenberg S, Sinha P (2009) Myeloid-derived suppressor cells: linking inflammation and cancer. J Immunol 182(8):4499–4506. https://doi.org/10.4049/jimmunol.0802740
267. Gmunder H, Eck HP, Droge W (1991) Low membrane transport activity for cystine in resting and mitogenically stimulated human lymphocyte preparations and human T cell clones. Eur J Biochem 201(1):113–117. https://doi.org/10.1111/j.1432-1033.1991.tb16263.x
268. Rodriguez PC, Quiceno DG, Zabaleta J, Ortiz B, Zea AH, Piazuelo MB, Delgado A, Correa P, Brayer J, Sotomayor EM, Antonia S, Ochoa JB, Ochoa AC (2004) Arginase I production in the tumor microenvironment by mature myeloid cells inhibits T-cell receptor expression and antigen-specific T-cell responses. Cancer Res 64(16):5839–5849. https://doi.org/10.1158/0008-5472.CAN-04-0465
269. Hoskin DW, Mader JS, Furlong SJ, Conrad DM, Blay J (2008) Inhibition of T cell and natural killer cell function by adenosine and its contribution to immune evasion by tumor cells (Review). Int J Oncol 32(3):527–535
270. Li J, Wang L, Chen X, Li L, Li Y, Ping Y, Huang L, Yue D, Zhang Z, Wang F, Li F, Yang L, Huang J, Yang S, Li H, Zhao X, Dong W, Yan Y, Zhao S, Huang B, Zhang B, Zhang Y (2017) CD39/CD73 upregulation on myeloid-derived suppressor cells via TGF-beta-mTOR-HIF-1 signaling in patients with non-small cell lung cancer. Oncoimmunology 6(6):e1320011. https://doi.org/10.1080/2162402X.2017.1320011
271. Li L, Wang L, Li J, Fan Z, Yang L, Zhang Z, Zhang C, Yue D, Qin G, Zhang T, Li F, Chen X, Ping Y, Wang D, Gao Q, He Q, Huang L, Li H, Huang J, Zhao X, Xue W, Sun Z, Lu J, Yu JJ, Zhao J, Zhang B, Zhang Y (2018) Metformin-induced reduction of CD39 and CD73 blocks myeloid-derived suppressor cell activity in patients with ovarian cancer. Cancer Res 78(7):1779–1791. https://doi.org/10.1158/0008-5472.CAN-17-2460

272. Noman MZ, Desantis G, Janji B, Hasmim M, Karray S, Dessen P, Bronte V, Chouaib S (2014) PD-L1 is a novel direct target of HIF-1alpha, and its blockade under hypoxia enhanced MDSC-mediated T cell activation. J Exp Med 211(5):781–790. https://doi.org/10.1084/jem.20131916
273. Hoechst B, Voigtlaender T, Ormandy L, Gamrekelashvili J, Zhao F, Wedemeyer H, Lehner F, Manns MP, Greten TF, Korangy F (2009) Myeloid derived suppressor cells inhibit natural killer cells in patients with hepatocellular carcinoma via the NKp30 receptor. Hepatology 50(3):799–807. https://doi.org/10.1002/hep.23054
274. Eisenblaetter M, Flores-Borja F, Lee JJ, Wefers C, Smith H, Hueting R, Cooper MS, Blower PJ, Patel D, Rodriguez-Justo M, Milewicz H, Vogl T, Roth J, Tutt A, Schaeffter T, Ng T (2017) Visualization of tumor-immune interaction – target-specific imaging of S100A8/A9 reveals pre-metastatic niche establishment. Theranostics 7(9):2392–2401. https://doi.org/10.7150/thno.17138
275. Shi H, Zhang J, Han X, Li H, Xie M, Sun Y, Liu W, Ba X, Zeng X (2017) Recruited monocytic myeloid-derived suppressor cells promote the arrest of tumor cells in the premetastatic niche through an IL-1beta-mediated increase in E-selectin expression. Int J Cancer 140(6):1370–1383. https://doi.org/10.1002/ijc.30538
276. Erler JT, Bennewith KL, Cox TR, Lang G, Bird D, Koong A, Le QT, Giaccia AJ (2009) Hypoxia-induced lysyl oxidase is a critical mediator of bone marrow cell recruitment to form the premetastatic niche. Cancer Cell 15(1):35–44. https://doi.org/10.1016/j.ccr.2008.11.012
277. Deng J, Liu Y, Lee H, Herrmann A, Zhang W, Zhang C, Shen S, Priceman SJ, Kujawski M, Pal SK, Raubitschek A, Hoon DS, Forman S, Figlin RA, Liu J, Jove R, Yu H (2012) S1PR1-STAT3 signaling is crucial for myeloid cell colonization at future metastatic sites. Cancer Cell 21(5):642–654. https://doi.org/10.1016/j.ccr.2012.03.039
278. Motallebnezhad M, Jadidi-Niaragh F, Qamsari ES, Bagheri S, Gharibi T, Yousefi M (2016) The immunobiology of myeloid-derived suppressor cells in cancer. Tumour Biol 37(2):1387–1406. https://doi.org/10.1007/s13277-015-4477-9
279. Kowanetz M, Wu X, Lee J, Tan M, Hagenbeek T, Qu X, Yu L, Ross J, Korsisaari N, Cao T, Bou-Reslan H, Kallop D, Weimer R, Ludlam MJ, Kaminker JS, Modrusan Z, van Bruggen N, Peale FV, Carano R, Meng YG, Ferrara N (2010) Granulocyte-colony stimulating factor promotes lung metastasis through mobilization of Ly6G+Ly6C+ granulocytes. Proc Natl Acad Sci U S A 107(50):21248–21255. https://doi.org/10.1073/pnas.1015855107
280. Chafe SC, Lou Y, Sceneay J, Vallejo M, Hamilton MJ, McDonald PC, Bennewith KL, Moller A, Dedhar S (2015) Carbonic anhydrase IX promotes myeloid-derived suppressor cell mobilization and establishment of a metastatic niche by stimulating G-CSF production. Cancer Res 75(6):996–1008. https://doi.org/10.1158/0008-5472.CAN-14-3000
281. Supuran CT, Winum JY (2015) Carbonic anhydrase IX inhibitors in cancer therapy: an update. Future Med Chem 7(11):1407–1414. https://doi.org/10.4155/fmc.15.71
282. Costa-Silva B, Aiello NM, Ocean AJ, Singh S, Zhang H, Thakur BK, Becker A, Hoshino A, Mark MT, Molina H, Xiang J, Zhang T, Theilen TM, Garcia-Santos G, Williams C, Ararso Y, Huang Y, Rodrigues G, Shen TL, Labori KJ, Lothe IM, Kure EH, Hernandez J, Doussot A, Ebbesen SH, Grandgenett PM, Hollingsworth MA, Jain M, Mallya K, Batra SK, Jarnagin WR, Schwartz RE, Matei I, Peinado H, Stanger BZ, Bromberg J, Lyden D (2015) Pancreatic cancer exosomes initiate pre-metastatic niche formation in the liver. Nat Cell Biol 17(6):816–826. https://doi.org/10.1038/ncb3169
283. Vadrevu SK, Chintala NK, Sharma SK, Sharma P, Cleveland C, Riediger L, Manne S, Fairlie DP, Gorczyca W, Almanza O, Karbowniczek M, Markiewski MM (2014) Complement c5a receptor facilitates cancer metastasis by altering T-cell responses in the metastatic niche. Cancer Res 74(13):3454–3465. https://doi.org/10.1158/0008-5472.CAN-14-0157
284. Hiratsuka S, Watanabe A, Aburatani H, Maru Y (2006) Tumour-mediated upregulation of chemoattractants and recruitment of myeloid cells predetermines lung metastasis. Nat Cell Biol 8(12):1369–1375. https://doi.org/10.1038/ncb1507
285. Ichikawa M, Williams R, Wang L, Vogl T, Srikrishna G (2011) S100A8/A9 activate key genes and pathways in colon tumor progression. Mol Cancer Res 9(2):133–148. https://doi.org/10.1158/1541-7786.MCR-10-0394
286. Giles AJ, Reid CM, Evans JD, Murgai M, Vicioso Y, Highfill SL, Kasai M, Vahdat L, Mackall CL, Lyden D, Wexler L, Kaplan RN (2016) Activation of hematopoietic stem/progenitor cells promotes immunosuppression within the pre-metastatic niche. Cancer Res 76(6):1335–1347. https://doi.org/10.1158/0008-5472.CAN-15-0204

287. Peinado H, Zhang H, Matei IR, Costa-Silva B, Hoshino A, Rodrigues G, Psaila B, Kaplan RN, Bromberg JF, Kang Y, Bissell MJ, Cox TR, Giaccia AJ, Erler JT, Hiratsuka S, Ghajar CM, Lyden D (2017) Pre-metastatic niches: organ-specific homes for metastases. Nat Rev Cancer 17(5):302–317. https://doi.org/10.1038/nrc.2017.6
288. Wang Y, Ding Y, Guo N, Wang S (2019) MDSCs: key criminals of tumor pre-metastatic niche formation. Front Immunol 10:172. https://doi.org/10.3389/fimmu.2019.00172
289. Yan HH, Jiang J, Pang Y, Achyut BR, Lizardo M, Liang X, Hunter K, Khanna C, Hollander C, Yang L (2015) CCL9 induced by TGFbeta signaling in myeloid cells enhances tumor cell survival in the premetastatic organ. Cancer Res 75(24):5283–5298. https://doi.org/10.1158/0008-5472.CAN-15-2282-T
290. Yan HH, Pickup M, Pang Y, Gorska AE, Li Z, Chytil A, Geng Y, Gray JW, Moses HL, Yang L (2010) Gr-1+CD11b+ myeloid cells tip the balance of immune protection to tumor promotion in the premetastatic lung. Cancer Res 70(15):6139–6149. https://doi.org/10.1158/0008-5472.CAN-10-0706
291. Binsfeld M, Muller J, Lamour V, De Veirman K, De Raeve H, Bellahcene A, Van Valckenborgh E, Baron F, Beguin Y, Caers J, Heusschen R (2016) Granulocytic myeloid-derived suppressor cells promote angiogenesis in the context of multiple myeloma. Oncotarget 7(25):37931–37943. https://doi.org/10.18632/oncotarget.9270
292. Kujawski M, Kortylewski M, Lee H, Herrmann A, Kay H, Yu H (2008) Stat3 mediates myeloid cell-dependent tumor angiogenesis in mice. J Clin Invest 118(10):3367–3377. https://doi.org/10.1172/JCI35213
293. Tartour E, Pere H, Maillere B, Terme M, Merillon N, Taieb J, Sandoval F, Quintin-Colonna F, Lacerda K, Karadimou A, Badoual C, Tedgui A, Fridman WH, Oudard S (2011) Angiogenesis and immunity: a bidirectional link potentially relevant for the monitoring of antiangiogenic therapy and the development of novel therapeutic combination with immunotherapy. Cancer Metastasis Rev 30(1):83–95. https://doi.org/10.1007/s10555-011-9281-4
294. Cui TX, Kryczek I, Zhao L, Zhao E, Kuick R, Roh MH, Vatan L, Szeliga W, Mao Y, Thomas DG, Kotarski J, Tarkowski R, Wicha M, Cho K, Giordano T, Liu R, Zou W (2013) Myeloid-derived suppressor cells enhance stemness of cancer cells by inducing microRNA101 and suppressing the corepressor CtBP2. Immunity 39(3):611–621. https://doi.org/10.1016/j.immuni.2013.08.025
295. Panni RZ, Sanford DE, Belt BA, Mitchem JB, Worley LA, Goetz BD, Mukherjee P, Wang-Gillam A, Link DC, Denardo DG, Goedegebuure SP, Linehan DC (2014) Tumor-induced STAT3 activation in monocytic myeloid-derived suppressor cells enhances stemness and mesenchymal properties in human pancreatic cancer. Cancer Immunol Immunother 63(5):513–528. https://doi.org/10.1007/s00262-014-1527-x
296. Li ZL, Ye SB, OuYang LY, Zhang H, Chen YS, He J, Chen QY, Qian CN, Zhang XS, Cui J, Zeng YX, Li J (2015) COX-2 promotes metastasis in nasopharyngeal carcinoma by mediating interactions between cancer cells and myeloid-derived suppressor cells. Oncoimmunology 4(11):e1044712. https://doi.org/10.1080/2162402X.2015.1044712
297. Ouzounova M, Lee E, Piranlioglu R, El Andaloussi A, Kolhe R, Demirci MF, Marasco D, Asm I, Chadli A, Hassan KA, Thangaraju M, Zhou G, Arbab AS, Cowell JK, Korkaya H (2017) Monocytic and granulocytic myeloid derived suppressor cells differentially regulate spatiotemporal tumour plasticity during metastatic cascade. Nat Commun 8:14979. https://doi.org/10.1038/ncomms14979
298. Ai L, Mu S, Wang Y, Wang H, Cai L, Li W, Hu Y (2018) Prognostic role of myeloid-derived suppressor cells in cancers: a systematic review and meta-analysis. BMC Cancer 18(1):1220. https://doi.org/10.1186/s12885-018-5086-y
299. Eriksson E, Wenthe J, Irenaeus S, Loskog A, Ullenhag G (2016) Gemcitabine reduces MDSCs, tregs and TGFbeta-1 while restoring the teff/treg ratio in patients with pancreatic cancer. J Transl Med 14(1):282. https://doi.org/10.1186/s12967-016-1037-z
300. Bayne LJ, Beatty GL, Jhala N, Clark CE, Rhim AD, Stanger BZ, Vonderheide RH (2012) Tumor-derived granulocyte-macrophage colony-stimulating factor regulates myeloid inflammation and T cell immunity in pancreatic cancer. Cancer Cell 21(6):822–835. https://doi.org/10.1016/j.ccr.2012.04.025
301. DeNardo DG, Brennan DJ, Rexhepaj E, Ruffell B, Shiao SL, Madden SF, Gallagher WM, Wadhwani N, Keil SD, Junaid SA, Rugo HS, Hwang ES, Jirstrom K, West BL, Coussens LM (2011) Leukocyte complexity predicts breast cancer survival and functionally regulates response to chemotherapy. Cancer Discov 1(1):54–67. https://doi.org/10.1158/2159-8274.CD-10-0028
302. Shojaei F, Wu X, Qu X, Kowanetz M, Yu L, Tan M, Meng YG, Ferrara N (2009) G-CSF-initiated myeloid cell mobilization and angiogenesis mediate tumor refractoriness to anti-VEGF therapy in mouse models.

Proc Natl Acad Sci U S A 106(16):6742–6747. https://doi.org/10.1073/pnas.0902280106
303. Sumida K, Wakita D, Narita Y, Masuko K, Terada S, Watanabe K, Satoh T, Kitamura H, Nishimura T (2012) Anti-IL-6 receptor mAb eliminates myeloid-derived suppressor cells and inhibits tumor growth by enhancing T-cell responses. Eur J Immunol 42(8):2060–2072. https://doi.org/10.1002/eji.201142335
304. Overall CM, Kleifeld O (2006) Tumour microenvironment – opinion: validating matrix metalloproteinases as drug targets and anti-targets for cancer therapy. Nat Rev Cancer 6(3):227–239. https://doi.org/10.1038/nrc1821
305. Ratner M (2014) Setback for JAK2 inhibitors. Nat Biotechnol 32(2):119. https://doi.org/10.1038/nbt0214-119a
306. Weiss JM, Subleski JJ, Back T, Chen X, Watkins SK, Yagita H, Sayers TJ, Murphy WJ, Wiltrout RH (2014) Regulatory T cells and myeloid-derived suppressor cells in the tumor microenvironment undergo Fas-dependent cell death during IL-2/alphaCD40 therapy. J Immunol 192(12):5821–5829. https://doi.org/10.4049/jimmunol.1400404
307. Ali K, Soond DR, Pineiro R, Hagemann T, Pearce W, Lim EL, Bouabe H, Scudamore CL, Hancox T, Maecker H, Friedman L, Turner M, Okkenhaug K, Vanhaesebroeck B (2014) Inactivation of PI(3)K p110delta breaks regulatory T-cell-mediated immune tolerance to cancer. Nature 510(7505):407–411. https://doi.org/10.1038/nature13444
308. Davis RJ, Moore EC, Clavijo PE, Friedman J, Cash H, Chen Z, Silvin C, Van Waes C, Allen C (2017) Anti-PD-L1 efficacy can be enhanced by inhibition of myeloid-derived suppressor cells with a selective inhibitor of PI3Kdelta/gamma. Cancer Res 77(10):2607–2619. https://doi.org/10.1158/0008-5472.CAN-16-2534
309. Iclozan C, Antonia S, Chiappori A, Chen DT, Gabrilovich D (2013) Therapeutic regulation of myeloid-derived suppressor cells and immune response to cancer vaccine in patients with extensive stage small cell lung cancer. Cancer Immunol Immunother 62(5):909–918. https://doi.org/10.1007/s00262-013-1396-8
310. Nefedova Y, Fishman M, Sherman S, Wang X, Beg AA, Gabrilovich DI (2007) Mechanism of all-trans retinoic acid effect on tumor-associated myeloid-derived suppressor cells. Cancer Res 67(22):11021–11028. https://doi.org/10.1158/0008-5472.CAN-07-2593
311. Khalil DN, Smith EL, Brentjens RJ, Wolchok JD (2016) The future of cancer treatment: immunomodulation, CARs and combination immunotherapy. Nat Rev Clin Oncol 13(5):273–290. https://doi.org/10.1038/nrclinonc.2016.25
312. Teachey DT, Bishop MR, Maloney DG, Grupp SA (2018) Toxicity management after chimeric antigen receptor T cell therapy: one size does not fit 'ALL'. Nat Rev Clin Oncol 15(4):218. https://doi.org/10.1038/nrclinonc.2018.19

Chapter 11

Rapid Production of Physiologic Dendritic Cells (phDC) for Immunotherapy

Douglas Hanlon, Olga Sobolev, Patrick Han, Alessandra Ventura, Aaron Vassall, Nour Kibbi, Alp Yurter, Eve Robinson, Renata Filler, Kazuki Tatsuno, and Richard L. Edelson

Abstract

Generation of large numbers of dendritic cells (DC) for research or immunotherapeutic purposes typically involves in vitro conversion of murine bone marrow precursors or human blood monocytes to DC via cultivation with supraphysiologic concentrations of cytokines such as GM-CSF and IL-4 for up to 7 days. Alternatively, our group has recently established a new approach, based on the underlying mechanism of action of a widely used cancer immunotherapy termed Extracorporeal Photochemotherapy (ECP). Our method of rapid and cytokine-free production of therapeutically relevant DC populations, leveraging the innate physiologic programs likely responsible for DC differentiation from blood monocytes in vivo, potentially offers a novel, inexpensive, and easily accessible source of DC for clinical and research uses. This approach involves ex vivo physiologic reprogramming of blood monocytes to immunologically tunable dendritic antigen-presenting cells, which we term "phDC," for physiological DC. To facilitate access and utilization of these new DC populations by the research community, in this chapter, we describe the use of a scaled-down version of the clinical ECP leukocyte-treatment device termed the Transimmunization (TI) chamber or plate, suitable for processing both mouse and human samples. We highlight the methodological sequences necessary to isolate mouse or human peripheral blood mononuclear cell (PBMC) from whole blood, and to expose those PBMC to the TI chamber for facilitating monocyte activation and conversion to physiological DC (phDC) through interaction with blood proteins and activated platelets under controlled flow conditions. We then provide sample protocols for potential applications of the generated DC, including their use as vaccinating antigen-presenting cells (APC) in murine in vivo antitumor models, and in human ex vivo T-cell stimulation and antigen cross-presentation assays which mimic clinical vaccination. We additionally highlight the technical aspects of loading mouse or human phDC with tumor-associated antigens (TAA) in the form of peptides or apoptotic tumor cells. We provide a simple and clinically relevant means to reprogram blood monocytes into functional APC, potentially replacing the comparatively expensive and clinically disappointing cytokine-derived DC which have previously dominated the dendritic cell landscape.

Key words Immunology, Cancer, Immunotherapy, Mouse models, Extracorporeal photochemotherapy, ECP, Cell reprogramming, Dendritic cells, DC

Douglas Hanlon and Olga Sobolev contributed equally to the work.

Samuel G. Katz and Peter M. Rabinovich (eds.), *Cell Reprogramming for Immunotherapy: Methods and Protocols*, Methods in Molecular Biology, vol. 2097, https://doi.org/10.1007/978-1-0716-0203-4_11,

1 Introduction

Extracorporeal photochemotherapy (ECP) is advantageous compared to widely used immunotherapies in three complementary ways: selectivity for pathogenic clones, favorable safety profile, and efficacy in both augmenting anticancer immunity and suppressing pathogenic autoreactivity [1]. All of these effects result from its efficient physiologic reprogramming of immunologically tunable dendritic antigen-presenting cells (DC), which are the master-switches of all antigen-specific T-cell responses.

As its name implies, ECP provides the clinical immunologist with two key advantages: outside-the-body production and sequestration of a fresh contingent of DC available for relevant antigen loading, and polarization of DC to either immunizing or tolerizing mode via platelet signaling and use of an exquisitely controllable ultraviolet-activated chemotherapeutic drug (8-methoxypsoralen or 8-MOP). Understanding of this therapy and its unique features will be aided by the succinct history of ECP's serendipitous discovery, its development into a therapy now regularly conducted at >350 university medical centers worldwide, and concomitant scientific investigations that led to the understanding of its mechanism of action. It is our hope that the elucidation of this mechanism, now uncovered to be intimately tied to the creation of physiologically derived DC populations directly and rapidly from blood monocytes [2], will facilitate the formulation of cost-effective and potent antigen-presenting cells (APC) for both basic scientific and clinical investigations.

ECP was first developed as a therapy for Cutaneous T-Cell Lymphoma (CTCL). CTCL is a malignancy of a subset of skin-homing CD4 T cells, referred to as "Cutaneous T cells", normally comprising about 10% of the circulating T-cell pool, distinguished by their recirculation between skin and blood and expression of a phenotypic marker, C LA, which contributes to their special ability to enter the skin papillary capillaries. Clonal expansion of malignant cells with skin-homing properties lead to a characteristic skin infiltration and lesions. Detailed investigation of CTCL cells, due to the advantage of analyzing a single pathogenic clone in each patient, has contributed to the recognition of CD4 as a marker of Th2 cells and the discovery of tissue homing phenomena in general. ECP's introduction as an FDA-approved anticancer immunotherapy is attributable to the ease of access of CTCL cells and our understanding of their biology.

In the process of developing a monoclonal antibody treatment for leukemic CTCL, it became desirable to first reduce the total body tumor burden in patients. Having already established that skin-infiltrating CTCL cells are in migratory equilibrium with those in the blood, we attempted to drastically reduce the number of malignant cells by passing patient blood extracorporeally, as a thin layer, through a device that allowed ultraviolet activation of 8-MOP, a naturally occurring plant-derived furocoumarin without biologic impact unless instantaneously converted into a DNA pyrimidine-crosslinking form by UVA exposure. Anticipating lethal damage to passaged CTCL cells, followed by their removal by the patient's reticuloendothelial system, we were astonished that the return to the patient of the damaged malignant cells led, in the best responders, to immunization against tumor antigens and selective immunotherapeutic elimination of the untreated CTCL cells. All participants, both responders and nonresponders, exhibited an excellent safety profile. In fact, the specificity and safety of that surprising result so far exceeded our expectations of the proposed monoclonal antibody therapy that we abandoned the latter and committed ourselves to an Odyssey of searching for the mechanistic underpinnings of that astounding ECP response.

That long and winding search for the mechanisms of ECP was motivated by recognition that the answer(s) might reveal insights with broad clinical and scientific implications. Along with clinical work, mechanistic understanding was enabled by the development of an experimental miniaturized ECP system, the "Transimmune" (TI) chamber or plate, suitable for both animal model studies and processing of small amounts of human blood samples [3]. We now know that the passage of patient monocytes, along with their T cells, through the ECP device, induces monocyte-to-DC maturation and tumor-antigen loading of the new DC. The sequence of events is as follows: (1) plasma fibrinogen is layered onto the plastic surfaces of the ECP device's cell passage chamber, (2) platelets adhere to the gamma chain of the immobilized fibrinogen, (3) the chamber-activated platelets instantaneously transpose preformed P selectin to the platelet external surface. This enables passaged monocytes, through their PSGL-1 (complementary to the platelet P-selectin) to transiently bind to the immobilized platelets and receive reprogramming signals to enter the DC differentiation pathway [4].This discovery most likely identified, for the first time, at least one major way that monocytes physiologically evolve into functional DC in vivo. We now recognize that this sequence provides the core reason for ECP's therapeutic efficacy and immune system partnership.

While the rest of the crosstalk between the platelets and the monocytes remains to be deciphered, and the full range of the properties and functional capabilities of the induced physiologic DC remain to be uncovered, several important conclusions are already clear. First, these ECP-derived DC internalize the 8-MOP-damaged CTCL cells and then process and present the array of CTCL tumor antigens to responding CD4 and CD8 T cells, which in turn collaborate in a clinically important, tumor-specific immune reaction. Second, a subset of new DC, themselves influenced by high exposures to ultraviolet-activated 8-MOP, are programmed to become potent immunosuppressive agents, also specifically directed to the antigens they have processed. Having converted into selectively tolerizing DC, they mediate the antigen-specific tolerance in the transplantation and GvHD ECP settings.

With the outlines of the mechanism underlying ECP's tight collaboration with the normal immune system coming into focus, the potential breadth of its clinical application is limited only by the availability of the relevant antigens, thereby conceptually encompassing various immunogenic cancers, transplantation reactions, autoimmunity, and microbial vaccination. While these goals may, at first consideration, appear absurdly ambitious, they are inherent in ECP's partnership with the immune system, which normally silently and smoothly accomplishes them all by itself. Armed with mechanistic knowledge, we can now reprogram ECP-induced DC to either a immunizing or tolerizing mode, thereby enhancing ECP's therapeutic impact in the desired clinical or experimental application. These possibilities are elaborated in the provided protocols.

This chapter focuses on the methodological sequences necessary to isolate mouse or human PBMC from whole blood, and to expose those PBMC to the "Transimmune" chamber to facilitate monocyte activation and conversion to physiological DC (phDC) through interaction with activated platelets under flow stress conditions. It then provides sample protocols for potential applications of the generated phDC, including their use as vaccinating APC in murine in vivo antitumor models, and in human ex vivo T-cell stimulation and antigen cross-presentation assays which mimic clinical vaccination. We additionally highlight the technical aspects of loading mouse or human phDC with tumor-associated antigens (TAA) in the form of peptides or 8-MOP/UVA-treated apoptotic tumor cells. It is our hope that the physiologically derived DC will afford a simple and clinically relevant way to reprogram blood monocytes into functional APC, potentially replacing the comparatively expensive and clinically disappointing cytokine-derived DC which has dominated the dendritic cell landscape for nearly two decades.

2 Materials

1. Wild-type male C57BL/6J mice at 4–6 weeks of age are used for the in vivo model of immunotherapeutic ECP. Male C57BL/6J mice at 10–12 weeks of age are used for autologous mouse serum production.
2. Rodent anesthesia system with active scavenging.
3. Isoflurane.

4. Cheek bleed GoldenRod lancets, 5 μm.
5. Plasticware: T75 (75 cm^2) tissue culture flasks, 12-well tissue culture plates, 96-well round-bottom tissue culture plates, 35 mm EasyGrip Petri Dishes, polypropylene 15- and 50-mL conical tubes, 1.5-mL conical tubes, 10-mL syringes (LUER-LokTip), 1-mL syringes (Slip tip), tissue culture scrapers, 25 G × 5/8 tumor injection needles, and 27 G × 1/2 retro-orbital injection needles.
6. Pipettes and tips: standard P10, P200, and P1000 pipettes and tips, tissue culture pipette-aid with 25-, 10-, and 5-mL serological pipettes.
7. Heparin 5000 U/mL (McKesson Packaging services, cat. Number 949512).
8. Lympholyte M (Cedarlane labs) murine peripheral blood mononuclear cell (PBMC) isolation medium for mouse protocol, Isolymph (CTL Scientific Supply Corp.) human PBMC isolation medium for human protocol.
9. Phosphate-buffered saline (1× DPBS without calcium or magnesium).
10. Ammonium-Chloride-Potassium (ACK) red cell lysis buffer.
11. Hemavet 950FS hematology counter (Drew Scientific, HV950FS) or other hematology counter for monitoring platelet numbers.
12. Heat-inactivated fetal bovine serum (FBS).
13. YUMM1.7 tumor cells (kindly provided by Prof. Marcus Bosenberg, Yale University School of Medicine, New Haven, CT).
14. YUMM1.7 cell culture medium: DMEM/F12 supplemented with 10% heat-inactivated fetal bovine serum (FBS), 1% penicillin/streptomycin, and 1% nonessential amino acids.
15. Trypsin-EDTA 0.25% (1×).
16. 8-Methoxypsoralen (8-MOP) (Therakos Inc., UVADEX 20 μg/mL, 10 mL).
17. UVA irradiator: The apparatus used in the experiments was developed by J&J for Prof. R. Edelson's laboratory, modeling the precise UVA exposure utilized in the Therakos UVAR XTS Photophoresis System (unfiltered bulb UVA wavelength of 320–370 nm delivering a total of 1.5 J/cm^2). Several machines are available in the laboratory on a collaborative basis; please contact Prof. R. Edelson for use of one. Alternative UVA irradiators are commercially available but have not been tested by us.
18. Laboratory UV face shield.

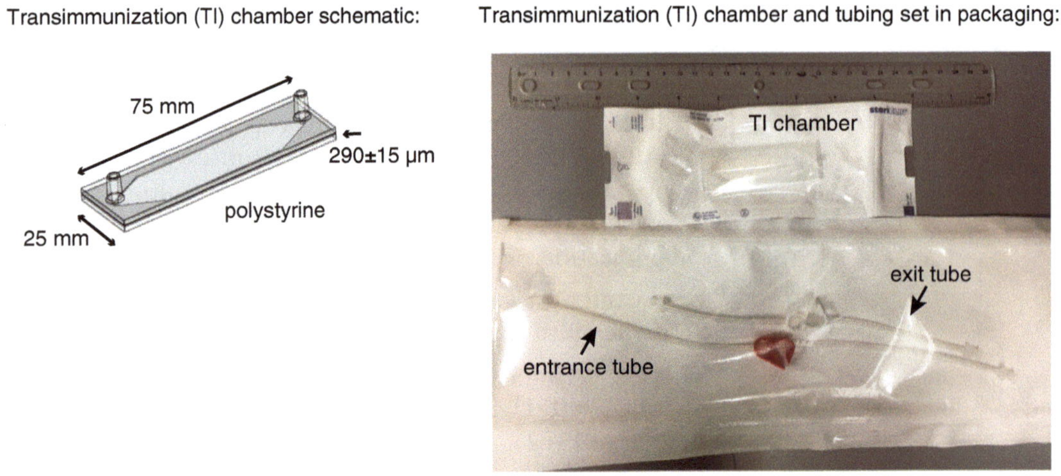

Fig. 1 Transimmunization (TI) chamber and tubing. Transimmunization (TI) or model ECP chamber/plate is represented schematically with the plate material and dimensions indicated, and photographed with its corresponding tubing sets in the original sterile packaging. The entrance and exit tubing is labeled

19. TI plate and tubing set: A miniaturized benchtop ECP flow plate for facilitating monocyte to DC conversion (Fig. 1). This device has been developed by Transimmune AG for Prof. R. Edelson's laboratory and is not yet commercially available. Please contact Prof. R. Edelson in order to obtain the device on a collaborative basis.
20. Programmable 2-channel syringe pump (New Era Pump Systems Inc., model NE-4000).
21. TI plate running platform: A holder for the TI plate, allowing for easy adjustment of TI plate angle as cells are flowing through it. This device has been developed by Transimmune AG for Prof. R. Edelson's laboratory, and is not yet commercially available. Please contact Prof. R. Edelson in order to obtain the device on a collaborative basis.
22. Autologous mouse serum: Blood collected from 10- to 12-week-old C57BL/6J mice by any institutionally approved method (e.g., eye bleed, tail vein bleed, cheek bleed, or terminal bleed-out by cardiac puncture), without any anticoagulants, is allowed to clot overnight at 4 °C. On the following day, the clotted blood is spun down at 3000 × *g* for 15 min, and the top serum-containing layer is collected. Serum can be used immediately to make the overnight PBMC culture medium (**item 22**), or preserved for future experiments. To preserve the serum, freeze in aliquots at −20 °C. Estimate collecting 1 mL blood for every 300 μL of serum needed; 300 μL of serum required for every treatment group of five mice in the experiment.

23. Mouse overnight PBMC culture medium: clear RPMI, 15% autologous mouse serum.
24. Human AB serum.
25. Human overnight PBMC culture medium: clear RPMI, 15% human AB serum.
26. Autologous mouse plasma for PBMC re-injection in the in vivo model of immunotherapeutic ECP: preparation described in Subheading 3.1, **step 3** "Optional."
27. Ag-specific T-cell line culture medium: AimV medium supplemented with 10% hAB serum (Lonza) and 600 IU/mL recombinant human IL-2 (Peprotech).
28. Human melanoma-specific TCR transgenic T-cell line (DMF5): The DMF5 line was created by transfection of patient-derived CD8+ T cells with a high-affinity TCR specific for the tumor-associated MART-$1_{(27-35)}$ epitope. This T-cell line was originally derived and expanded for autologous T-cell transfer therapy within the Rosenberg group, NCI Surgery Branch, and were a kind gift of John R.Wunderlich, Surgery Branch Cell Prep Core. Cells were expanded and cryopreserved within the Edelson group in the Department of Dermatology, Yale University. Cell lines could potentially be distributed to other investigators with permission of Dr. Wunderlich and/or NCI Surgery Branch.
29. Human HPV-associated TCR transgenic T cells: CD8+ T cell lines were originally derived from patent TIL and expanded to test reactivity against HPV-associated tumors by the Hinrichs group, NCI Clinical Research Center, and were a kind gift of Christian Hinrichs. E6- and E7-specific lines were expanded and cryopreserved within the Edelson group in the Department of Dermatology, Yale University. Cell lines could potentially be distributed to other investigators with permission of Dr. Hinrichs and/or NCI Clinical Research Center.
30. Other human melanoma and HPV-reactive CD8+ T-cell lines: A limited number of commercially available sources of cryopreserved tumor-specific and pathogen-specific T-cell lines, including those reactive against HPV-associated E6 or E7 oncoproteins are available from Astarte Biologics. These lines can be expanded in the presence of IL-2 and cryopreserved in multiple aliquots to allow reproducible T-cell stimulation assays.
31. Human IFNγ ELISA kit.
32. Human Monocyte Isolation Kit.
33. Human CD8+ T-cell Isolation kit.
34. Human coculture medium: RPMI medium supplemented with 10% hAB serum (Lonza) or autologous serum if available,

nonessential amino acids (100×, Sigma-Aldrich), sodium pyruvate (100×, Invitrogen), MEM vitamin solution (100×, Invitrogen), 2 mercaptoethanol (100×, Invitrogen), and 10 μM ciprofloxacin (Serologicals Proteins).

35. BD Biosciences Fixation/Permeabilization Solution Kit with BD GolgiStop and appropriate antihuman IFNγ Ab. In the absence of appropriate tetramer(s) for evaluating DC-driven CD8+ T-cell stimulation and replication, cocultured CD8+ T cells can be analyzed for intracellular cytokine production using this commercially available intracellular cytokine staining kit.

3 Methods

DC reprogramming with model ECP for immunotherapy.

3.1 Preparation of PBMC from Mouse Anticoagulated Blood

1. Using the cheek bleed lancets, collect blood by cheek bleed (*see* **Note 1**) from C57BL/6 mice, using institutional guidelines to determine maximal collection amount permitted. Blood should be collected into a 15-mL conical tube containing 5000 U/mL heparin, with the calculation of 10% final heparin content (i.e., if collecting 1 mL of blood, preadd to the tube 100 μL of 5000 U/mL heparin). Mix tube periodically during blood collection to ensure proper anticoagulation. For mouse source and blood pooling guidelines, please *see* **Note 2**.
2. Set up one 15-mL conical tube with 4-mL Lympholyte M for each 15-mL conical tube of blood (up to 4 mL) collected. Slowly layer blood on top of Lympholyte, and spin the tube for 35–40 min at 700 × *g* without brake.
3. Harvest each lymphocyte buffy coat into a clean 15-mL conical tube, fill to 15 mL with PBS, and spin for 10 min at 250 × *g* in a standard tissue culture centrifuge to collect the PBMC.

   Optional (for Subheading 3.7): If preparing cells for the mouse in vivo tumor assay, before collecting the buffy coat, first remove the top clear layer above the buffy coat (autologous mouse plasma, Subheading 2, **item 26**) into a clean 15-mL conical tube and reserve at 4 °C for use the following day, leaving 0.5–1 mL liquid remaining above buffy coat.
4. Remove most of media with a pipette (do not decant as it is a soft pellet) and resuspend the PBMC pellet by flicking.
5. Add 2 mL of ACK red cell lysis buffer to the pellet, incubate on ice for 10 min, then fill to 15 mL with PBS, and spin for 10 min at 250 × *g* in a standard tissue culture centrifuge to collect the PBMC.
6. Remove most of media with a pipette (do not decant, as it is a soft pellet) and resuspend the PBMC pellet by flicking.

7. Resuspend pellet in 1 mL of PBS. Count cells and perform a platelet count on the sample (*see* **Note 3a**). Platelet counts can be done manually using a hemocytometer utilizing the central erythrocyte grid, or with an automated analyzer such as a Hemavet Multispecies Hematology Analyzer.
8. Spin the tube for 10 min at 250 × *g* in a standard tissue culture centrifuge to collect the PBMC, remove most of media with a pipette (do not decant, as it is a soft pellet), add 300 μL of FBS, and resuspend the PBMC by pipetting.
9. At this point of the protocol, several different options are possible (*see* **Note 4**).

3.2 Preparation of PBMC from Human Anticoagulated Blood

1. Collect blood from consented human donors according to institutional guidelines as per approved study protocol. Blood can be collected into appropriately sized syringes containing 5000 U/mL heparin so that the final amount of heparin when blood is added is 25 U/mL (i.e., if collecting 50 mL of blood, preadd 0.25 mL of 5000 U/mL heparin to a 60 cc syringe). Mix syringe periodically during blood collection to ensure proper anticoagulation.
2. Set up a 50-mL conical tube with 15 mL Isolymph for each 35 mL blood collected. Slowly, layer blood on top of Isolymph, to ensure minimal mixing of the layers. Spin the tube for 35–40 min at 700 × *g* without brake.
3. Leaving 4–5 mL of plasma just above the gradient, remove the plasma (*see* **Note 3b**). Then harvest the buffy coat and remaining plasma transferring up to two lymphocyte buffy coats into a clean 50-mL conical tube, fill to 50 mL with PBS, and spin for 10 min at 250 × *g* in a standard tissue culture centrifuge to collect the PBMC.
4. Remove media with a pipette or vacuum aspiration (do not decant, as it is a soft pellet) and resuspend the PBMC pellet by flicking.
5. Resuspend each pellet in 5 mL of PBS, combining tubes. Count cells, taking into consideration that most human donors will have between 1 and 1.75×10^6 PBMC per 1 mL of blood and diluting accordingly prior to counting. Perform a platelet count on the sample (*see* **Note 3b**). Platelet counts can be done manually using a hemocytometer utilizing the central erythrocyte grid, or with an automated analyzer such as a Hemavet Multispecies Hematology Analyzer.
6. Spin the tube for 10 min at 250 × *g* in a standard tissue culture centrifuge to collect the PBMC, remove media with a pipette or vacuum aspiration (do not decant, as it is a soft pellet), and resuspend the cells by pipetting at 8.33 million per mL in FBS.

3.3 Optional: Preparation of Mouse Tumor Cells as an Antigen Source

1. Grow tumor cells of interest in the appropriate medium to 60–70% confluence. For example, culture freshly thawed YUMM1.7 cells (Subheading 3.7) in YUMM1.7-cell culture medium under standard tissue culture conditions (37 °C, 5% CO_2).
2. Collect cells at 60–70% confluence. For YUMM1.7 tumor cells (Subheading 3.7), add enough 0.25% Trypsin-EDTA to cover the bottom of the cell culture plate or flask (e.g., 4 mL for a T75 flask). Rest the cell culture vessels at room temperature for 3–4 min, occasionally gently tapping on the bottom or sides of the vessel to help detach the cells. Add 1 mL of FCS per 4 mL of Trypsin-EDTA to stop the reaction once most of the cells have detached. Collect the cells by pipetting into a 15-mL conical tube. Rinse the flask with PBS and add the rinse to the same collection tube. Fill the tube to 15 mL with PBS. Spin down for 10 min at 250 × *g* in a standard tissue culture centrifuge to collect the tumor cells and resuspend in a small volume of FBS.
3. Take out a small aliquot and count the cells. When using tumor cells as antigen source, the aim is to have a 1:1 final ratio of PBMC to tumor cells. Therefore, calculate total tumor cell number, and adjust it to match the PBMC count obtained in Subheading 3.1, **step 8**, in an equal 300 μL FBS volume (ex. if PBMC are at 2.5×10^6 cells/300 μL, then adjust the tumor cell amount and volume to also be 2.5×10^6 cells/300 μL).
4. With tissue culture hood lights off, add 8-MOP to the tumor cells. The optimal amount of 8-MOP and UVA will differ for every tumor cell line, and should be titrated beforehand (*see* **Note 5**). For YUMM1.7 tumor cells (Subheading 3.7), to do so, add 8-MOP for the final concentration 100 ng/mL. Mix well, wrap the tube in foil to protect it from light, and transfer to a standard incubator. Incubate at 37 °C for 20 min.

   Caution: 8-MOP is a photoactivatable carcinogen. Handle with care, use appropriate protective gear, and dispose of materials that come into contact with 8-MOP appropriately.
5. Immediately after placing the tumor cells for incubation, turn on the UVA light source (*see* **Note 6**). Standard UVA lamps need 15–20 min to reach full emission potency and equilibrate.

   Caution: UVA light is carcinogenic. When working with a UVA light source, do not allow exposure to skin or eyes, use appropriate protective gear such as lab coat with long sleeves, gloves, and UV-protective face shield.
6. Precoat a 12-well tissue culture plate by adding 1 mL of FBS to as many wells as you will need (prepare one well for every 300 μL of cells to be treated). Incubate the FBS-filled plate in a 4 °C refrigerator for 20 min, and then remove FBS from wells by pipetting.

7. Remove the tube of cells (**step 4**) from the incubator, and add 300 μL of 8-MOP-treated cells per well to the precoated 12-well plate (**step 6**). Expose the plate, with the lid off, to the prewarmed UVA light source for the desired time, which should be individually pretitrated for every tumor cell line of interest (*see* **Note 5**). For YUMM1.7 tumor cells (Subheading 3.7), expose the plate to 4 J/cm^2 UVA irradiation.
8. Switch off the UVA light source, remove the plate, and individually collect each 300 μL cell aliquot into a clean 1.5-mL conical tube, carefully swirling and rinsing down the wells to recover as many cells as possible. Proceed to Subheading 3.5, **step 3** (Fig. 2).

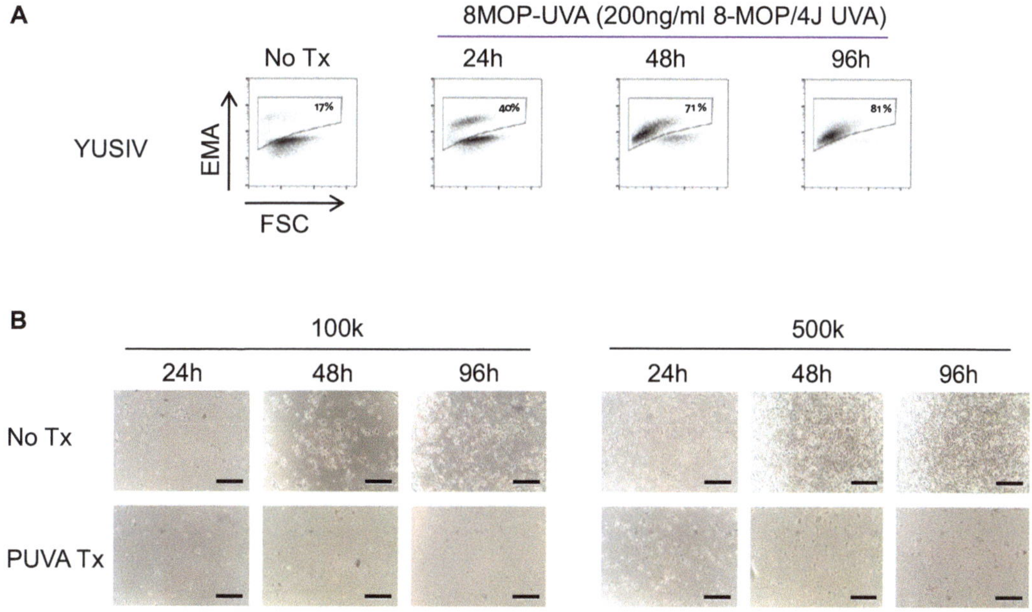

Fig. 2 phDC in vivo immunotherapy assay using tumor cellular antigens in mouse YUMM1.7 syngeneic melanoma model. (**a**) Schematic description of YUMM1.7 Transimmunization treatment experimental workflow. Animals are inoculated subcutaneously (s.c.) with syngeneic tumor cells; animals with palpable tumors are treated twice weekly by blood draw, isolation of PBMC from blood, PBMC flow passage through the autologous platelet-coated TI plate in the presence of 8-MOP/UVA-treated tumor cells, followed by PBMC and tumor cell coincubation overnight, and reinjection of coincubated cells intravenously (i.v.) into tumor-bearing animals. Tumor volume is measured throughout the experiment. (**b**) YUMM1.7 tumor volume over time plotted for C57BL/6 mice inoculated with 1×10^5 YUMM1.7 tumor cells, and receiving either six transimmunization treatments (black lines), or six control treatments (gray lines). Data are from a representative YUMM1.7 Transimmunizaton experiment, with each line showing tumor growth for an individual mouse. (**c**) Cumulative data over nine independent YUMM1.7 experiments. To adjust for interexperimental variation, in the cumulative data, the individual tumor volumes from each experiment were normalized to the average final volume of PBS control tumors for that experiment, and mean relative tumor volumes across all experiments were then calculated. (**b**, **c**) "PBS control" mice in all experiments were bled on the same schedule as the experimental animals, but received six sterile PBS reinfusions. Error bars represent SEM, *p*-values calculated for each time point using Sidak's multiple comparisons test; $^{**}p = 0.0013$; $^{****}p < 0.0001$. This figure has been modified from [3]

3.4 Optional: Preparation of Human Tumor Cells as an Antigen Source

1. Human tumor cell harvest, exposure to 8-MOP and UVA irradiation and monitoring of tumor cell death is carried out essentially identically to the procedures described for murine tumors in Subheading 3.3. Confirmation of suitable levels of apoptosis induction requires a tumor cell kill curve to be carried out for every individual cell line to be used as an antigen source (*see* **Note 5**). The results of such a kill curve are shown visually in Fig. 3 for the human melanoma cell line YUSIV at the ideal apoptosis-inducing dose of 200 ng/mL 8-MOP and 4 J UVA light.

3.5 Mouse or Human PBMC Treatment with Model ECP

1. Remove the TI tubing sets and TI plate (Fig. 1) from packaging and orient yourself. The longer tube with the red clamp is the "entrance" tube, while the shorter one with the white clamp is the "exit" tube.
2. Precoat the TI tubing sets with FBS. To do so, use a 10-mL syringe to fill the entrance tube with FBS, and close the clamp before detaching the syringe. Repeat with the exit tube. Return the FBS-filled tubes to the original packaging.
3. Ensure the total volume of cells to be treated is 600 μL in FBS. If making phDC without cellular antigens, add 300 μL of FBS to 300 μL of mouse or human PBMC obtained in Subheading 3.1 or 3.2, respectively. If making phDC loaded

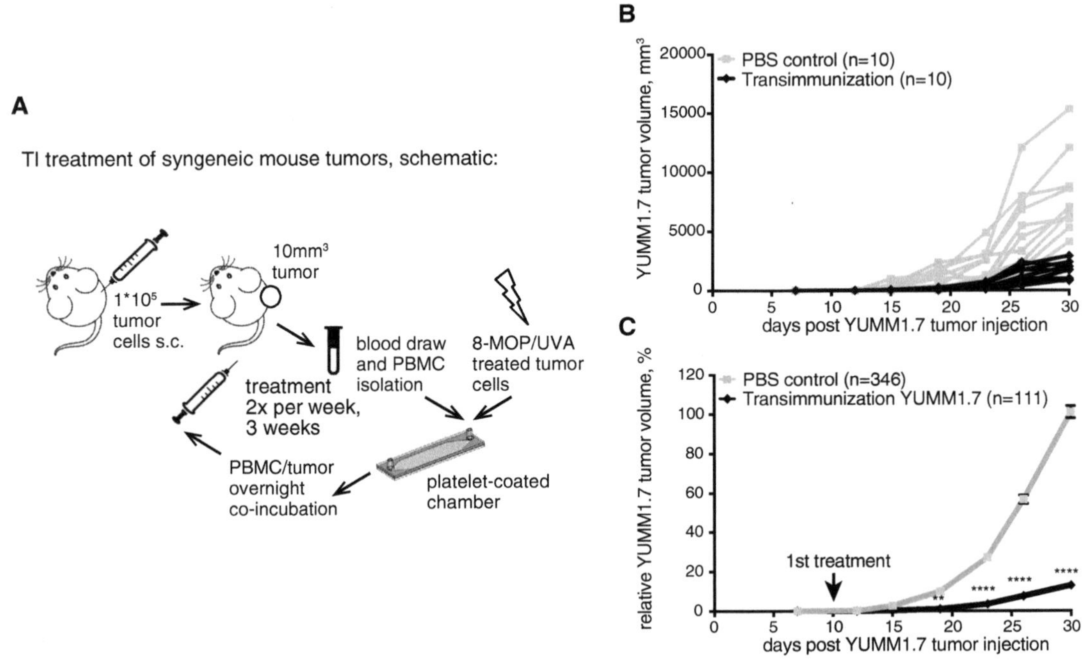

Fig. 3 Induction of cell death in YUSIV human melanoma cell line following 8-MOP/UVA treatment. Patient-derived melanoma cell line YUSIV was tested for induction of cell death and proliferation inhibition upon 8-MOP/UVA treatment and was analyzed by (**a**) ethidium monoazide bromide (EMA) staining, with gated box indicating nonviable cells and (**b**) image analysis of untreated and treated cells over 96 h of culture. Cells were seeded at 100k or 500k cells per well in 6-well plates and imaged at 10× magnification on 24, 48, and 96 h after initial plating. Scale bars indicate 500 μm. All data are representative of at least three replicates

with cellular antigens, ex. 8-MOP/UVA-treated tumor cells, at this point, combine 300 μL of PBMC obtained in Subheading 3.1 or 3.2 with 300 μL of 8-MOP/UVA-treated tumor cells from Subheading 3.3 or 3.4, respectively. Mix the cells by flicking the tube.

4. Precoat the TI plate. To do so, remove a TI plate from packaging. Using a P1000 pipette, take up the 600 μL of cells to be treated. Insert the P1000 pipette tip firmly into either one of the two TI plate ports. Holding the TI plate at a gentle upward angle (~45°, port with the P1000 tip inserted lower than the unoccupied port), depress the pipette plunger and slowly fill the plate, taking care to avoid creating bubbles. When the plate is filled and liquid just reaches the opposite port (~450 μL), place the plate flat, and remove the P1000 pipette tip from the port, without releasing the plunger. Return the remainder of the cells (~150 μL) to the original 1.5-mL conical tube.
5. Move the FBS-filled tubes (**step 2**), the filled TI plate, and the 1.5-mL conical tube with the remaining cells into a 37 °C incubator, and incubate for 1 h.
6. Remove all components from the incubator. Remove the FBS-filled tubing from packaging, release the clamps, and drain the FBS by gravity. With the TI plate flat, depress the plunger of a P1000 pipette and insert it firmly into one of the TI plate ports. Holding the TI plate at a gentle upward angle (~45°, port with the P1000 tip inserted lower than the unoccupied port), release the pipette plunger and slowly fill the pipette tip, removing cells from the newly coated plate. Add the recovered cells (~450 μL) back into the 1.5-mL conical tube with the remainder (~150 μL).
7. Connect the FBS precoated entrance and the exit tubing to the precoated TI plate. Secure the plate into the TI plate running platform (*see* **Note 7**). Connect the exit tubing to a clean 1.5-mL conical collection tube.
8. Turn on the syringe pump and adjust flow rate to 0.09 mL/min, but do not start the pump yet. Using a 1-mL syringe, draw up the 600 μL of cells in FBS, taking care to avoid bubbles in the syringe. Attach the syringe to the entrance tubing and slowly load the cells until fully loaded. Close the red clamp, detach the 1-mL syringe from the entrance tubing, and attach the entrance tubing to the syringe pump. Tilt the TI plate at a ~30° toward the syringe pump so that it will fill "uphill."
9. Release the red clamp and start the syringe pump. If an air bubble/etc. impedes the initial flow, gently flick tubing or plate with your finger to start the cells moving. Once the TI plate is completely filled, tilt it in the opposite direction, ~30° away from the syringe pump, and let the cells empty "downhill" from it into the exit tubing and the collection tube. Once all

the cells are in the collection tube, stop the pump. At 0.09 mL/min, 600 μL should take about 6.5 min.

10. To wash the TI plate, disconnect the entrance tubing from the syringe pump, and use a syringe to load ~500 μL FBS into the entrance tubing (completely filling the entrance tube, but not entering the TI plate). Close the red clamp, disconnect the syringe, and connect the entrance tubing to the syringe pump. Set the syringe pump to 0.49 mL/min, but do not start the pump yet. Tilt the TI plate at a ~30° toward the syringe pump so that it will fill "uphill."
11. Release the red clamp and start the syringe pump. Once the TI plate is completely filled, tilt it in the opposite direction, ~30° away from the syringe pump to let the cells empty "downhill," and flick or tap the top of the plate firmly (can use fingers or a small tool such as a pen or pencil), as it empties to help detach any cells that may have adhered to the plate. When all liquid has moved from the plate to the exit tubing, turn up the pump speed to 2.49 mL/min and let it run until all the liquid is in the collection tube. Stop the pump, disconnect, and discard the TI plate and tubing, and reserve the collection tube.
12. The collection tube, now comprising the plate-passed cells and the FBS wash, should have a total volume of ~1.2 mL. Spin the tube in a benchtop microfuge at 250 × *g* for 8 min. Remove the supernatant gently with a P1000 pipette.
13. At this point of the protocol, several different options are possible (*see* **Note 8**).

3.6 Mouse or Human PBMC Overnight Culture

1. Resuspend each tube of TI-treated cells in 2 mL of mouse or human overnight PBMC culture medium. Plate the 2 mL of cells in culture medium in 35 mm EasyGrip petri dishes (not tissue culture treated) and culture overnight under standard tissue culture conditions (37 °C, 5% CO_2). For additional experimental possibilities at this point of the protocol, please *see* **Note 9**.
2. On the following day, detach any adherent cells from the bottom of the petri dish with a tissue culture scraper, rotating the dish while scraping to ensure even cell collection. Collect the cells in a 15-mL tube. Add 2 mL PBS to the dish and repeat the scraping step, collecting the cells into the same tube. Rinse the 35 mm dish with 1 mL PBS and add the rinse to the same tube.
3. Spin the collection tube for 10 min at 250 × *g* in a standard tissue culture centrifuge to collect the cells. Carefully pipette off supernatant and resuspend the pellet by flicking the tube. Do not decant, as the pellet is soft and can be lost.
4. The TI-treated phDC cells are now ready to be used in your assay of interest. Subheadings 3.7–3.9 below describe some of the murine and human assays that we have employed.

3.7 phDC in vivo Immunotherapy Assay Using Tumor Cellular Antigens in Mouse YUMM1.7 Syngeneic Melanoma Model

This method describes one possible application of mouse phDC, as in vivo immunotherapy for syngeneic tumors using the YUMM1.7 melanoma model. Figure 2 gives an overview of the basic experimental design and shows representative data. More on this assay system can be found in our recent publication [5].

1. To implant syngeneic YUMM1.7 melanoma tumors into the flank of mice, culture freshly thawed YUMM1.7 cells in YUMM1.7 cell culture medium under standard tissue culture conditions (37 °C, 5% CO_2). Use cells for injection after one cell passage.
2. Collect YUMM1.7 cells at 60–70% confluence by adding enough Trypsin-EDTA to cover the bottom of the cell culture plate or flask (e.g., 4 mL for a T75 flask). Rest the cell culture vessels at room temperature for 3–4 min, gently tapping on the bottom or sides of the vessel occasionally to help detach the cells. Add 1 mL of FCS per 4 mL of Trypsin-EDTA to stop the reaction once most of the cells are detached. Collect the cells by pipetting into a 15-mL conical tube. Rinse the flask with PBS and add the rinse to the same collection tube. Fill the tube to 15 mL with PBS. Spin down for 10 min at 250 × *g* in a standard tissue culture centrifuge to collect the tumor cells.
3. Resuspend YUMM1.7 tumor cells in PBS at 1×10^6 cells/mL. Inject 100 μL (1×10^5 YUMM1.7 tumor cells) subcutaneously into the right flanks of anesthetized (e.g., 3–5% Isoflurane inhalant gas, anesthesia confirmed by lack of hind paw pinch reflex) recipient 4–6-week-old wild-type male C57BL/6J mice.
4. Monitor tumor volume via bi-weekly measurement of perpendicular tumor diameters and height using a caliper. Calculate YUMM1.7 tumor volume as tumor length × width × height/2. Initiate immunotherapy when YUMM1.7 tumors first become palpable, usually day 7–10 post tumor implantation.
5. Divide mice into control and experimental groups. For instance, a simple YUMM1.7 experiment might have ten tumor-bearing mice, five of which are treated with phDC immunotherapy (experimental), and five with PBS (control). Each mouse receives six treatments over 3 weeks, with two treatments per week. Blood is collected from all ten mice on Monday and Thursday, and "control" blood either pooled with "experimental" or discarded (*see* **Note 2**). Pooled blood from up to ten mice is processed through one TI plate and cultured overnight. The cell therapy product is returned only to the five experimental mice on the following day, Tuesday or Friday, while the control mice receive a PBS injection (*see* **Note 10**).

6. On treatment days (e.g., Mondays and Thursdays), collect blood from tumor-bearing mice, and process it to obtain PBMC by following the protocol in Subheading 3.1 (remember to collect and reserve autologous plasma in **step 3**). Prepare 8-MOP/UVA-treated apoptotic YUMM1.7 tumor cells by following the YUMM1.7 protocol in Subheading 3.3. Combine the PBMC and the 8-MOP/UVA-treated YUMM1.7 tumor cells (*see* **Note 11**) and pass them through one TI plate per 5–10 mice bled, as described in Subheading 3.5. Incubate the plate-passed cells overnight following the mouse protocol in Subheading 3.6, **step 1**.
7. On cell return days (e.g., Tuesdays and Fridays), collect the cells incubated overnight by following the mouse protocol in Subheading 3.6, **steps 2** and **3**. Resuspend the cell pellet in ~120 μL of autologous plasma (prepared in Subheading 3.1, **step 3**) per mouse to be treated (e.g., if blood used to make the cellular product was collected from 5 experimental mice ±5 control mice, then resuspend the pellet in 5 × 120 μL = 600 μL autologous plasma). Inject 100 μL cells per mouse into the retro-orbital plexus of appropriately anesthetized (e.g., 3–5% Isoflurane inhalant gas, anesthesia confirmed by lack of hind paw pinch reflex) tumor-bearing experimental mice. If so desired, reinject control tumor bearing mice with 100 μL of autologous plasma alone, or of PBS.
8. Repeat the treatment (**steps 6** and 7) twice a week for 3 weeks, for a total of 6 treatments. Monitor tumor volume throughout by measuring perpendicular tumor diameters and height using a caliper. Terminate the experiment when control tumor volumes reach maximal size permitted by institutional guidelines and protocols.

3.8 Human phDC ex vivo Immunotherapy Assays Using Tumor-Associated Peptide Antigens or Whole Tumor Cellular Antigens in Human T Cell Co-culture Model

1. To determine potency of phDC versus alternative APC sources, such as GM-CSF/IL-4-induced monocyte-derived DC (moDC), for stimulation of CD4 or CD8+ T cells, an investigator may load DC populations with defined tumor-associated antigens (Ags) in the form either of known HLA allele-fitting peptides, or of tumor cells expressing the antigenic protein of interest (*see* **Note 12**).
2. For peptide Ag assays, derive human phDC as described (Subheadings 3.2 and 3.5, with no added cellular Ags) and during overnight culture period (Subheading 3.6), add HLA-matched form-fitting or "long-peptide" (LP) Ag, such as melanoma-associate MART-1 peptides [6–8] directly to the 2-mL culture dish. Peptides are routinely solubilized and stored at −20 °C in DMSO at 1000× concentration (10 mg/mL) and then added to cultures at 2.0 μL/2 mL for 10 μg/mL final concentration.

3. For assays with whole human tumor/other cell lines expressing the Ag of interest, generate human phDC as described (Subheading 3.2 and 3.5) with the addition of apoptotic tumor cells (Subheading 3.4) to the PBMC population prior to exposure to the TI chamber.
4. For either peptide or cellular Ag-loaded phDC population, T-cell coculture assays must be titrated to determine the APC/T cell ratio which maximizes Ag presentation to CD4-restricted T cells or cross-presentation to CD8-restricted T cells.
5. T-cell lines are routinely cultured directly from LN2 stocks for 48 h prior to utilization in media supplemented with hIL-2 (Subheading 2, **item 27**).
6. For the majority of T-cell stimulation assays, our laboratory uses a 2:1 PBMC-derived APC:T cell ratio (numerically 200,000 PBMC: 100,000 T cells) cocultured in TC-treated 96-well round bottom plates for 72 h at 37 °C; or, in the case of purified monocytes as APC, a 20,000 monocyte: 100,000 T-cell ratio (reflecting the monocyte % within PBMC at approximately 10% for most human donors). Potential titration of cell ratios (e.g., 1:10, 1:5, 1:2, and 1:1 for both APC and T-cell components) and coculture times (for example 24, 48, 72, and 96 h) should be carried out for each T-cell type utilized in any new stimulation assay. Examples of human T-cell lines we have utilized in both peptide-based and tumor-based stimulation assays include: (a) DMF5 human TCR transgenic T-cell lines specific for melanoma-associated differentiation Ags such as MART-1 (derived originally from patient antitumor CD8+ T cells from TIL ([9]; Subheading 2, **item 28**), (b) HPV-16-reactive TCR transgenics similarly derived targeting E6 or E7 epitopes ([10]; Subheading 2, **item 29**), or (c) commercially available HPV-16 restricted T-cell lines expanded from normal donors (Subheading 2, **item 30**).
7. Post coculture, the 96-well plates are centrifuged at 700 × *g* for 10 min at 4 °C to firmly pellet cells, the supernatants are harvested, and compared for IFNγ production as a measure of T-cell stimulatory capacity utilizing a commercially available ELISA kit (Subheading 2, **item 31**).

3.9 phDC ex vivo Expansion of Autologous T Cells from Normal Donor or Cancer Patient Blood

1. To utilize phDC as stimulatory APC for expansion of circulating T-cell populations accessed from human blood, derive phDC as described (Subheadings 3.2 and 3.5, with no added cellular Ags) and during a coculture period with purified CD8+ T cells (Subheading 2, **item 33**), add HLA-matched form-fitting or "long-peptide" (LP) Ag, such as melanoma-associated MART-1 peptides (Subheading 3.8, **step 2**), to phDC or comparator APC population.

2. To directly compare phDC to comparator populations such as cytokine-derived DC at identical APC concentrations, purify blood monocytes from PBMC (*see* **Note 13**) either prior to exposure to the "Transimmune" plate system (after Subheading 3.2), or directly after but prior to T-cell coculture. To purify blood monocytes, use a commercially available "untouched" monocyte isolation kit (Subheading 2, **item 32**).
3. Purify "untouched" CD8+ T cells from human peripheral blood for coculture with phDC by purifying PBMC (Subheading 3.2) and then using a commercially available CD8+ isolation kit (Subheading 2, **item 33**). Use autologous, not just HLA-matched CD8+ T cells, whenever possible (*see* **Note 14**).
4. For DC/CD8+ cocultures, plate CD8+ T cells and phDC or comparator APC in 96-well round-bottomed plates in coculture medium (Subheading 2, **item 34**). Keeping the CD8+ cell number constant at 0.5 × 10^6 cells/well, set up cocultures at varying APC:T cell ratios (e.g., 1:5, 1:10, 1:20, 1:40, achieving the ratios by changing the number of APC added per well) in the presence of relevant tumor or pathogen-specific antigenic peptide, protein, or phagocytic substrate (as an example, 10 μM of MART-1 HLA-A2-restricted $MART_{27L}$ long peptide was used in Fig. 4).

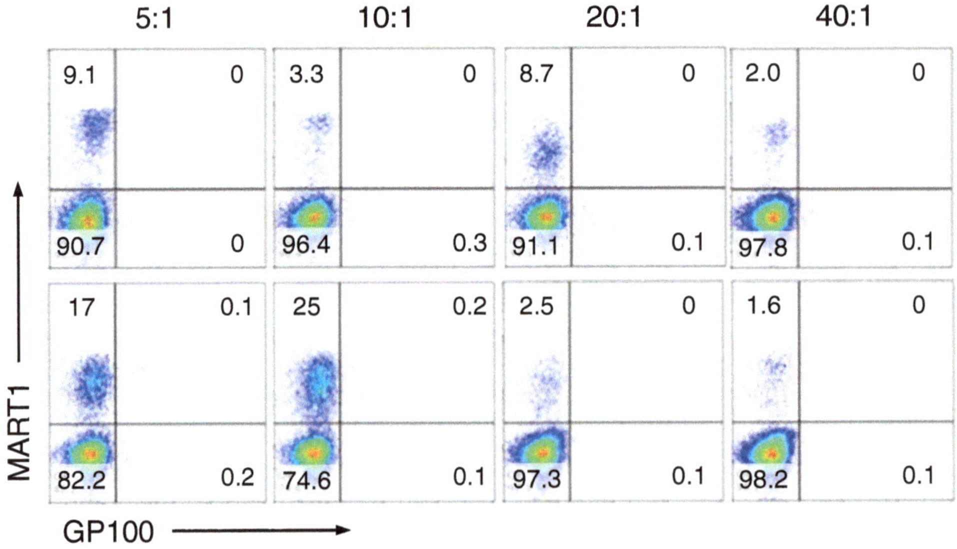

Fig. 4 Expansion of peripheral blood MART-1-specific CD8+ T cells following coculture with phDC. Cocultures were set up between traditional GM-CSF/IL-4 DC (top) and monocyte-derived phDC (bottom) and autologous $CD8^+$ T cells at four different T:APC ratios (5:1, 10:1, 20:1, and 40:1). Thirteen days following coculture, nonadherent T cells were collected and stained for tetramer positivity. Gp100 was used as a negative control tetramer. Representative flow cytometry profiles of $CD8^+$ T cells stained for gp100 and MART tetramers are shown. This figure has been modified from [8]

5. In parallel, set up negative control wells, including non-Ag-loaded APC, APC loaded with irrelevant peptide, and CD8+ T cells alone in the presence or absence of peptide.
6. At day 3 of coculture, change medium by gently pipetting half of the medium out of the wells, and replacing it with fresh coculture medium. Add 12.5 IU/mL IL-2 and 5 ng/mL IL-7 cytokines to all wells containing T cells.
7. Between days 9 and 11 of coculture, change medium by gently pipetting half of the medium out of the wells, and replacing it with fresh coculture medium containing IL-2 and IL-7 at the same concentrations as above.
8. On day 13 of coculture, harvest floating nonadherent cells and spin for 10 min at 250 × *g* in a standard tissue culture centrifuge and wash pellet 1× in RPMI. Stain with an appropriate nonsterically hindering anti-CD8 Ab as well as relevant tumor or pathogen epitope-specific tetramer or dextramer if available (e.g., 1:25 dilution MART-1 PE-conjugated dextramer (Immudex) and, as a negative control, 1:25 dilution gp100 APC-conjugated dextramer (Immudex) were used in Fig. 4). Alternatively, in the absence of appropriate tetramer readout, CD8+ T cells can instead be stained intracellularly for T-cell stimulation-associated cytokines, such as IFNγ (Subheading 2, **item 35**) or IL-2.
9. Assay specific T-cell expansion by FACS analysis of dextramer-positive (or activation cytokine-positive) CD8+ T cells in experimental vs. control cocultures.

4 Notes

1. We use the cheek bleed method routinely when preparing PBMC for the mouse in vivo tumor assay (Subheading 3.7) because it is gentle enough to allow for repeated bleeding (the mice are bled twice a week during treatment) while sparing the eyes, important because the manufactured therapeutic phDC are returned to the animal via retro-orbital injection. We therefore recommend cheek bleeds as the blood collection method for manufacture of phDC intended for the in vivo assay (Subheading 3.7). Other institutionally approved methods (tail bleed, eye bleed, cardiac puncture, etc.) may be appropriate when collecting large amounts of blood for other purposes and/or from different mouse models, but have not been tested or compared by us, and are therefore left to the investigators' discretion.
2. Mouse source and blood pooling depend on the intended use of the manufactured phDC. If collecting blood for the in vivo tumor assay (Subheading 3.7), in each 15-mL tube, pool

blood form 5–10 C57BL/6 tumor-bearing experimental mice that are 4–6 weeks of age at the time of the experiment. At the same time, to ensure equivalent treatment of all animals, bleed the corresponding 5–10 PBS control mice, either adding their blood in the same 15-mL tube if you want a higher therapeutic PBMC yield, or discarding it. The treated cells will be returned only to the experimental mice. We have performed treatments with both the combined experimental/control PBMC, and with experimental PBMC alone, without observing any major difference in antitumor therapeutic efficacy. If collecting PBMC for other application (in vitro phenotypic assays, coculture assays, etc.), then collect 2–4 mL of blood (~1×10^7 PBMC) from assay-appropriate mouse source per 15-mL conical tube, using as many 15-mL conical tubes as needed to reach the total PBMC number required for the assay.

3. Platelets are a very important component of this protocol, as they are necessary to obtain phDC.
   (a) Mouse samples should have a platelet/white blood cell (WBC) ratio of ~20:1; if the ratio is 1 or below, please discard the sample and repeat the PBMC isolation, paying particular attention to the anticoagulant, centrifugation conditions, and red blood cell lysis step. If the problem persists, try omitting the red blood cell lysis step, as it depletes some platelets along with the red cells.
   (b) Human samples should have a platelet/white blood cell (WBC) ratio of ~25–30:1. For human platelet-containing PBMC, no RBC lysis step is necessary, and the majority of the platelets can be co-retained with the PBMC layer by simply including the 4–5 mL of plasma above the PBMC interphase. This 4–5 mL just above the buffy coat is rich in platelets. If one is going to platelet-deplete the PBMC, this layer could be discarded along with the other plasma.
4. At this point, many experimental options are possible. For instance, the PBMC mixture can be depleted of component cell types (platelets, monocytes, T cells, etc.) to test their contribution; we have successfully and informatively conducted such experiments in both mouse and human PBMC [3, 5], using Miltenyi and EasySep cell isolation/depletion kits following manufacturers' instructions. Alternatively, extra cells can be added to the PBMC, such as 8-MOP/UVA-treated tumor cells as a source of antigen (Subheadings 3.3 and 3.4), which we have also successfully tested with both mouse and human tumor cell lines [3, 5].
5. Each tumor cell line has its own unique sensitivity to killing by 8-MOP/UVA. Therefore, an 8-MOP/UVA killing curve

should be determined in advance for each tumor cell line of interest. To do so, incubate tumor cells of interest with varying doses of 8-MOP, and then expose them to varying doses of UVA irradiation (Subheading 3.3, **steps 4–8** mouse, Subheading 3.4 human). UVA light intensity should be measured and normalized from any newly utilized light source; we monitor UVA bulb output using an International Light IL1700 Research Radiometer equipped with a XRD140A low-profile probe. The probe is placed above a "blank" well of the tissue culture dish utilized for irradiation, with the well containing identical cell media (traditionally 1× PBS or clear tissue culture media) as cells being irradiated. UVA dose is then calculated from output readings according to individual manufacturer's instructions. Plate the 8-MOP/UVA-treated cells and monitor their viability by any desired method (microscopy, flow cytometry, colony formation assay) over a period of 1 week–10 days. Select the lowest 8-MOP/UVA dose that ensures 100% tumor cell death within that time period. The results of a sample kill curve for the human melanoma cell line YUSIV is shown visually in Fig. 3. As a guideline, in our hands, the optimal doses for murine cell lines include 100 ng/mL 8-MOP with 4 J/cm^2 UVA for YUMM1.7 melanoma, 200 ng/mL 8-MOP with 4 J/cm^2 UVA for B16OVA melanoma, and 600 ng/mL 8-MOP with 7 J/cm^2 UVA for MC38 colon carcinoma cells. Doses used for human tumor cell lines include 200 ng/mL 8-MOP with 4–8 J/cm^2 UVA for squamous cell carcinoma cell lines, and 200 ng/mL 8-MOP with 4 J/cm^2 UVA for primary melanoma cell lines from the Yale biobank collection.

6. UVA light sources vary in their output intensity, which decays over time with use. Before using a new UVA light source, calibrate the UVA emission from where the cells will be irradiated, and from that, calculate cell exposure time. Recalibrate UVA emission and exposure time every 6 months. As reference, our in-house UVA light source delivers 4 J/cm^2 in approximately 9–10 min of cell exposure.
7. The TI plate running platform both secures the TI plate, and allows it to be easily manipulated and tilted in the desired directions during the procedure. However, the plate can also be tilted by raising one end onto a small flat object, such as the edge of a lab book, or edge of a tissue culture dish or flask.
8. At this point, the PBMC passed through the TI plate contains monocytes that have been activated by interaction with platelets under flow to enter the phDC activation pathway and to become immuno-stimulatory antigen-presenting cells. To instead obtain immunosuppressive cells, the TI plate-passed PBMC can be resuspended in 300 μL of FBS and treated with

8-MOP/UVA, following **steps 4–8** of the Subheading 3.3 and using 100 ng/mL 8-MOP in **step 4** and an exposure of 4 J/cm^2 in **step 7**.

9. During the overnight incubation, the activated mouse or human monocytes within plate-passed PBMC upregulate activation markers [3] and also take up antigen. This is a good point for adding noncellular antigens, such as peptides, proteins, or nanoparticles, if the intent is to create antigen-specific phDC.
10. This treatment regimen was developed for the specific growth kinetics of YUMM1.7 tumors in C57BL/6 J mice. It may need to be titrated for other murine in vivo tumor systems with slower or faster tumor growth rates—respectively increasing or reducing the number of treatments.
11. In this experimental setup, the 8-MOP/UVA-treated apoptotic tumor cells provide to the phDC the antigens that determine the antitumor immune specificity. It is therefore very important to use the same tumor cells both for establishing cancer in the mouse, and for treating that cancer with model ECP. We have demonstrated this specificity by showing that 8-MOP/UVA-treated YUMM1.7 tumor cells protect mice against YUMM1.7, but not against MC38 tumors, while, conversely, 8-MOP/UVA-treated MC38 cells are uniquely protective against MC38, but not against YUMM1.7 tumors [3].
12. Always remember to choose an APC blood donor who has the identical HLA restriction as the reactive CD8+ T cells you will be using as stimulation partners. For example, the tumor-specific T cells mentioned in our human T-cell stimulation assays (Subheading 2, **items 28–30**) are all restricted to HLA-A2-specific epitopes, so your chosen human blood donor must also be HLA-A2.
13. Although monocyte percentages within PBMC vary significantly between human donors, a good rule-of-thumb is start with approximately ten-fold more PBMC than monocytes you expect to need for any coculture experiments (i.e., a 10% monocyte yield). Also expect a 10–15% total cellular loss over MACs-based isolation columns and plan starting cell numbers accordingly.
14. When utilizing purified donor CD8+ T cells, or even CD8+ TCR transgenics, you may generate yourself from transfection of donor T cells in T-cell stimulation assays; it may be advisable to use the same donor as your source of both T cells and APC for phDC creation. In our experience, although adequately purified CD8+ T-cell populations should react extremely specifically against a particular HLA-restricted

epitope displayed on your APC populations, the possibility of minor histocompatibility mismatches or cross-reactivity to related Ags (particularly when using pathogen-specific T-cell readouts) can lead to spurious, nonspecific stimulation results.

References

1. Raval JS, Ratcliffe NR (2018) Extracorporeal photopheresis and personalized medicine in the 21st century: the future's so bright! J Clin Apher 33(4):461–463
2. Edelson RL (2014) Mechanistic insights into extracorporeal photochemotherapy: efficient induction of monocyte-to-dendritic cell maturation. Transfus Apher Sci 50:322–329
3. Ventura A et al (2018) Extracorporeal photochemotherapy drives monocyte-to-dendritic cell maturation to induce anticancer immunity. Cancer Res 78:4045–4058
4. Durazzo TS, Tigelaar RE, Filler R, Hayday A, Girardi M, Edelson RL (2014) Induction of monocyte-to-dendritic cell maturation by extracorporeal photochemotherapy: initiation via direct platelet signaling. Transfus Apher Sci 50:370–378
5. Ventura A, Vassall A, Yurter A, Robinson E, Filler R, Hanlon D, Meeth K, Ezaldein H, Girardi M, Sobolev O, Bosenberg MW, Edelson RL (2019) Novel protocol for generating physiologic immunogenic dendritic cells. J Vis Exp 147. https://doi.org/10.3791/59370
6. Chauvin JM, Larrieu P, Sarrabayrouse G, Prévost-Blondel A, Lengagne R, Desfrançois J, Labarrière N, Jotereau F (2012) HLA anchor optimization of the melan-A-HLA-A2 epitope within a long peptide is required for efficient cross-priming of human tumor-reactive T cells. J Immunol 188(5):2102–2110
7. Hong E, Usiskin IM, Bergamaschi C, Hanlon DJ, Edelson RL, Justesen S, Pavlakis GN, Flavell RA, Fahmy TM (2016) Configuration-dependent presentation of multivalent IL-15:IL-15Rα enhances the antigen-specific T cell response and anti-tumor immunity. J Biol Chem 291(17):8931–8950
8. Kibbi N, Sobolev O, Girardi M, Edelson RL (2016) Induction of anti-tumor CD8 T cell responses by experimental ECP-induced human dendritic antigen presenting cells. Transfusion Apher Sci 55(1):146–152
9. Johnson LA, Heemskerk B, Powell DJ Jr et al (2006) Gene transfer of tumor-reactive TCR confers both high avidity and tumor reactivity to nonreactive peripheral blood mononuclear cells and tumor-infiltrating lymphocytes. J Immunol 177:6548–6559
10. Draper L et al (2015) Targeting of HPV-16+ epithelial cancer cells by TCR gene engineered T cells directed against E6. Clin Cancer Res 21(19):4431–4439

Chapter 12

Reprogramming Exosomes for Immunotherapy

Qinqin Cheng, Xiaojing Shi, and Yong Zhang

Abstract

Exosomes are nanosized vesicles secreted by nearly all types of cells and play important roles in intercellular communication. Given their unique and important pharmacological properties, exosomes have been emerging as a new class of cell-free therapeutics. Herein, we describe exosomes developed against epidermal growth factor receptor (EGFR), a key factor in epithelial malignancies, involved in enhanced tumor growth, invasion, and metastasis. The exosomes are genetically modified for displaying two distinct types of monoclonal antibodies on the exosome surface, resulting in novel *s*ynthetic *m*ultivalent *a*ntibodies *r*e*t*argeted *exo*some*s* (SMART-Exos) that can simultaneously target tumor-associated human EGFR and T-cell surface CD3 receptor. By redirecting and activating T cells toward attacking EGFR-expressing cancer cells, the designed SMART-Exos exhibit highly potent and specific antitumor activity. In this chapter, the methodologies are outlined for generating and using SMART-Exos for cancer immunotherapy.

Key words Exosome, Immunotherapy, Genetic engineering, Monoclonal antibody, EGFR, CD3, Cancer, T cell, Isolation, Flow cytometry, Immunoblot, Confocal microscopy, Cytotoxicity

1 Introduction

Exosomes are nanoscale membranous vehicles secreted by nearly all types of cells and are found in a wide range of tissues and body fluids [1, 2]. By carrying various forms of membrane proteins and soluble cargos including proteins, lipids, and nucleic acids, exosomes function as important mediators for intercellular communication through modulating signaling pathways in target cells and/or transferring membrane receptors and cargos to recipient cells [3, 4]. Compared with conventional nanoparticles, naturally occurring exosomes are expected to possess excellent biocompatibility, high efficiency for therapeutic delivery, and low immunogenicity [5]. Notably, exosomes are characterized by abundant membrane proteins, critical for their biological functions and therapeutic applications. For instance, tetraspanin CD9 highly expressed on the exosome surface can promote direct membrane fusion with

Samuel G. Katz and Peter M. Rabinovich (eds.), *Cell Reprogramming for Immunotherapy: Methods and Protocols*, Methods in Molecular Biology, vol. 2097, https://doi.org/10.1007/978-1-0716-0203-4_12,

target cells, which facilitate cytosolic delivery of therapeutics [6]. By triggering a "don't eat me" signal, transmembrane protein CD47 is shown to protect exosomes from clearance by circulating monocytes and macrophages [7]. Given their strong therapeutic potential, exosomes have been emerging as a promising form of nanomedicine. A considerable number of preclinical and clinical studies indicate that exosome-based therapeutics display excellent efficacy for treatment of a variety of human diseases, including cancer, neurodegenerative diseases, liver diseases, ischemic diseases, and immune disorders [8–13].

Currently, different methods are established for loading exosomes with various types of endogenous and exogenous agents for therapeutic delivery [5, 14, 15]. Herein, we develop a new approach for reprogramming exosomes to modulate cellular immunity in a controlled and directed manner. Through genetically displaying two distinct types of monoclonal antibodies on the exosome surface using the transmembrane (TM) domain of human platelet-derived growth factor receptor (PDGFR), a novel class of exosomes is generated, namely *s*ynthetic *m*ultivalent *a*ntibodies *ret*argeted *exo*some*s* (SMART-Exos) [16]. By targeting T-cell surface CD3 and cancer cell-associated epidermal growth factor receptor (EGFR), the resulting SMART-Exos not only show tight binding to both EGFR-expressing triple-negative breast cancer (TNBC) cells and T cells but also induce potent and specific killing of EGFR-positive TNBC cells in the presence of nonactivated human peripheral blood mononuclear cells (hPBMCs) (*see* **Note 1**). In this chapter, the procedures are elaborated for generation and characterization of SMART-Exo for cancer immunotherapy. It is likely that these methods can be easily adopted for other disease models by utilizing different types of functional antibodies.

2 Materials

2.1 Plasmids

1. Mammalian expression vector pDisplay.
2. Anti-EGFR antibody-PDGFR TM domain fusion plasmid: hemagglutinin (HA)-αEGFR-PDGFR TM domain.
3. Anti-CD3 antibody-PDGFR TM domain fusion plasmid: HA-αCD3-PDGFR TM domain.
4. Anti-EGFR/CD3 antibody-PDGFR TM domain fusion plasmid: HA-αEGFR-αCD3-PDGFR TM domain.
5. Anti-CD3/EGFR antibody-PDGFR TM domain fusion plasmid: HA-αCD3-αEGFR-PDGFR TM domain.
6. ZymoPURE II Plasmid Maxiprep Kit.

2.2 Cell Culturing and Transfection

1. Expi293F cell line.
2. Expi293 expression medium.

3. Opti-Minimal Essential Medium (Opti-MEM) I reduced serum medium.
4. ExpiFectamine 293 transfection reagent.
5. 125-mL polycarbonate Erlenmeyer flasks.

2.3 Exosome Isolation

1. Dulbecco's phosphate-buffered saline (DPBS) (1×).
2. Benchtop centrifugation instrument: Heraeus Megafuge 40R Refrigerated Centrifuge; rotor: TX-750 Swinging Bucket Rotor; tube: 50-mL centrifuge tubes.
3. Floor centrifugation instrument: J2–21 Floor Model Centrifuge (Beckman Coulter, Indianapolis, IN, USA); rotor: JA-17 Fixed-Angle Aluminum Rotor; tube: Nalgene Oak Ridge high-speed polypropylene copolymer centrifuge tubes.
4. Ultracentrifugation instrument: Optima L-80 XP Ultracentrifuge (Beckman Coulter); rotor: Type 70 Ti Fixed-Angle Titanium Rotor, tube: polycarbonate bottle with cap assembly.
5. Pierce Coomassie (Bradford) Protein Assay Kit.
6. Synergy H1 Hybrid Multi-Mode Microplate Reader.

2.4 Immunoblots

1. Phosphate-buffered saline (PBS): 8 g/L NaCl, 0.2 g/L KCl, 1.78 g/L $Na_2HPO_4{\cdot}2H_2O$ and 0.27 g/L KH_2PO_4, pH 7.4, filtered with a 0.22-μm disc filter.
2. 4–20% ExpressPlus-PAGE gels.
3. NuPAGE LDS Sample Buffer (4×).
4. Trans-Blot SD Semi-Dry Transfer Cell.
5. ExpressPlus-PAGE gel running buffer: Tris base 6.06 g/L, MOPS 10.46 g/L, SDS 1 g/L, EDTA 0.3 g/L, add deionized water to 1 L.
6. Immunoblot transfer buffer: 5.82 g/L Tris base, 2.93 g/L glycine, 200 mL/L methanol, add deionized water to 1 L.
7. Immun-Blot PVDF membrane.
8. Membrane wash buffer: PBS-T (PBS with 0.1% Tween-20).
9. Blocking solution: dissolve 5% BSA in PBS-T.
10. Primary antibodies: anti-HA antibody (clone: 2-2.2.14, Thermo Fisher Scientific), anti-CD63 antibody (clone: H5C6, BioLegend, San Diego, CA, USA), anti-CD81 antibody (clone: 1.3.3.22, Thermo Fisher Scientific), and anti-CD9 antibody (clone: D8O1A, Cell Signaling Technology).
11. Secondary antibodies: anti-mouse IgG-HRP and anti-rabbit IgG-HRP.
12. SuperSignal West Pico PLUS chemiluminescent substrate.
13. ChemiDoc Touch Imaging System.

2.5 Flow Cytometry Analysis

1. Jurkat cell line.
2. MDA-MB-468 cell line.
3. RPMI-1640 medium with 10% fetal bovine serum (FBS).
4. DPBS, 1× (Mediatech).
5. 0.05% Trypsin-EDTA (1×).
6. LSR II Flow Cytometer.
7. Anti-HA antibody (clone: 2-2.2.14, Thermo Fisher Scientific).
8. Alexa Fluor 488-labeled goat anti-mouse IgG H&L antibody.
9. Sample wash buffer: DPBS with 2% FBS.

2.6 Confocal Microscopy of Cell–Cell Cross-Linking Mediated by SMART-Exos

1. Jurkat cell line.
2. MDA-MB-468 cell line.
3. RPMI-1640 medium with 10% FBS.
4. MitoSpy Red.
5. Carboxyfluorescein succinimidyl ester (CFSE).
6. Leica SP8 confocal laser scanning microscope.
7. 24-well cell culture plates (sterile, flat bottom).
8. 12-mm circular coverslips (glass).
9. Microslides (25 × 75 mm, glass).
10. Fluoromount.

2.7 Cytotoxicity Induced by SMART-Exos

1. Jurkat cell line.
2. MDA-MB-468 cell line.
3. MDA-MB-453 cell line.
4. RPMI-1640 medium with 10% FBS.
5. Human peripheral blood mononuclear cells (hPBMCs) from HemaCare.
6. 3-(4,5-dimethylthiazol-2-yl)-2,5-diphenyltetrazolium bromide (MTT) solution: dissolve 5-mg MTT in 1-mL PBS. Sterilize by filtration with a 0.22-μm disc filter.
7. Lysis buffer (20% SDS in 50% dimethylformamide, pH 4.7).
8. Synergy H1 Hybrid Multi-Mode Microplate Reader (BioTek).
9. 96-well plates (sterile, flat bottom).

3 Methods

3.1 Plasmid Construction and Preparation

1. Use the pDisplay vector for constructing SMART-Exos expression plasmids, which contains a human PDGFR TM domain for displaying proteins of interest on the mammalian cell surface. Synthetic genes encoding scFv fragments of

αEGFR cetuximab and αCD3 UCHT1 antibodies could be purchased from Integrated DNA Technologies, Inc. (Coralville, IA), in which a $(GGGGS)_3$ linker is inserted between V_H and V_L regions. To generate single polypeptide encoding dual scFv antibodies, perform overlap extension polymerase chain reactions (PCR) by placing a $(GGGGS)_3$ flexible linker between two individual scFv fragments. Arrange the orientations of variable regions for each scFv or dual scFv antibody as follows: $V_{H\text{-}\alpha EGFR}$-$V_{L\text{-}\alpha EGFR}$ (αEGFR scFv), $V_{L\text{-}\alpha CD3}$-$V_{H\text{-}\alpha CD3}$ (αCD3 scFv), $V_{H\text{-}\alpha EGFR}$-$V_{L\text{-}\alpha EGFR}$-$V_{L\text{-}\alpha CD3}$-$V_{H\text{-}\alpha CD3}$ (αEGFR/αCD3 scFv), and $V_{L\text{-}\alpha CD3}$-$V_{H\text{-}\alpha CD3}$-$V_{H\text{-}\alpha EGFR}$-$V_{L\text{-}\alpha EGFR}$ (αCD3/αEGFR scFv). Add a hemagglutinin (HA)-tag at the N terminus of the scFv or dual scFv antibody during PCR amplification. Insert the resulting DNA fragments between the N-terminal signal peptide and the TM domain of PDGFR using flanking restriction sites BglII and SalI (Fig. 1). Confirm the generated expression vectors by DNA sequencing.

2. Prepare transfection-grade plasmids for the sequence-verified expression constructs by using endotoxin-free buffers and ZymoPURE II Plasmid Maxiprep Kits (ZYMO Research).

3.2 Transient Transfection

Perform transfection by following the guidelines of Expi293 expression system. Below is a brief outline of the method for transfection of 30 mL of Expi293F cell culture.

1. On the day prior to transfection (Day -1), determine cell density and viability (*see* **Note 2**). To transfect the cells on the following day, split the Expi293F cell culture to a final density of 2×10^6 viable cells/mL with fresh prewarmed Expi293 expression medium and incubate the cells on an orbital shaker (125 rpm/min) for 16–24 h in a 37 °C incubator with a humidified atmosphere of 8% CO_2 in air.

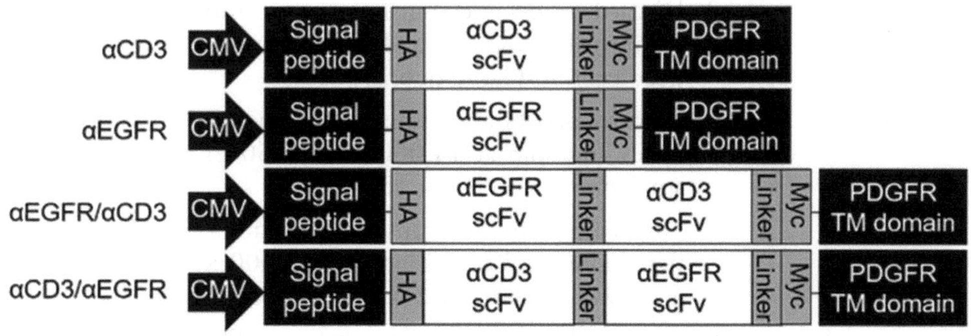

Fig. 1 Molecular designs of antibody-PDGFR TM domain fusions for the generation of SMART-Exos. Each fusion protein contains an N-terminal HA epitope tag and flexible $(GGGGS)_3$ linkers

2. On the next day (Day 0), determine cell density and viability (*see* **Note 3**) and resuspend cells in fresh prewarmed Expi293 expression medium. For each 30 mL of transfection, add appropriate volume of cell suspension with a total of 7.5×10^7 cells to a sterile 125-mL polycarbonate Erlenmeyer flask and bring up the volume to 25.5 mL by adding fresh prewarmed Expi293 expression medium. Return the cells to the CO_2 incubator.
3. Prepare plasmid DNA/ExpiFectamine 293 complexes. Dilute 30-μg plasmid DNA (*see* **Note 4**) with 1.5-mL Opti-MEM I reduced serum medium. Mix gently. Dilute 80-μL ExpiFectamine 293 reagent with 1.5-mL Opti-MEM I reduced serum medium. Mix gently and allow to incubate at room temperature (RT) for 5 min (*see* **Note 5**). Add the diluted plasmid DNA to the diluted ExpiFectamine 293 reagent. Mix gently.
4. Incubate plasmid DNA/ExpiFectamine 293 complexes at RT for 20–30 min, and then slowly transfer the solution to the shaker flask from **step 2**. Swirl the flask gently during addition. Return the cells to the CO_2 incubator.
5. On the first day after transfection (Day 1, 18–22 h posttransfection), add 150-μL ExpiFectamine 293 transfection enhancer 1 and 1.5-mL ExpiFectamine 293 transfection enhancer 2 to the shaker flask. Gently swirl the flask during addition. Return the cells to the CO_2 incubator (*see* **Note 6**).
6. Day 3 and Day 6: Collect exosome-containing cell culture media via centrifugation (*see* **Note 7**).

3.3 Exosome Isolation

Purify the expressed SMART-Exos from the collected culture media through differential centrifugation.

1. Collect cell cultures containing expressed SMART-Exos.
2. Centrifuge at $100 \times g$ for 10 min at 4 °C to remove residual cells (*see* **Note 8**).
3. Discard pellets and transfer supernatants to clean 50-mL centrifuge tubes and spin at $4000 \times g$ for 30 min at 4 °C to remove dead cells and large debris.
4. Discard pellets and transfer supernatants to clean centrifuge tubes and spin at $14{,}000 \times g$ for 50 min at 4 °C to remove small debris and larger vesicles.
5. Discard pellets and transfer supernatants to clean ultracentrifuge tubes and centrifuge at $371{,}000 \times g$ (60,000 rpm) in a Type 70 Ti rotor for 2 h at 4 °C (*see* **Note 9**).
6. Discard the supernatants, carefully refill the tubes with PBS to wash the pellets. Repeat this step once.

7. Resuspend the pellet in PBS (50–200 μL) and transfer to 1.7-mL Eppendorf tubes.
8. Vortex for 20 min at RT and then centrifuge at 14,000 × *g* for 15 min at 4 °C to remove precipitation. Repeat the centrifugation step once.
9. Collect the supernatants and filter with 0.22-μm filters (*see* **Note 10**).
10. Measure protein concentrations using the Coomassie (Bradford) protein assay kits.
11. Store SMART-Exos at 4 °C or on ice prior to immediate use or at −80 °C for long-term storage (*see* **Note 11**).

3.4 Immunoblots

1. Reduce the isolated SMART-Exos (4 μg of protein in total) with 10-mM dithiothreitol in 1 × LDS sample buffer at 95 °C for 5 min (*see* **Note 12**).
2. Load the sample into each well (up to 40 μL per well for 4–20% ExpressPlus-PAGE gels).
3. Run the gel at 150 V for 50 min at RT.
4. Transfer samples to PVDF membranes at 15 V for 35 min at RT (*see* **Note 13**).
5. Wash the PVDF membrane briefly in wash buffer.
6. Block the PVDF membrane in PBS-T with 5% BSA for 1 h at RT.
7. Incubate with appropriate primary antibodies in PBS-T at RT.
8. Wash three times with PBS-T for a total of 30 min.
9. Incubate with appropriate HRP conjugated secondary antibodies for 1 h at RT.
10. Wash three times with PBS-T for a total of 30 min.
11. Visualize the immunoblots using the ChemiDoc Touch Imaging System upon the addition of SuperSignal West Pico PLUS chemiluminescent substrate (Fig. 2).

3.5 Flow Cytometry

1. Harvest cells and wash twice in PBS with 2% FBS at 400 × *g* for 5 min at 4 °C.
2. Discard the supernatants and incubate the cells (0.2 million) with exosomes (0.1 mg/mL) in PBS with 2% FBS for 30 min at 4 °C (*see* **Note 14**).
3. Wash twice with PBS containing 2% FBS.
4. Discard the supernatants and incubate with the anti-HA antibody for 30 min at 4 °C.
5. Wash twice with PBS containing 2% FBS.
6. Discard the supernatants and incubate with the Alexa Fluor 488-labeled goat anti-mouse IgG H&L antibody for 30 min at 4 °C.

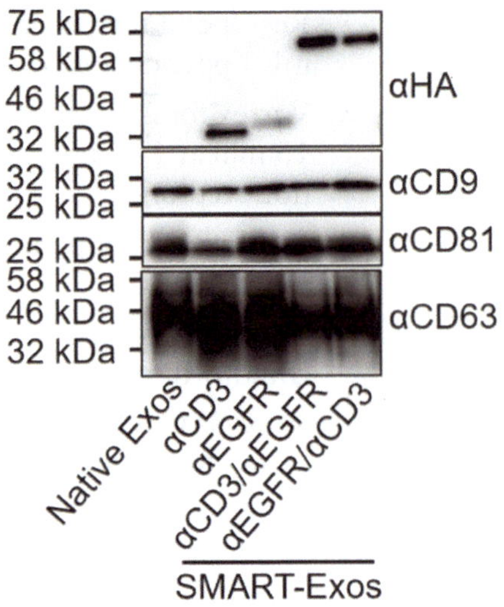

Fig. 2 Immunoblot analysis of native exosomes and the generated SMART-Exos using antibodies specific for HA to detect expression of scFv antibodies and for CD9, CD81, and CD63 exosomal marker proteins. A total of 4 μg exosomes (by total protein concentration) for each sample was loaded for immunoblot analysis. (Adapted with permission from *J. Am. Chem. Soc.* 2018, 140, 48, 16,413–16,417. Copyright (2018) American Chemical Society)

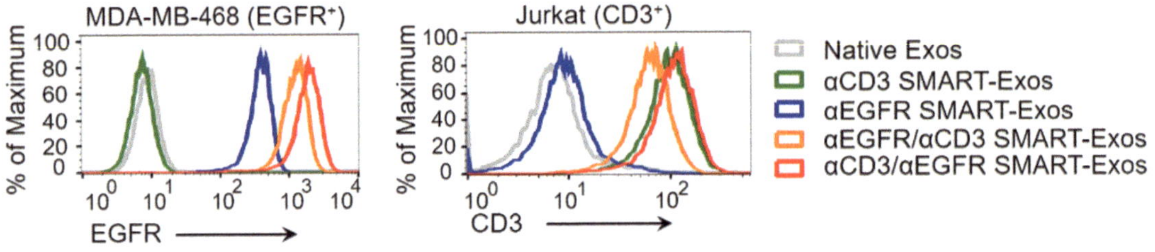

Fig. 3 Flow cytometric analysis of the binding of the generated SMART-Exos to MDA-MB-468 cells ($EGFR^+$ $CD3^-$) and Jurkat cells ($EGFR^-$ $CD3^+$). (Adapted with permission from *J. Am. Chem. Soc.* 2018, 140, 48, 16,413–16,417. Copyright (2018) American Chemical Society)

7. Wash twice with PBS containing 2% FBS.
8. Analyze samples using an LSR II Flow Cytometer (Fig. 3).

3.6 Confocal Microscopy of Cell–Cell Cross-Linking Mediated by SMART-Exos

1. Incubate MDA-MB-468 cells and Jurkat cells with 200 nM MitoSpy Red and 5 μM CFSE at 37 °C for 20 min, respectively (*see* **Note 15**).
2. Wash twice with PBS to remove free dyes (*see* **Note 16**).
3. Incubate Jurkat cells (6×10^4) with α-CD3/α-EGFR SMART-Exos (0.1 mg/mL) or a mixture (1:1) of αCD3 SMART-Exos and αEGFR SMART-Exos in 100 μL PBS for 30 min at 4 °C.
4. Wash Jurkat cells with 1 mL of cold PBS to remove free exosomes.

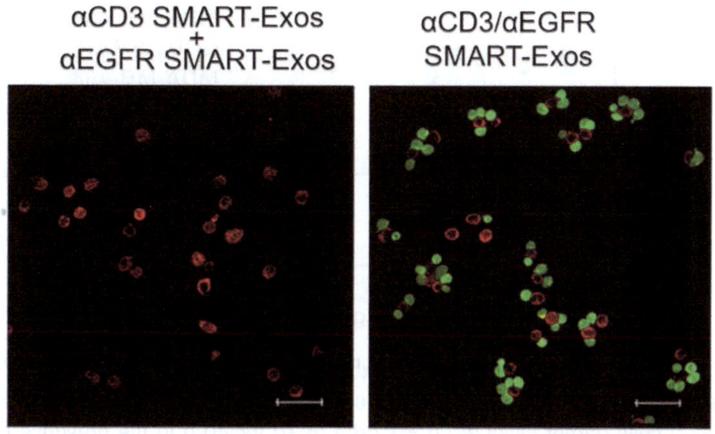

Fig. 4 Confocal microscopic analysis of the cross-linking of Jurkat and MDA-MB-468 cells induced by SMART-Exos. Fluorescently labeled target cancer cells (red) and Jurkat cells (green) were mixed at a 1:3 ratio and incubated for 5 h in the presence of αCD3/αEGFR SMART-Exos or a mixture (1:1) of αCD3 SMART-Exos and αEGFR SMART-Exos. Nonbound cells were washed away by PBS. Scale bars: 50 μm. (Adapted with permission from *J. Am. Chem. Soc.* 2018, 140, 48, 16,413–16,417. Copyright (2018) American Chemical Society)

5. Mix Jurkat cells (6×10^4) and MDA-MB-468 cells (2×10^4) in 500 μL of RPMI-1640 medium with 10% FBS.
6. Add cell mixtures into clear bottoms of 24-well plates with 12-mm coverslips at the bottoms and incubate for 5 h at 37 °C (*see* **Note 17**).
7. Gently wash the cells four times with PBS to remove unattached Jurkat cells.
8. Carefully remove the coverslips from 24-well plates and mount onto slides with fluoromount, then image with a Leica SP8 confocal laser scanning microscope using FITC (for CFSE) and rhodamine (for MitroSpy Red) filters (Fig. 4).

3.7 Cytotoxicity Induced by SMART-Exos

In this section, we will use a colorimetric MTT assay to measure viability.

1. Mix cancer cells (target cells) and nonactivated hPBMCs (effector cells) with a ratio of 1:10, then add the cell mixtures into the clear bottoms of 96-well plates (*see* **Note 18**).
2. Prepare SMART-Exos in DPBS with concentrations ranging from 100 μg/mL to 1 pg/mL with a ten-fold serial dilution.
3. Add 10 μL SMART-Exo dilution into corresponding wells and use 10 μL DPBS as a control.
4. Incubate the 96-well plates for 40 h at 37 °C.
5. Wash cells twice with PBS to remove hPBMC suspensions.

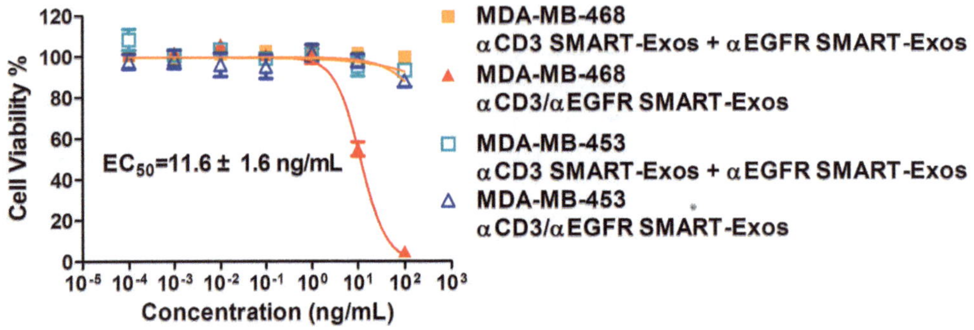

Fig. 5 In vitro cytotoxicity of the αCD3/αEGFR SMART-Exos for MDA-MB-468 ($EGFR^+$) or MDA-MB-453 ($EGFR^-$) cells. Nonactivated hPBMCs (effector cells) were incubated with MDA-MB-468 ($EGFR^+$) and MDA-MB-453 ($EGFR^-$) cells (target cells) at an E:T ratio of 10 for 40 h in the presence of αCD3/αEGFR SMART-Exos or a mixture (1:1) of αCD3 SMART-Exos and αEGFR SMART-Exos. After removing hPBMCs suspensions, cell viabilities of target cells were measured through MTT assays. (Adapted with permission from *J. Am. Chem. Soc.* 2018, 140, 48, 16,413–16,417. Copyright (2018) American Chemical Society)

6. Add 10 μL MTT solution with 90 μL culture medium to each well and incubate for 4 h at 37 °C.
7. Add 100 μL of lysis buffer into each well and incubate for 3 h at 37 °C.
8. Measure absorbance at 570 nm using a BioTek Synergy H1 Hybrid Multi-Mode Microplate reader.
9. Calculate cytotoxicity as the percentage of cell viability = $[(\text{absorbance}_{\text{experimental}} - \text{absorbance}_{\text{spontaneous average}})/(\text{absorbance}_{\text{maximal viability average}} - \text{absorbance}_{\text{spontaneous average}})] \times 100$ (Fig. 5).

4 Notes

1. The αCD3/αEGFR SMART-Exos are expected to induce formation of immunological synapses between T cells and EGFR-expressing tumor cells, leading to the activation of possibly both CD4 and CD8 effector T cells to release cytolytic proteins and inflammatory cytokines for potent anticancer activity, which is independent of antigen presentation by MHC class I molecules or costimulation.
2. Determine cell number and viability using the trypan blue dye exclusion method. Cell viability should be >95% to proceed with transfection.
3. To start transfection, the final cell density should be at least 2.5×10^6 cells/mL with >95% viability.

4. To ensure sterility, DNA should be filtered through a 0.22-μm filter prior to use. A total plasmid DNA of 1.0 μg per mL of culture volume to be transfected is appropriate for most proteins.
5. Longer incubation time may result in decreased performance.
6. Enhancer 1 and enhancer 2 can be mixed together just prior to additions to the cell culture.
7. Cells or media (if recombinant protein is secreted) may be harvested beginning at approximately 48 h posttransfection and assayed for recombinant protein expression. Optimal time for harvesting SMART-Exos depends on the expression levels and stability of the fusion proteins on the exosome surface. For antibody protein fusions, 3–5 days are typical harvest times.
8. Keep all the centrifugation steps at 4 °C to retain the maximum biological activities of proteins expressed on the exosome surface.
9. It is recommended that the supernatants be centrifuged at 371,000 × *g* for 2 h to maximize the yields of exosomes. Ultracentrifugation at lower speeds (e.g., 100,000 × *g*) for 2 h or longer is also acceptable for this step.
10. For cell-based assays, the supernatants should be sterilized through filtration with 0.22-μm disc filters. In our experience, there is only a very small percentage of vesicles larger than 200 nm in diameter after the ultracentrifugation step.
11. Exosomes can be stored on ice or at 4 °C for up to 6 h or at −80 °C for up to 6 months.
12. Using LDS sample buffer alone allows successful extraction of tetraspanins. However, it could be difficult to extract certain transmembrane proteins from exosomes, which requires optimization of the sample buffer for more efficient extraction, such as using RIPA buffer. Reducing exosomes with DTT can help detect HA-tagged fusion protein and CD9 exosomal marker. But CD81 and CD63 require nonreducing electrophoresis for immunoblotting, possibly due to the disulfide bond-dependent epitopes for their respective antibodies.
13. The transfer time is dependent on the size of target protein.
14. Keep all steps at 4 °C to avoid the interference of exosome uptakes by cells. Native exosomes isolated from nontransfected Expi293F cells should also be included as controls.
15. For CFSE staining, resuspend cells at 10–100 × 10^6 cells/mL in the 5 μM CFSE working solution in PBS. For MitoSpy Red staining, prepare the working solution with incomplete

culture medium and a concentration at 50–250 nM is recommended, if labeling mitochondria for live cell imaging.

16. Before the wash steps, quench the staining by adding five times of the original staining volume of cell culture medium with 10% FBS.
17. Incubation at 37 °C for 5 h allows adherent MDA-MB-468 cells to attach to the bottoms of 24-well plates.
18. Each well should contain 1×10^4 cancer cells and 1×10^5 non-activated hPBMCs.

Acknowledgments

This work was supported, in part, by University of Southern California School of Pharmacy Start-Up Fund for New Faculty, University of Southern California Ming Hsieh Institute for Engineering Medicine for Cancer, American Association of Pharmaceutical Scientists (AAPS) Foundation New Investigator Grant (to Y.Z.), STOP CANCER Research Career Development Award (to Y.Z.), PhRMA Foundation Research Starter Grant in Translational Medicine and Therapeutics (to Y.Z.), P30CA014089 to the USC Norris Comprehensive Cancer Center, and P30DK048522 to the USC Research Center for Liver Diseases.

References

1. Thery C, Zitvogel L, Amigorena S (2002) Exosomes: composition, biogenesis and function. Nat Rev Immunol 2:569–579
2. Yanez-Mo M, Siljander PR, Andreu Z, Zavec AB, Borras FE, Buzas EI, Buzas K, Casal E, Cappello F, Carvalho J, Colas E, Cordeiro-da Silva A, Fais S, Falcon-Perez JM, Ghobrial IM, Giebel B, Gimona M, Graner M, Gursel I, Gursel M, Heegaard NH, Hendrix A, Kierulf P, Kokubun K, Kosanovic M, Kralj-Iglic V, Kramer-Albers EM, Laitinen S, Lasser C, Lener T, Ligeti E, Line A, Lipps G, Llorente A, Lotvall J, Mancek-Keber M, Marcilla A, Mittelbrunn M, Nazarenko I, Nolte-'t Hoen EN, Nyman TA, O'Driscoll L, Olivan M, Oliveira C, Pallinger E, Del Portillo HA, Reventos J, Rigau M, Rohde E, Sammar M, Sanchez-Madrid F, Santarem N, Schallmoser K, Ostenfeld MS, Stoorvogel W, Stukelj R, Van der Grein SG, Vasconcelos MH, Wauben MH, de Wever O (2015) Biological properties of extracellular vesicles and their physiological functions. J extracell Ves 4:27066
3. Colombo M, Raposo G, Thery C (2014) Biogenesis, secretion, and intercellular interactions of exosomes and other extracellular vesicles. Annu Rev Cell Dev Biol 30:255–289
4. Jiang XC, Gao JQ (2017) Exosomes as novel bio-carriers for gene and drug delivery. Int J Pharm 521:167–175
5. Armstrong JP, Holme MN, Stevens MM (2017) Re-engineering extracellular vesicles as smart nanoscale therapeutics. ACS Nano 11:69–83
6. van Dongen HM, Masoumi N, Witwer KW, Pegtel DM (2016) Extracellular vesicles exploit viral entry routes for cargo delivery. MMBR 80:369–386
7. Kamerkar S, LeBleu VS, Sugimoto H, Yang S, Ruivo CF, Melo SA, Lee JJ, Kalluri R (2017) Exosomes facilitate therapeutic targeting of oncogenic KRAS in pancreatic cancer. Nature 546:498–503
8. Johnsen KB, Gudbergsson JM, Skov MN, Pilgaard L, Moos T, Duroux M (2014) A comprehensive overview of exosomes as drug delivery vehicles – endogenous nanocarriers for targeted cancer therapy. Biochim Biophys Acta 1846:75–87

9. Cooper JM, Wiklander PB, Nordin JZ, Al-Shawi R, Wood MJ, Vithlani M, Schapira AH, Simons JP, El-Andaloussi S, Alvarez-Erviti L (2014) Systemic exosomal siRNA delivery reduced alpha-synuclein aggregates in brains of transgenic mice. Mov Disord 29:1476–1485
10. Oh K, Kim SR, Kim DK, Seo MW, Lee C, Lee HM, Oh JE, Choi EY, Lee DS, Gho YS, Park KS (2015) In vivo differentiation of therapeutic insulin-producing cells from bone marrow cells via extracellular vesicle-mimetic nanovesicles. ACS Nano 9:11718–11727
11. Fais S, O'Driscoll L, Borras FE, Buzas E, Camussi G, Cappello F, Carvalho J, Cordeiro da Silva A, Del Portillo H, El Andaloussi S, Ficko Trcek T, Furlan R, Hendrix A, Gursel I, Kralj-Iglic V, Kaeffer B, Kosanovic M, Lekka ME, Lipps G, Logozzi M, Marcilla A, Sammar M, Llorente A, Nazarenko I, Oliveira C, Pocsfalvi G, Rajendran L, Raposo G, Rohde E, Siljander P, van Niel G, Vasconcelos MH, Yanez-Mo M, Yliperttula ML, Zarovni N, Zavec AB, Giebel B (2016) Evidence-based clinical use of nanoscale extracellular vesicles in nanomedicine. ACS Nano 10:3886–3899
12. Gyorgy B, Hung ME, Breakefield XO, Leonard JN (2015) Therapeutic applications of extracellular vesicles: clinical promise and open questions. Annu Rev Pharmacol Toxicol 55:439–464
13. Lener T, Gimona M, Aigner L, Borger V, Buzas E, Camussi G, Chaput N, Chatterjee D, Court FA, Del Portillo HA, O'Driscoll L, Fais S, Falcon-Perez JM, Felderhoff-Mueser U, Fraile L, Gho YS, Gorgens A, Gupta RC, Hendrix A, Hermann DM, Hill AF, Hochberg F, Horn PA, de Kleijn D, Kordelas L, Kramer BW, Kramer-Albers EM, Laner-Plamberger S, Laitinen S, Leonardi T, Lorenowicz MJ, Lim SK, Lotvall J, Maguire CA, Marcilla A, Nazarenko I, Ochiya T, Patel T, Pedersen S, Pocsfalvi G, Pluchino S, Quesenberry P, Reischl IG, Rivera FJ, Sanzenbacher R, Schallmoser K, Slaper-Cortenbach I, Strunk D, Tonn T, Vader P, van Balkom BW, Wauben M, Andaloussi SE, Thery C, Rohde E, Giebel B (2015) Applying extracellular vesicles based therapeutics in clinical trials - an ISEV position paper. J Extracell Ves 4:30087
14. Luan X, Sansanaphongpricha K, Myers I, Chen H, Yuan H, Sun D (2017) Engineering exosomes as refined biological nanoplatforms for drug delivery. Acta Pharmacol Sin 38:754–763
15. Bell BM, Kirk ID, Hiltbrunner S, Gabrielsson S, Bultema JJ (2016) Designer exosomes as next-generation cancer immunotherapy. Nanomedicine 12:163–169
16. Cheng Q, Shi X, Han M, Smbatyan G, Lenz HJ, Zhang Y (2018) Reprogramming exosomes as nanoscale controllers of cellular immunity. J Am Chem Soc 140:16413–16417

Chapter 13

Nanoparticles for Immune Cell Reprogramming and Reengineering of Tumor Microenvironment

Ketki Bhise, Samaresh Sau, Rami Alzhrani, and Arun K. Iyer

Abstract

Nanoparticles in cancer therapy have garnered significant attention in the past few decades. Cancer immunotherapy, which is aptly called "the new-generation cancer therapy," is slowly making remarkable strides in the improvement of patient outcome and longevity. Taken together, nanoparticles in immune therapy have the potential to offer advantages of both nanoparticles and immune therapy on a single platform.

Key words Cancer, Macrophage polarization, Immune therapy, Immune cell reprogramming, Nanoparticles

1 Introduction

The tumor microenvironment is a complex dynamic system and is an important factor that dictates the therapeutic patient outcome. Monocytes and macrophages are cells derived from the myeloid cell lineage. The premature cells reside in the bone marrow, where upon maturation monocytes are released in the bloodstream. The circulating monocytes enter into tissues and differentiate into the resident tissue macrophages. Upon response to certain environmental signals, the macrophages further differentiate into either of the two activated forms which ultimately dictate the tumor proliferation. Resident tissue macrophages can polarize into either M1 (classically activated) or M2 (alternatively activated) types. Broadly speaking, M1 macrophages are antitumorigenic, whereas M2 macrophages are protumorigenic. The M1 macrophages are activated by lipopolysaccharide and IFN-γ and secrete TNFα, IL-12, iNOS, and ROS which have tumor disruptive functions. They also promote Th1 responses. On the other hand, the tumor-promoting M2 macrophages are activated by IL-4, IL-13, IL-10, and glucocorticoid hormones. They release IL-10, promoting Th2 immune response.

Samuel G. Katz and Peter M. Rabinovich (eds.), *Cell Reprogramming for Immunotherapy: Methods and Protocols*, Methods in Molecular Biology, vol. 2097, https://doi.org/10.1007/978-1-0716-0203-4_13,

M2 macrophages are responsible for proliferation, angiogenesis, and remodeling of extracellular matrix. Tumor-associated macrophages (TAMs) have M2-like phenotype whose presence has been linked to a poor patient outcome [1–3]. Recent studies have shown the improvement in cancer chemotherapy based on remodeling or reeducating the tumor microenvironment to elicit antitumorigenic immune responses. Nanoparticulate drug delivery systems have been widely studied to be effective platforms to modulate the tumor microenvironment [4–9]. Such an approach to reengineer the malfunctioning tumor microenvironment has attracted the attention of new-generation cancer therapeutics.

This article briefly describes some of the molecular biology techniques that are used to study the interaction of nanoparticles with cells of the immune system. One section describes the protocol to study the effect of nanoparticles on activated macrophages. Cancer and macrophage cells are grown together in vitro, and the macrophages are activated to either M1 or M2 subtypes by action of cytokines. While there are many ways in which this protocol can be used to suit ones needs, an example would be to study the differential uptake of nanoparticles by M1- or M2-activated macrophages and to establish whether the protumorigenic M2 macrophages are reeducated to the antitumorigenic M1 macrophages on treatment with drug-loaded nanoparticles.

In animal models, once the subject is treated with anticancer drug-loaded nanoparticles, it would be of interest to see the influence they have on immune cells by means of immunohistochemistry and tissue staining. The protocol for the same is discussed in the second section.

Finally, we have written a protocol in the third section that talks about immune cell activation on nanoparticle treatment.

2 Materials

2.1 Macrophage Polarization

1. 6-Well Transwell inserts 0.4 μM.
2. 6-Well culture plates.
3. RAW264.7 cells, maintained in 5% CO_2 at 37 °C in Dulbecco's Minimum Essential Medium supplemented with 10% fetal bovine serum (FBS) and 1% penicillin/streptomycin (Pen/Strep) antibiotic (*see* **Note 1**).
4. Cancer cell lines, for example, A498, HeLa, and SKOV3. The cells are maintained and passaged in their respective media.
5. 15-ng/ml IFN-γ (*see* **Note 2**).
6. 100-ng/ml lipopolysaccharide (LPS) (*see* **Note 2**).
7. 20-ng/ml IL-4 recombinant protein (*see* **Note 2**).

8. Nanoparticles to stimulate the activity of polarized macrophages. The nanoparticle can be of any category, either polymer or lipid, or a combination of both (*see* **Note 3**). As an example, we have performed this experiment with polymer nanoparticles.
9. Dialysis bag, MWCO 2 kDa.
10. Lyophilizer or freeze dryer.
11. Isopropanol.
12. TRIzol.
13. OneStep RT-PCR kit (QIAGEN).
14. Primers for RT-PCR.
15. Caspase-Glo 3/7 assay system (Promega).
16. Nanodrop spectrophotometer machine or Quant-iT™ RNA assay kit.

2.2 Nanoparticle H-MnO$_2$-PEG/C&D Synthesis

1. Doxorubicin hydrochloride—optimized at different feeding ratios with Ce6, from 20 to 50 mg.
2. mPEG-5 K-NH$_2$—50 mg.
3. $KMnO_4$—300 mg.
4. Chlorine e6 (Ce6)—optimized at different feeding ratios with doxorubicin hydrochloride.
5. Ultrasonicator (Vendor VWR™).
6. SiO_2—40 mg.

2.3 Immune Cell Activation Study

1. Syngeneic 4 T1 tumor-bearing mice.
2. ELISA assay kit. Antibodies to IL-10, IL-12p40, IFN-γ, TNFα.
3. Anti-CD-206-FITC, anti-CD11b-PE, anti-F4/80-Alexa Fluor 647, anti-CD3-FITC, anti-FoxP3-PE, anti-CD4-PerCP, anti-CD8-APC, anti-CD3-APC, anti-CD8-PE.
4. RBC lysis buffer.
5. Flow cytometer.

2.4 Immuno-histochemistry and Tissue Sectioning

Prepare all solutions using deionized water. All solutions are stored at room temperature unless otherwise specified.

1. Blocking buffer (TBSA): 5% bovine serum albumin (BSA), 0.1% triton X-100. Store the buffer at 4 °C.
2. Xylene.
3. Phosphate buffered saline (PBS).
4. Ethyl alcohol.
5. Hydrogen peroxide.
6. DAPI.

7. Primary antibodies against the protein of interest, and secondary antibody, for example, horseradish peroxidase (HRP)-conjugated secondary antibody. Secondary antibody can be used at dilutions tenfold higher than the corresponding primary antibodies.
8. Working concentrations of primary antibodies: anti-TNFα (M1-macrophage marker)—1:100–1:200. Prior to beginning IHC protocol, heat-mediated antigen retrieval is required. Anti-CXCL-10 (M1-macrophage marker)—Concentration of antibody is assay dependent. Anti-CCL22 (M2-macrophage marker)—Concentration of antibody is assay dependent. Anti-CD206 (M2-macrophage marker)—1:50–1:500. Anti-CD3 (cytotoxic T-lymphocyte marker)—1:150. Anti-CD8 (cytotoxic T-lymphocyte marker)—1:100–1:500. Prior to beginning IHC protocol, heat-mediated antigen retrieval with citrate buffer pH 6 is required.
9. Fluorescent microscope with 40× objective (EVOS™ FL Auto Imaging System, Thermo Fisher Scientific, Waltham, MA, USA).

3 Methods

3.1 Macrophage Polarization

1. Grow RAW264.7 cells to 80% confluency and seeded in the Transwell insert at a density of 0.3 million/well.
2. Incubate at 37 °C overnight for adherence, then treat either with IFN-γ and LPS (for M1-macrophage polarization) or IL-4 (for M2-macrophage polarization) over a period of 12 h to overnight. Treat in triplicates, with separate plates for M1- and M2-polarized macrophages.
3. Plate cancer cells in the 6-well plates and allow to adhere to the surface.
4. Place inserts containing RAW264.7 cells over the 6-well plates (*see* **Note 4**).
5. To load drug (C4.16) into polymer nanoparticles, add 100 mg of conjugated polymer styrene maleic anhydride-D-alpha-tocopheryl polyethylene glycol succinate (SMA-TPGS) to 100 ml of water.
6. Dissolve 30 mg of drug C4.16 in 1 ml of DMSO.
7. Add solution from **step 6** dropwise to solution from **step 5** under stirring with a magnetic stir bar. Add 40-mg 1-Ethyl-3-(3-dimethylaminopropyl)carbodiimide (EDC).
8. Adjust the pH of the suspension to 5.0 and keep stirring for 30 min. Next, increase the pH to 11 and keep stirring for another 30 min. Gradually decrease the pH to 7.8–8.0.
9. Add suspension to 2-kDa cut-off dialysis bag.

10. Remove unentrapped drug and DMSO by dialyzing against 1-l deionized water for 4–5 h.
11. Collect drug-loaded nanoparticles from the dialysis bag.
12. Lyophilize the nanoparticles to obtain fine powder. Switch on the lyophilizer and allow it to reach the maximum vacuum and minimum condenser temperature. Freeze the liquid nanoparticle suspension completely by cooling in a mixture of dry ice in acetone. Plug in the frozen sample to the lyophilizer and allow the freeze drying to proceed. Care should be taken that the frozen sample does not liquefy during the process.
13. Resuspend nanoparticles in 1 ml of water.
14. After 12-h incubation, add either drug-loaded nanoparticles or the free drug. Concentration of free drug should be equivalent to the drug concentration in the nanoparticle and depends on the sensitivity of each cell type.
15. Use Caspase-Glo 3/7 according to the manufacturer's instructions in cancer cells to measure the extent of apoptosis-mediated cancer cell death [10].
16. For the macrophages, add TRIzol according to the manufacturer's instructions.
17. Precipitate RNA by adding 0.5-ml of isopropanol to each ml of TRIzol.
18. Incubate for 10 min on ice.
19. Centrifuge at 12,000 × *g* for 10 min at 4 °C to attain RNA pellet.
20. Remove isopropanol solution.
21. Add 1 ml of 75% of ethanol (per 1-ml TRIzol) to wash the RNA pellet and resuspend in RNAse-free water.
22. Quantify RNA using either Nanodrop spectrophotometer machine or Quant-iT™ RNA assay kit (*see* **Note 5**).
23. Evaluate markers for M1 (CD86, iNOS) and M2 (CD206, Arginase 1) by RT-PCR using QIAGEN® OneStep RT-PCR kit (*see* **Note 6**).
24. Evaluate cDNA product from RT-PCR by agarose gel to detect modulation of macrophage markers. For successful therapy, there should be an upregulation of M1-macrophage markers and downregulation of M2-macrophage markers.

3.2 Nanoparticle H-MnO_2-PEG/C&D Synthesis for Immune Cell Activation [11, 12]

1. Adjust to pH 10 with 20-ml anhydrous ethanol by ca. 28 wt% ammonia. Sonicate for 10 min. Add 4-ml tetraethyl orthosilicate (TEOS) and vigorously stir for 20 h. Centrifuge at 3500 rpm = 1,164 × *g* radius of the rotor is 84.841 mm for 30 min, followed by washing thrice with deionized water and twice by ethanol to make monodispersed silica nanoparticles ($sSiO_2$). The $sSiO_2$ will function as a hard template of the hollow nanoshell.

2. Add dropwise 10 ml of aqueous $KMnO_4$ (300 mg/ml) to the suspension of 40-mg $sSiO_2$ prepared in **step 1**, under ultrasonication.
3. Incubate the $KMnO_4/sSiO_2$ for 6 h under ultrasonication.
4. Centrifuge at 20,814 × *g* for 10 min to pellet mesoporous MnO_2-coated $sSiO_2$.
5. Dissolve MnO_2-coated $sSiO_2$ in 10 ml of 2 M Na_2CO_3 solution.
6. Mix and incubate at 60 °C for 12 h.
7. Centrifuge mixture at 20,814 × *g* for 10 min and remove supernatant.
8. Resuspend pellet in 10-ml water and repeat centrifugation.
9. Repeat **step 8**, three times.
10. Mix a 5-mL H-MnO_2 solution (2 mg/mL) from **step 9** with 10 mL of cationic polymer poly-allylamine hydrochloride (PAH) solution (5 mg/mL) under ultrasonication (*see* **Note 7**).
11. Stir the mixture from **step 10** for 2 h with a magnetic stir bar.
12. Centrifuge suspension at 20,814 × *g* for 10 min and remove supernatant.
13. Resuspend pellet in 10-ml water to obtain a H-MnO_2/PAH solution.
14. Add 10-mL anionic polymer poly-acrylic acid (PAA) (5 mg/mL) under ultrasonication.
15. Stir for 2 h with a magnetic stir bar.
16. Centrifuge suspension at 20,814 × *g* for 10 min and remove supernatant.
17. Resuspend pellet in 10-ml water.
18. Add and mix 50-mg $mPEG_{5000}$-NH_2.
19. Ultrasonicate for 30 min.
20. Add 15-mg EDC.
21. Stir for 12 h with a magnetic stir bar.
22. Centrifuge mixture at 20,814 × *g* for 10 min and remove supernatant.
23. Resuspend pellet in 10-ml water and repeat centrifugation.
24. Repeat **step 21**, three times.
25. Resuspend H-MnO_2-PEG pellet in water at 0.2 mg/ml.
26. Add doxorubicin (DOX) and photosensitizer chlorine e6 (Ce6) at drug: MnO_2 feeding ratios of 0.5, 1, 2, 4, and 6 by wt/wt%.
27. Incubate for 12 h at room temperature, stirring in dark to yield H-MnO_2-PEG/C&D that is ready for use [11] (*see* **Notes 8** and **9**).

3.3 Immune Cell Activation Study (Fig. 1)

1. Inject intravenously syngeneic 4 T1 tumor-bearing mice with PBS, free Ce6 + Dox, H-SiO_2-PEG-C&D, H-MnO_2-PEG/Ce6, or H-MnO_2-PEG/C&D (dose of MnO_2 = 10 mg/kg, SiO_2 = 25 mg/kg, Ce6 = 4.7 mg/kg, and Dox = 4.5 mg/kg), respectively.
2. To achieve photodynamic therapy response from Ce6, irradiate a 660-nm laser on the tumor with power of 5 mW/cm^2 for 1 h.
3. At day 5 of postphotodynamic irradiation, collect all tumors, mince into small pieces, and homogenize in PBS using a glass homogenizer to get a single-cell suspension.
4. Lyse tumor pieces with red blood cells (RBC) lysis buffer to remove red blood cells.
5. Aliquot 100-μl cell suspension and add the following antibodies in their respective dilutions (noted as per manufacturer's recommendations).
6. Add 0.125-μg anti-CD11b-PE, 0.4-μg anti-F4/80-AlexaFluor 647, and 10-μg anti-CD206-FITC to define polarization of macrophages (*see* **Note 10**).
7. Add 0.25-μg anti-CD4-PerCP, 0.125-μg anti-CD8-APC, 0.25-μg anti-CD3-FITC, and 1-μg anti-FoxP3-PE to define the regulatory T cells (*see* **Note 11**).
8. Add 0.25-μg anti-CD3-APC and 0.25-μg anti-CD8-PE to evaluate cytotoxic T lymphocyte (CTL) infiltration.
9. Incubate for 30 min at 4 °C in dark.
10. Determine IL-10, IL-12p40, interferon gamma (IFN-γ), and tumor necrosis factor alpha (TNFα) cytokine levels using an ELISA assay on the supernatant of tumor lysates of control and treated mice (*see* **Note 12**).

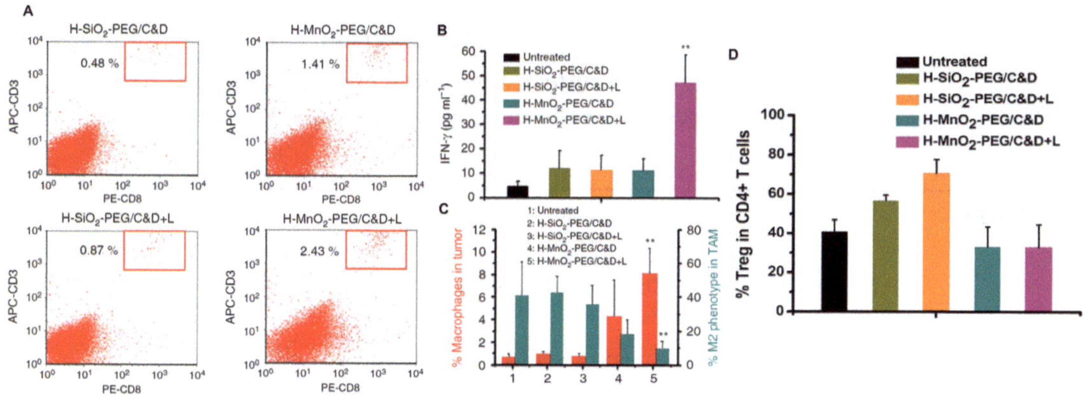

Fig. 1 (**a**) Activation of CD8+ T cell in nanoparticles (H-Mn022-PEG/C&D) with combination of photodynamic light (L) treatment as compared to control. (**b**) Subsequently significant increase of T-cell activation cytokine, IFN-g. (**c**) Increase of M1-macrophage infiltration and decrease of tumorigenic M2-macrophages in H-Mn022-PEG/C&D + L-treated tumor. (**d**) Decrease of Foxp3+ Treg cell in H-Mn022-PEG/C&D + L-treated tumor supports the resurrection of anti-tumor immune response. (Figure was adapted from ref. 11)

3.4 Immunohistochemistry and Tissue Sectioning

Immunohistochemistry (IHC) is a technique that is used to detect an expression of specific protein of interest on the cells or tissues. IHC can be done either by direct detection via adding a labeled primary antibody or by indirect detection by adding an unlabeled primary antibody and then labeled secondary antibody. IHC can be divided into three categories depending on the way of sample processing: IHC-formalin fixed, paraffin embedded, and IHC frozen. Here, we discuss the protocol of staining tissues on a paraffin-embedded glass slide. Such slides can be preserved for a long duration of time, which is an added advantage. Perform all the following steps at room temperature:

1. Wash slide 3× in xylene for 5 min.
2. Wash slide 2× in 100% ethyl alcohol for 3 min.
3. Wash slide 1× in 95% ethyl alcohol for 3 min.
4. Wash slide 1× in 70% ethyl alcohol for 3 min.
5. Wash slide 1× in 50% ethyl alcohol for 3 min.
6. Wash slide gently using distilled water for 5 min.
7. Quench endogenous peroxidase via using 3% hydrogen peroxide in PBS for 5 min.
8. Wash the slide 3× with PBS, 5 min each.
9. Block the slide using TPBS for 1 h.
10. After 1-h blocking, add primary antibody in TBSA overnight (*see* **Note 13**).
11. Wash the slide 3× with PBS, 5 min each.
12. Add secondary antibody diluted in TBSA, secondary antibody should be diluted according the manufacturer's instructions.
13. Add DAPI dye to stain the nucleolus of cells or tissue, the final concentration of DAPI used is 1 μg/ml.
14. Mount the slides with the coverslip.
15. Image the slide using florescence microscopy to study the expression of the protein of interest.

4 Notes

1. RAW264.7 cells multiply rapidly usually after the first two passages.
2. All dilutions should be done in the cell culture media.
3. The nanoparticles are loaded with drug and treated at a dose sufficient to induce apoptosis of cancer cells, while ensuring minimal side effects to normal cells. The quantity of drug loaded in the nanoparticle vesicle is determined by a suitable analytical technique, more precisely high-performance liquid chromatography (HPLC).

4. To better understand this, visualize a dual-chambered system with cancer cells in the bottom chamber and RAW264.7 cells in the upper chamber. Since the inserts have a porous membrane of 0.4-μM size cutoff, there will be an exchange of media and cytokines between both the chambers, provided the volume of media in the well is sufficient enough to be in contact with cells in both the chambers. This step is referred to as "cell co-culture" and is responsible for cross-talk between cancer and immune cells.
5. For RNA purity, the ratio of absorbance of RNA at 260 and 280 nm (A260/A280) assesses the RNA and DNA purity. For DNA, the accepted ratio (A260/A280) is ~1.8, while the accepted ratio for RNA is ~2.0. A lower ratio may be a sign of impurity present in either the RNA or DNA samples.
6. This kit contains a unique combination of Omniscript® and Sensiscript® Reverse Transcriptases having high affinity to RNA template. These transcriptases provide highly specific reverse transcription ranged from 1 pg to 2 μg.
7. This procedure allows to enhance the physiological stability. H-MnO_2 nanoshells are modified with polyethylene glycol (PEG_{5000}) through a layer-by-layer (LBL) polymer-coating method.
8. The nanocell is fabricated with mPEG-5K-NH_2 to improve the plasma stability. Transmission electron microscopy analysis indicates the hollow-shaped morphology and acidic pH responsive Dox degradation of H-MnO_2-PEG/C&D. The use of MnO_2 is used to increase the oxygenation in tumor environment that could reverse the tumor hypoxia [11].
9. Antitumor efficacy of H-MnO_2-PEG/C&D may be improved by combination with a programmed death-ligand 1 (PD-L1) checkpoint inhibitor. Antitumor immune cell activation is measured in treatment with H-MnO_2-PEG/C&D and photodynamic light (L) irradiation (Fig. 2).
10. In flow cytometry, CD11b+F4/80+ positive stained can be defined as M1-macrophage phenotype and CD11b+F4/80+CD206+ cells can be defied as M2 macrophages.
11. Tumor cell suspension can be gated with CD4+ staining and then CD4+ cells can be analyzed to find Treg cells with anti-FoxP3-PE+ and anti-CD25+ staining [13]. Tregs usually represent approximately 5–10% of the total CD4+ T-cell population.
12. Decrease of IL-10 is an indicator of inhibition of tumorigenic M2-macrophage function, and increase of IL-12 supports the upmodulation of tumoricidal M1-macrophage function. Activated CD8+ T cells secrete tumoricidal cytokines like TNFα and IFN-γ.

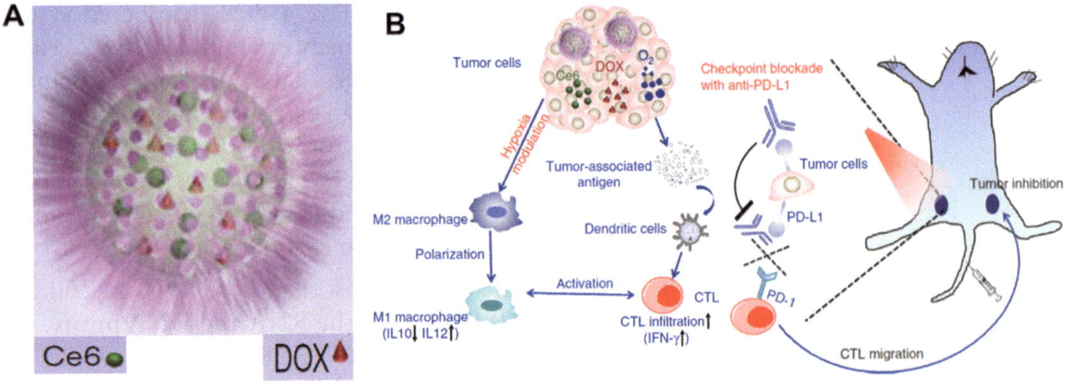

Fig. 2 (**a**) Development of hollow manganese dioxide (H-MnO_2) nanoparticles and the subsequent dual-drug loading of photodynamic agent chlorine e6 (Ce6), and a chemotherapy drug doxorubicin (Dox), namely H-MnO_2-PEG/C&D. (**b**) The proposed mechanism of antitumor immune responses induced by H-MnO_2-PEG/C&D in combination with anti-PD-L1 therapy. (Figure was adopted and revised from ref. 11)

13. Primary antibody should be diluted according the manufacture instruction. If multiple antibodies will be used, different excitation/emission (different colors) of secondary antibodies should be considered.

Acknowledgments

The authors acknowledge partial support for this work by Wayne State University startup funding. A.K.I. acknowledges a grant from the Department of Defense (DoD) CDMRP KCRP Idea Development Award (#W81XWH-18-1-0471) as an Early Career Investigator. S.S. and A.K.I. acknowledge a grant support from the Burroughs Wellcome Trust. K.B. acknowledges Thomas C. Rumble University Graduate Fellowship and Graduate Research Fellowship from the Wayne State University. R.A. acknowledges the scholarship support from College of Pharmacy at Taif University and Saudi Arabian Cultural Mission (SACM).

References

1. Liu CY et al (2013) M2-polarized tumor-associated macrophages promoted epithelial-mesenchymal transition in pancreatic cancer cells, partially through TLR4/IL-10 signaling pathway. Lab Investig 93:844–854
2. Zanganeh S et al (2016) Iron oxide nanoparticles inhibit tumour growth by inducing pro-inflammatory macrophage polarization in tumour tissues. Nat Nanotechnol 11:986–994
3. Xie Z et al (2017) Immune cell-mediated biodegradable theranostic nanoparticles for melanoma targeting and drug delivery. Small 13:1–10
4. Alsaab HO et al (2017) PD-1 and PD-L1 checkpoint signaling inhibition for cancer immunotherapy: mechanism, combinations, and clinical outcome. Front Pharmacol 8:1–15
5. Sau S et al (2018) Multifunctional nanoparticles for cancer immunotherapy: a ground-

breaking approach for reprogramming malfunctioned tumor environment. J Control Release 274:24–34
6. Bhise K et al (2018) Combination of vancomycin and cefazolin lipid nanoparticles for overcoming antibiotic resistance of MRSA. Materials (Basel) 11(7):pii: E1245
7. Bhise K, Kashaw SK, Sau S, Iyer AK (2017) Nanostructured lipid carriers employing polyphenols as promising anticancer agents: quality by design (QbD) approach. Int J Pharm 526:506–515
8. Bhise K et al (2017) Nanomedicine for cancer diagnosis and therapy: advancement, success and structure-activity relationship. Ther Deliv 8:1003–1018
9. Sau S et al (2018) Abstract 4660: tumor multicomponent targeting nanoparticle library for personalized cancer therapy and imaging. Cancer Res 78:4660–4660
10. Alsaab HO et al (2018) Tumor hypoxia directed multimodal nanotherapy for overcoming drug resistance in renal cell carcinoma and reprogramming macrophages. Biomaterials 183:280–294
11. Yang G et al (2017) Hollow MnO_2as a tumor-microenvironment-responsive biodegradable nano-platform for combination therapy favoring antitumor immune responses. Nat Commun 8:902. https://doi.org/10.1038/s41467-017-01050-0
12. Huang P et al (2011) Biomaterials folic acid-conjugated silica-modified gold nanorods for X-ray/CT imaging-guided dual-mode radiation and photo-thermal therapy. Biomaterials 32:9796–9809
13. Zampell JC et al (2012) CD4+ cells regulate fibrosis and lymphangiogenesis in response to lymphatic fluid stasis. PLoS One 7:e49940. https://doi.org/10.1371/journal.pone.0049940

Chapter 14

CRISPR/Cas9 Gene Targeting in Primary Mouse Bone Marrow-Derived Macrophages

Will Bailis

Abstract

CRISPR-Cas9 technology allows for rapid, targeted genome editing at nearly any loci with limited off-target effects. Here, we describe a method for using retroviral transduction to deliver single-guide RNA to primary bone marrow-derived macrophages. This protocol allows for high-throughput reverse genetics assays in primary immune cells and is also compatible with retroviral systems for transgene expression.

Key words CRISPR/Cas9, Gene targeting, Macrophage, Retrovirus, Bone marrow, Mice

1 Introduction

Genomic manipulation and reverse genetics hold tremendous value for elucidating the function of undercharacterize loci in organismal genomes. While gene editing has long been tractable in viruses and prokaryotes [1–3], there has been a continuous effort to develop increasingly robust and efficient techniques for use on eukaryotic genomes. These endeavors first culminated with the groundbreaking work of Capecchi and Smithies demonstrating that exogenous DNA could be targeted into the eukaryotic genome through a process called homologous recombination [4–6]. Though powerful, these methods were inefficient, with low-frequency integration events, and off-target events could be observed at rates similar to the desired on-target integrations [4, 7]. The discovery that introducing double-stranded DNA breaks into the genome could increase the efficiency of recombination soon led to the use of rare cutting endonuclease enzymes, like I-SceI, to enhance genome engineering efforts; however, these techniques were limited by the low frequency of endonuclease recognition sites in target genomes [8, 9].

Samuel G. Katz and Peter M. Rabinovich (eds.), *Cell Reprogramming for Immunotherapy: Methods and Protocols*, Methods in Molecular Biology, vol. 2097, https://doi.org/10.1007/978-1-0716-0203-4_14,

In an effort to introduce greater flexibility in selecting target sites for gene editing, researchers generated fusion proteins by combining zinc finger DNA-binding motifs that display programmable binding site selectability with the DNA cleavage domain of the FokI endonuclease, called zinc finger nucleases (ZFN) [10–12]. By combining six or seven zinc fingers that each recognize 3-bp DNA motifs, ZNFs have the capacity to specifically target any 18- to 21-bp DNA sequence [13]. Similar to ZFNs, transcription activator-like effector (TALE) proteins from Xanthomonas bacteria have also been fused to the endonuclease domain of FokI to generate TALENs, which benefit from the single-base pair recognition site of TALE proteins, providing even greater flexibility compared to the 3-bp recognition site in ZFNs [14, 15].

While both ZFNs and TALENs offer the ability to perform gene editing on a genome scale, both techniques require engineering and cloning of a new set of proteins for each target site desired, limiting both throughput and scalability for these technologies. This significant barrier to large-scale genome editing has recently been overcome by the discovery of clustered regularly interspaced short palindromic repeat (CRISPR) DNA sequences in bacteria that encoded RNAs that guide a single endonuclease enzyme (Cas9) to specific sequences of infecting bacteriophage genomes [16–20]. Seminal work by the Doudna and Charpentier groups, as well as others, demonstrated that the bacterial CRISPR locus could be engineered into a single-guide RNA (sgRNA) and used to target Cas9 to loci for genome engineering in both prokaryotic and eukaryotic genomes [21–26]. As CRISPR engineering only requires a single endonuclease protein and the unique sequence of a sgRNA requires designing only a 20-bp sequence, the method is remarkably efficient and highly scalable. Indeed, much like shRNA, CRISPR techniques lend themselves to both targeted, single-gene approaches and large-pooled libraries with whole-genome coverage [27–29]. Moreover, any cell type or source can be used, providing that both the Cas9 endonuclease and a sgRNA are expressed. To this end, transgenic animal models expressing a Cas9 transgene have been developed to enable ready genome editing in primary cells and tissues [30]. This chapter covers a retroviral sgRNA delivery method compatible with Cas9 transgenic animals for genome engineering of primary bone marrow-derived macrophages.

2 Materials

2.1 Cell Culture and Retroviral Production

1. HEK293T cells (ATCC).
2. Incubator 37 °C.
3. 0.25% trypsin/EDTA solution.
4. DMEM media (no additives).

5. DMEM10 media (DMEM, 10% FBS, 2-mM L-glutamine, 100-U/ml penicillin, 100-μg/ml streptomycin sulfate).
6. 6-well flat-bottomed tissue culture plate.
7. Phosphate-buffered saline (PBS).
8. LipoD293 (SignaGen) or comparable transfection reagent.
9. Retroviral sgRNA expression vector (*see* **Note 1**).
10. pCL-ECO (Addgene, 12371) or equivalent packaging vector.
11. 1.5-ml microfuge tubes.

2.2 Bone Marrow Harvest and Bone Marrow-Derived Macrophage (BMDM) Culture

1. Ethanol 70%.
2. Forceps.
3. Scissors, surgical.
4. 3- and 10-ml syringes.
5. Needles, 25G.
6. Rosa26-Cas9 knockin on B6J mice (The Jackson Laboratory, 026179), 6–10 weeks old.
7. 10-mm Petri dish.
8. M-CSF (Peprotech), alternatively L929 cell supernatant can be used.
9. DMEM10 medium.
10. ACK (ammonium–chloride–potassium) lysing buffer.
11. CO_2 incubator 37 °C.
12. Centrifuge (4000 × *g*) and rotors to hold 15- and 50-ml tubes.
13. 15- and 50-ml disposable polystyrene conical centrifuge tubes with screw cap.
14. 1.5-ml microcentrifuge tubes.
15. 12-well flat-bottom culture plate (nontissue culture treated).

3 Methods

Cary out all procedures using sterile technique. Other than the tissue harvest from mice, all steps should be carried out in a sterilized biosafety cabinet.

3.1 Retroviral Supernatant Production

1. Culture and passage HEK293T cells in DMEM10 media, maintaining them at under 70% confluency.
2. Once cells have been grown to the desired number (passage 3–12 only), remove culture media, wash cells with 1× PBS, and add trypsin/EDTA solution to detach cells.
3. Collect HEK293T suspension into a 50-ml conical tube and add DMEM10 at 4× of the trypsin/EDTA volume.

4. Pellet HEK293T cells at 450 × *g* for 5 min.
5. Resuspend the cell pellet and count viable cells by trypan blue exclusion.
6. Split confluent HEK293T cells into 6-well plates at 5 × 10^5 cells per well in 2-ml DMEM10.
7. Place cells into the 37 °C incubator and culture overnight.
8. The next day the cells should be 75–90% confluent, before proceeding to transfection.
9. Remove 1 ml of media from each well of HEK293T cell cultured, so ~1 ml remains.
10. According to the manufacturer's protocol for the transfection reagent used (here for LipoD293), bring plain DMEM and LipoD293 to room temperature.
11. For each well of transfected HEK293T cells, prepare 1-μg retroviral plasmid and 0.5-μg pCL-ECO vector in 50-μl plain DMEM, mix by pipetting or briefly vortexing.
12. Dilute 6 μl of LipoD293 transfection reagent in 50 μl of plain DMEM.
13. Add diluted transfection reagent 1:1 to the DNA mixture dropwise and then mix by pipetting or briefly vortexing.
14. Incubate the mixture for 10–20 min at room temperature.
15. Add transfection mixture dropwise to each well of HEK293T cells cultured.
16. Return cells to the 37 °C incubator.
17. 24 h after adding the transfection mixture, prewarm fresh DMEM10 to 37 °C.
18. Carefully remove all media from the cultured HEK293T cells, making sure not to detach cells, and replace with 1 ml of fresh, prewarmed DMEM10 (it is best to do this one well at a time to limit stress on the cells).
19. Return cells to the 37 °C incubator.
20. 24 h after replacing the culture media, harvest supernatants from the HEK293T cultures and transfer to a 15-ml conical tube.
21. Spin supernatants at 3000 × *g* for 10 min to clear cell debris and aliquot at 1 ml of supernatant in 1.5-ml microfuge tubes.
22. Supernatant can be used immediately for transduction or placed on dry ice and then frozen at −80 °C for long-term storage.

3.2 Bone Marrow Isolation, BMDM Preparation, and Retroviral Transduction

1. Sacrifice one mouse by cervical dislocation or approved methods.
2. Use sterile forceps and surgical scissors throughout the protocol and sterilize the skin of the mice with 70% ethanol.
3. At the top of each hind leg, cut the skin and pull it toward the foot, exposing the muscle.

4. From the feet toward the torso, cut the muscle back from the bone using a surgical scissors.
5. Remove the hind legs by cutting between the end of the femur and pelvis and place them into a sterile 15-ml conical containing sterile, ice-cold 1× PBS (5–10 ml).
6. Transfer the bones into a sterile 10-mm Petri dish and remove the excess muscle from the bone using a surgical scissors or blade.
7. Separate the tibia from the femur at the joint and cut the ends of the bones off using the surgical scissors.
8. Using a 10-ml syringe containing ice-cold 1× PBS and a 25G needle, insert the needle into the bone cavity and flush out the marrow.
9. Repeat **step 8** until all the marrow is flushed out (bones will become opaque and white, with little red remaining).
10. Collect the supernatant containing the flushed marrow and transfer over a 70-μm nylon cell strainer placed on top of a 50-ml conical tube.
11. Using the back-end of a 3-ml syringe plunger, crush the bone marrow fragments on the nylon cell strainer, rinsing the plunger and the filter 3× with 10 ml of ice-cold, 1× PBS.
12. Centrifuge the filtered cell suspension at 450 × *g* for 10 min at 4 °C.
13. Remove the supernatant and disrupt the cell pellet with 3-ml ACK lysis buffer; incubate at room temperature for 5 min to allow for red blood cell lysis.
14. Add 20-ml DMEM10 to the cell suspension at 450 × *g* for 10 min at 4 °C.
15. Resuspend the cell pellet in 10-ml DMEM10 and count cells by trypan blue exclusion.
16. Adjust media volume with DMEM10 to bring the cell density to 1×10^6 cells/ml and add M-CSF to a final concentration of 10 ng/ml.
17. Plate 1 ml of the cell suspension per well in a 12-well tissue culture plate and place the plate in the 37 °C incubator and culture overnight.
18. 24 h after establishing the bone marrow culture, collect retroviral supernatants and add M-CSF at 10 ng/ml (for frozen retroviral supernatant, allow to thaw on ice first).
19. Add 1 ml of retroviral supernatant to the 1 ml of culture media on the bone marrow cells and return the plate to the 37 °C incubator (*see* **Note 2**).
20. (Optional) Cells can be sorted 48 h after the addition of retroviral supernatant, based on retroviral vector reporter gene expression, if a pure population of cells is required.

21. At day 3 of culture, carefully detach the cultured cells from the 12-well plate and collect them into a 15-ml conical.
22. Centrifuge the cell suspension at 450 × *g* for 10 min at 4 °C.
23. Resuspend the cell pellet in DMEM10 containing and count the cells using trypan blue exclusion.
24. Replate the cells at ~1 × 10^6 cells in 10 ml of DMEM10 containing 10 ng/ml M-CSF in a 10-mm Petri dish.
25. Cells are ready for confirmation of knockout (*see* **Notes 3** and **4**) and use at days 7–12 of culture for downstream assays.

4 Notes

1. There are multiple resources available for designing sgRNA for any given target DNA sequence. For design of individual sgRNA, more information can be found at: https://zlab.bio/guide-design-resources and http://www.rgenome.net/. Predesigned pooled CRISPR libraries and sequence information for sgRNA are also readily available: https://www.addgene.org/crispr/libraries/.
2. While the method described above can be used for sgRNA expressing retroviral vectors, the same protocols are compatible for the retroviral transduction of BMDMs with any retroviral vector and can be used for gene expression studies as well.
3. For protein-coding genes, the efficiency of knockout effects from sgRNA expression is dependent on the rate of protein turnover, the preexisting pool of mRNA for the target, and the rate of mRNA turnover for the gene. Thus, the kinetics of knockout are unique to each target and must be empirically evaluated. When available, the use of an antibody for flow cytometry, Western blot, or microscopy to confirm knockout is preferred. Although the use of one sgRNA only yields local in-del mutations rather than loss of the entire loci or whole exons of a gene, it is often possible to check for loss of transcript after sgRNA expression using real-time quantitative PCR, when antibodies are not available. Finally, the surveyor nuclease assay can be used to assess Cas9 nuclease activity at the target site; however, for the reasons listed above, the kinetics of DNA damage at target loci and protein loss are not always tightly correlated.
4. While computations methods exist for estimating the likelihood of a given sgRNA for generating a frame-shift mutation in a given gene, there currently is no way to predict whether a given sgRNA will efficiently result in the loss of a gene product without empirical testing. Moreover, while the off-target sites available to a given sgRNA can be determined

computationally, there is always the possibility for some off-target effects with any genome editing technique. For these reasons, it is highly recommended that researchers design and evaluate at least three sgRNA per target as well as use a non-targeting sgRNA or a sgRNA targeting an irrelevant gene to ensure that efficient gene product loss can be achieved and to control for off-target effects and the potential for DNA damage to influence the observed phenotype.

References

1. Danna K, Nathans D (1971) Specific cleavage of simian virus 40 DNA by restriction endonuclease of Hemophilus influenzae. Proc Natl Acad Sci U S A 68(12):2913–2917
2. Kelly TJ Jr, Smith HO (1970) A restriction enzyme from Hemophilus influenzae. II. J Mol Biol 51(2):393–409
3. Smith HO, Wilcox KW (1970) A restriction enzyme from Hemophilus influenzae. I Purification and general properties. J Mol Biol 51(2):379–391
4. Capecchi MR (1989) Altering the genome by homologous recombination. Science 244(4910):1288–1292
5. Smithies O, Gregg RG, Boggs SS, Koralewski MA, Kucherlapati RS (1985) Insertion of DNA sequences into the human chromosomal beta-globin locus by homologous recombination. Nature 317(6034):230–234
6. Thomas KR, Folger KR, Capecchi MR (1986) High frequency targeting of genes to specific sites in the mammalian genome. Cell 44(3):419–428
7. Lin FL, Sperle K, Sternberg N (1985) Recombination in mouse L cells between DNA introduced into cells and homologous chromosomal sequences. Proc Natl Acad Sci U S A 82(5):1391–1395
8. Rouet P, Smih F, Jasin M (1994) Introduction of double-strand breaks into the genome of mouse cells by expression of a rare-cutting endonuclease. Mol Cell Biol 14(12):8096–8106
9. Rudin N, Sugarman E, Haber JE (1989) Genetic and physical analysis of double-strand break repair and recombination in Saccharomyces cerevisiae. Genetics 122(3):519–534
10. Bibikova M, Carroll D, Segal DJ, Trautman JK, Smith J, Kim YG, Chandrasegaran S (2001) Stimulation of homologous recombination through targeted cleavage by chimeric nucleases. Mol Cell Biol 21(1):289–297. https://doi.org/10.1128/MCB.21.1.289-297.2001
11. Kim YG, Cha J, Chandrasegaran S (1996) Hybrid restriction enzymes: zinc finger fusions to Fok I cleavage domain. Proc Natl Acad Sci U S A 93(3):1156–1160
12. Porteus MH, Baltimore D (2003) Chimeric nucleases stimulate gene targeting in human cells. Science 300(5620):763. https://doi.org/10.1126/science.1078395
13. Urnov FD, Rebar EJ, Holmes MC, Zhang HS, Gregory PD (2010) Genome editing with engineered zinc finger nucleases. Nat Rev Genet 11(9):636–646. https://doi.org/10.1038/nrg2842
14. Boch J, Scholze H, Schornack S, Landgraf A, Hahn S, Kay S, Lahaye T, Nickstadt A, Bonas U (2009) Breaking the code of DNA binding specificity of TAL-type III effectors. Science 326(5959):1509–1512. https://doi.org/10.1126/science.1178811
15. Moscou MJ, Bogdanove AJ (2009) A simple cipher governs DNA recognition by TAL effectors. Science 326(5959):1501. https://doi.org/10.1126/science.1178817
16. Barrangou R, Fremaux C, Deveau H, Richards M, Boyaval P, Moineau S, Romero DA, Horvath P (2007) CRISPR provides acquired resistance against viruses in prokaryotes. Science 315(5819):1709–1712. https://doi.org/10.1126/science.1138140
17. Garneau JE, Dupuis ME, Villion M, Romero DA, Barrangou R, Boyaval P, Fremaux C, Horvath P, Magadan AH, Moineau S (2010) The CRISPR/Cas bacterial immune system cleaves bacteriophage and plasmid DNA. Nature 468(7320):67–71. https://doi.org/10.1038/nature09523
18. Ishino Y, Shinagawa H, Makino K, Amemura M, Nakata A (1987) Nucleotide sequence of the iap gene, responsible for alkaline phosphatase isozyme conversion in Escherichia coli, and identification of the gene product. J Bacteriol 169(12):5429–5433
19. Jansen R, Embden JD, Gaastra W, Schouls LM (2002) Identification of genes that are associated with DNA repeats in prokaryotes. Mol Microbiol 43(6):1565–1575

20. Senoo-Matsuda N, Yasuda K, Tsuda M, Ohkubo T, Yoshimura S, Nakazawa H, Hartman PS, Ishii N (2001) A defect in the cytochrome b large subunit in complex II causes both superoxide anion overproduction and abnormal energy metabolism in Caenorhabditis elegans. J Biol Chem 276(45):41553–41558. https://doi.org/10.1074/jbc.M104718200
21. Cong L, Ran FA, Cox D, Lin S, Barretto R, Habib N, Hsu PD, Wu X, Jiang W, Marraffini LA, Zhang F (2013) Multiplex genome engineering using CRISPR/Cas systems. Science 339(6121):819–823. https://doi.org/10.1126/science.1231143
22. Deltcheva E, Chylinski K, Sharma CM, Gonzales K, Chao Y, Pirzada ZA, Eckert MR, Vogel J, Charpentier E (2011) CRISPR RNA maturation by trans-encoded small RNA and host factor RNase III. Nature 471(7340):602–607. https://doi.org/10.1038/nature09886
23. Gasiunas G, Barrangou R, Horvath P, Siksnys V (2012) Cas9-crRNA ribonucleoprotein complex mediates specific DNA cleavage for adaptive immunity in bacteria. Proc Natl Acad Sci U S A 109(39):E2579–E2586. https://doi.org/10.1073/pnas.1208507109
24. Jinek M, Chylinski K, Fonfara I, Hauer M, Doudna JA, Charpentier E (2012) A programmable dual-RNA-guided DNA endonuclease in adaptive bacterial immunity. Science 337(6096):816–821. https://doi.org/10.1126/science.1225829
25. Jinek M, East A, Cheng A, Lin S, Ma E, Doudna J (2013) RNA-programmed genome editing in human cells. elife 2:e00471. https://doi.org/10.7554/eLife.00471
26. Mali P, Yang L, Esvelt KM, Aach J, Guell M, DiCarlo JE, Norville JE, Church GM (2013) RNA-guided human genome engineering via Cas9. Science 339(6121):823–826. https://doi.org/10.1126/science.1232033
27. Shalem O, Sanjana NE, Hartenian E, Shi X, Scott DA, Mikkelson T, Heckl D, Ebert BL, Root DE, Doench JG, Zhang F (2014) Genome-scale CRISPR-Cas9 knockout screening in human cells. Science 343(6166):84–87. https://doi.org/10.1126/science.1247005
28. Wang T, Wei JJ, Sabatini DM, Lander ES (2014) Genetic screens in human cells using the CRISPR-Cas9 system. Science 343(6166):80–84. https://doi.org/10.1126/science.1246981
29. Wu X, Scott DA, Kriz AJ, Chiu AC, Hsu PD, Dadon DB, Cheng AW, Trevino AE, Konermann S, Chen S, Jaenisch R, Zhang F, Sharp PA (2014) Genome-wide binding of the CRISPR endonuclease Cas9 in mammalian cells. Nat Biotechnol 32(7):670–676. https://doi.org/10.1038/nbt.2889
30. Platt RJ, Chen S, Zhou Y, Yim MJ, Swiech L, Kempton HR, Dahlman JE, Parnas O, Eisenhaure TM, Jovanovic M, Graham DB, Jhunjhunwala S, Heidenreich M, Xavier RJ, Langer R, Anderson DG, Hacohen N, Regev A, Feng G, Sharp PA, Zhang F (2014) CRISPR-Cas9 knockin mice for genome editing and cancer modeling. Cell 159(2):440–455. https://doi.org/10.1016/j.cell.2014.09.014

Chapter 15

Genome-Wide CRISPRi/a Screening in an In Vitro Coculture Assay of Human Immune Cells with Tumor Cells

Jialing Zhang, Stephan S. Späth, and Samuel G. Katz

Abstract

Cell-based immunotherapy has achieved preclinical success in certain types of cancer patients, with a few approved cell-based products for clinical use. These achievements revitalized the field of cell engineering/immunotherapy and brought attention to the opportunities that cell-based immunotherapeutics can offer to patients. On the other hand, obvious indications emphasize the need for a better understanding of the biological mechanisms involved in the immune response. This knowledge may not only ameliorate safety and efficacy, but also determine the possibilities and limitations in use of immune cell engineering for cancer treatment, and facilitate developing novel immunotherapeutic strategies. Recently developed technology based on CRISPR-dCas9 has an immense potential to systematically uncover genetic mechanisms by identifying subsets of essential genes involved in interactions of cancer cells with the immune system. This chapter will present a reliable and reproducible general protocol for the application of genome-wide sgRNA gene-editing tools in the recently established two-cell type co-culture, consisting of immune cells as effectors and cancer cells as targets, utilizing CRISPRi/a-dCas9-based technology.

Key words CRISPRi/a, Genome-wide screening, Two-cell type coculture assay, Cell-based immunotherapy

1 Introduction

The ongoing development of precise molecular, cell-manipulation tools is opening up a wide prospect for specific cell-based therapeutic approaches. In particular, recent applications of the CRISPR/Cas9 system have rapidly advanced the landscape of genome engineering in the fields of cancer biology and therapeutic gene-editing applications in human disease treatment [1]. Multiple ongoing clinical trials are utilizing CRISPR/Cas9 to knockout (KO) PD-1 from T cells of patients with lung cancer (NCT02793856) [2], prostate cancer (NCT02867345), bladder cancer (NCT02863913), esophageal cancer (NCT03081715), leukemia and lymphoma (NCT03166878, NCT03398967), renal cell cancer

Samuel G. Katz and Peter M. Rabinovich (eds.), *Cell Reprogramming for Immunotherapy: Methods and Protocols*, Methods in Molecular Biology, vol. 2097, https://doi.org/10.1007/978-1-0716-0203-4_15,

(NCT02867332), and others (NCT03044743, NCT03399448). These studies are the first proof of concept of the application of CRIPSR/Cas9 technology in human cancer treatment. CRISPR screens have also been widely adopted to conduct comprehensive genome-wide scale gene disruption to facilitate the discovery of essential and/or drug resistance genes in various cancer types [1, 3]. Subsequent modifications of the CRISPR/Cas9 system, through the establishment of CRISPR interference (CRIPSRi) and activation (CRISPRa) tools (Fig. 1), facilitate controlling transcript levels of endogenous genes [4, 5].

CRISPR screens allow the identification of genes, which influence a specific phenotype in an unbiased fashion. Hence, the chosen approach requires a number of technical and rational considerations.

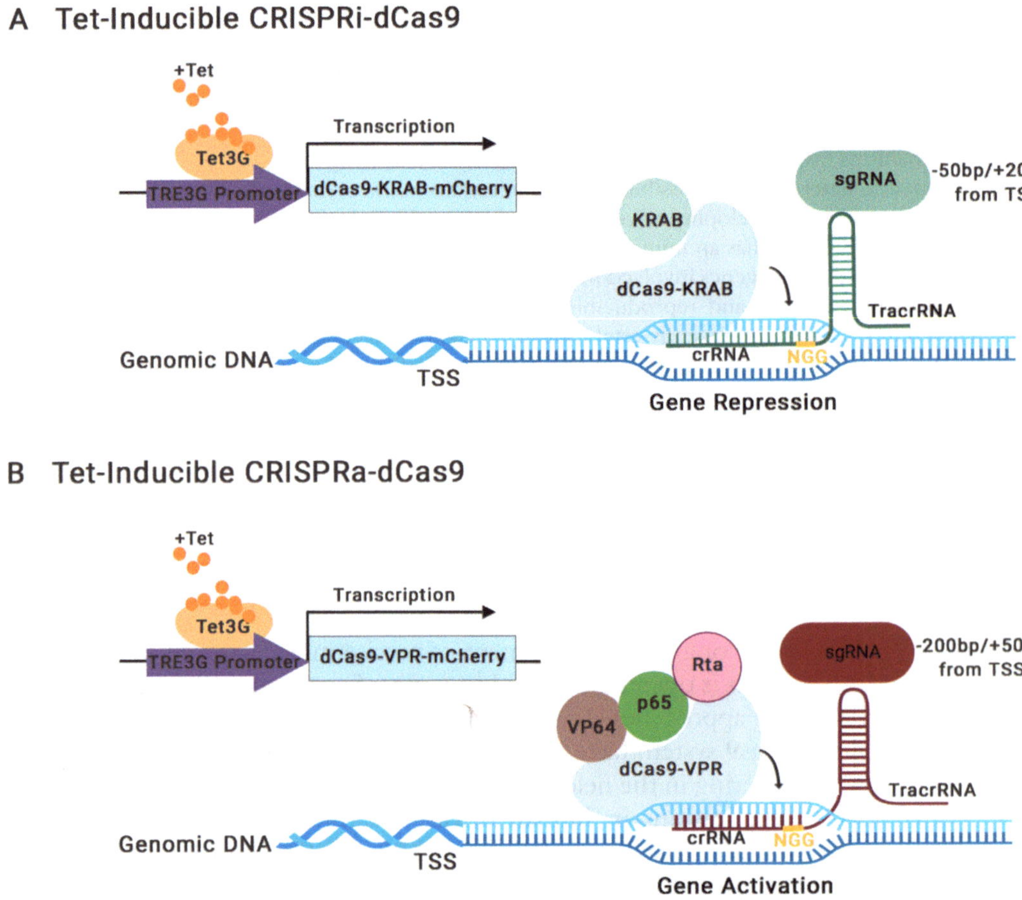

Fig. 1 Basic principles of the CRISPRi/a-dCas9 system. The necessary components and basic mechanism of CRISPRi-dCas9 (gene repression) (**a**) and CRISPRa-dCas9 (gene activation) (**b**) are shown. Important functional components are color coded for easy identification. The KRAB domain is the main component resulting in gene repression, while VP64-RelA-RtaAD (VRP)-linked domains result in gene activation. This system is based on the Tet-inducible system, utilizing the Tet-On-3G domain and the addition of Tetracycline, for proper functioning

1. The Cas9 enzyme is natively an RNA-guided DNA endonuclease. The most typical outcome after a double-strand DNA break is repair via the nonhomologous end-joining (NHEJ) pathway, which often introduces small insertions or deletions [6]. Several modified genetic screening platforms for mammalian cells based on this pathway, also known as CRISPR nuclease (CRISPRn), have been established [1]. A mutant form of Cas9 (dCas9) was engineered to inactivate the nuclease activity of Cas9, without affecting its ability to bind DNA through sgRNA targeting [7]. This in turn sterically prevents transcription factors and RNA polymerase from accessing DNA promoter regions. This tool can be used to either inhibit or activate transcriptional activity as CRISPRi and CRISPRa approaches, respectively [8] (Fig. 1). CRISPRi is based on the fusion of dCas9 to a transcriptional repressor domain (e.g., Krüppel-associated box (KRAB) domain), which promotes heterochromatin-mediated gene silencing in mammalian-derived cells [7] (Fig. 1a). CRISPRa constructs recruit multiple activator domains that are directly fused to dCas9. There are currently several CRISPRa approaches, including the VPR approach, via a direct fusion of multiple activator domains (e.g., VP64, p65, and RTA (VPR)) to dCas9 [9] (Fig. 1b); the SunTag approach, via a protein scaffold consisting of VP64, which is fused to superfolder GFP (sfGFP) and an antibody single-chain variable fragment (scFv), which is able to target the transcriptional activator GCN4 epitope [7]; and the SAM approach, via an RNA scaffold that consists of MS2 RNA hairpins, which bind to dimers of MS2 coat protein (MCP) fused to p65 and HSF1 transcriptional activation domains [10]. Genome scale screens, based on CRISPRn, CRISPRi, and CRISPRa technologies, have already been successfully tested in pooled screens [1].
2. The sgRNA expression construct library pools are stably integrated into the genomes of mammalian cells, via lentiviral transduction and designed to target on specific pathway components and/or genomes on a wide scale, utilizing 5–10 sgRNAs to target each gene. Although the system is very reliable, it is recommended to optimize and validate optimal infection conditions. Although the number of sgRNAs per gene can be varied (at least 3–5 sgRNA/per gene), it is recommended to design 5–10 sgRNAs/per gene in a primary screen, in order to achieve more reliable results for specific gene targeting. There are numerous resources available that can assist in sgRNA design, such as:
   (a) https://omictools.com/crispr-grna-design-tool-tool
   (b) https://portals.broadinstitute.org/gpp/public/analysis-tools/sgrna-design-crisprai

CRISPRa and CRISPRi pooled libraries generally select a representative transcription start site (TSS) for a gene (Fig. 1a, b). In addition to genome-wide sgRNA pools, sgRNA subsets that target genes according to their biofunctional theme are also commercially available (e.g., Kinases Screen, Cancer and Apoptosis Screen, Membrane Proteins Screen, and Mitochondrial Proteins Screen). An alternative approach is to customize the CRISPRi/a library pools, which can be done in-house or through commercial vendors. The relative abundance of each sgRNA in a given experimental cell pool can be quantified by comparing the frequencies of cells expressing sgRNA at an initial time point, compared to those at a later time point [1]. It is important to consider certain limitations regarding the end point of the utilized cell viability screens, and many mechanisms of genetic interactions between perturbation/depletion and cell state specific gene expression changes over time have been noted [11–13]. The isolation of cells at multiple time points may provide clearer distinctions, but it is also associated with a higher cost.

3. In a pooled genetic CRISPR-based screen, lentiviral delivery of dCas9 and the sgRNA often utilize a drug resistance gene (e.g., puromycin, hygromycin, neomycin, blasticidin) or fluorescence-reporter gene tag (e.g., GFP, BFP, mCherry) to select for pure target cell populations. In general, puromycin treatment results in quicker elimination of nonresistant cells, while neomycin or hygromycin treatments may take weeks, with the appropriate treatment doses varying across selected cell types [14].
4. The most crucial step in the general screening strategy is to deliver the dCas9 gene prior to sgRNA pools. In most cases, the delivery dCas9 and/or sgRNA pools (being inserted in expression plasmids, carrying either drug resistance and/or fluorescent protein-coding genes) can be accomplished by several well-established approaches, with lentiviral delivery often achieving the best transfection efficiencies [1, 8, 15]. During the next step, it is absolutely crucial to confirm dCas9 presence and stability by evaluating its expression levels, both at genetic (e.g., RNA by RT-qPCR) and protein levels (e.g., Western blot). Utilization of a pre-established dCas9 expressing cell line, prior to the introduction of sgRNA pools, can significantly improve the quality of library screening and further provides a universal background during the screening procedure.
5. Ultimately, the critical step in a pooled CRISPR-based screen is the isolation of genomic DNA. Many commercial kits and homemade protocols are readily available, although there is no one-size-fits-all solution, due to variations in yield and quality across different cell types. The quantification of library abundance, as a read-out of library screens, requires a two-step PCR

amplification step on the isolated genomic DNA, prior to subjecting the PCR product to next-generation sequencing.

Presented below is a reliable and reproducible general protocol for genome-wide sgRNA screening in a two-cell type in vitro co-culture assay, consisting of immune cells as effectors and cancer cells as targets, utilizing the recently established CRISPRi/a-dCas9-based technology. This protocol is also compatible with Tet-inducible CRISPRi/a-dCas9 system for gene-editing applications in either all types of immune or cancer cells.

2 Materials

2.1 Equipment

1. Thermocycler (Bio-Rad T100™, temperature range: 0–100 °C).
2. Bench centrifuge or microcentrifuge.
3. Spectrophotometer.
4. CO_2 incubator (temperature range 37 °C).
5. Microbiological culture hood.
6. Temperature-regulated shaker (temperature range covering 37 °C).
7. Electroporator and electroporating cuvettes.
8. Large gel system.
9. QuBit fluorometer.
10. Bioanalyzer 2100.
11. Access to an Illumina sequencing facility (preferably HiSeq 4000 sequencer, Illumina).

2.2 Plasmid Vectors

1. pHR-TRE3G-KRAB-dCas9-mCherry and/or pHR-UCOE-TRE3G-dCas9-VPR-mCherry.
2. pHR-Tet3G.
3. pU6-sgRNA-Ef1α-Puro-T2A-BFP.
4. hCRISPRi sgRNA library pool and/or hCRISPRa sgRNA library pool.
5. Lentiviral packaging plasmids: pCMV-dR8.91 and pMD2-G.

2.3 Cells and Culture Media

1. HEK293T cell line.
2. Effector cells (e.g., NK cells, T cells, TILs).
3. Target cells (e.g., cancer cells, primary or cell line).
4. Complete HEK 293 T cell culture medium (DMEM, 10% FBS, 1× Pen-Strep).
5. Cell culture medium for effector and target cells.

2.4 Supplies and Reagents

1. Pipette tips (RNAse and DNAse free).
2. Sterile culture dish (15 cm^2).
3. Sterile 6-well plates.
4. Sterile 1.7-ml tubes.
5. Sterile 15-ml falcon tubes.
6. Sterile T25 flasks.
7. Centrifuge tubes (Max volume: 1.5 ml; RNase and DNase free).
8. PCR tubes or 96-well PCR plate (Max volume: 50 μl; RNAse and DNAse free).
9. X-tremeGENE HP DNA transfection reagent.
10. Polybrene (final working concentration: 8 μg/ml).
11. DEAE dextran sulfate (final working concentration: 6 μg/ml).
12. Puromycin (final working concentration: 1 or 2 μg/ml).
13. Tetracycline (stock in DMSO: 10 mg/ml, final working concentration: 500 ng to 1 ug/ml).
14. 2× annealing buffer (200-mM potassium acetate; 60-mM HEPES-KOH, pH 7.4).
15. 1× PBS.
16. DMSO.
17. Designed oligonucleotides (Table 1).
18. Restriction enzymes: BstXI and BlpI.
19. 10× CutSmart buffer.
20. Ultrapure agarose.
21. Electrophoresis buffer (10× TAE or 10× TBE).
22. DNA loading dye (30% glycerol, 0.25% bromophenol blue, and 0.25% xylene cyanol FF).
23. SYBR Green I nucleic acid gel stain-10,000×.
24. Premade LB-Agar plates supplemented with ampicillin (100 μg/ml).
25. *DH5α* chemically competent cells and/or MegaX DH10B electrocompetent cells.
26. Lenti-X concentrator.
27. Genomic DNA extraction kit (e.g. Macherey Nagel Blood Kit).
28. NucleoSpin Blood Kits.
29. Quick-gRNA MidiPrep Kit.
30. SbfI-HF restriction enzyme.
31. Gel extraction kit.
32. Phusion high-fidelity DNA polymerase.

Table 1
Designed oligonucleotides for insertion of gRNA into pU6-sgRNA plasmid

Primer	Sequence (5'-3')
5′ Forward	TTGNNNNNNNNNNGTTTAAGAGC
3′ Reverse	CTTGTTGNNNNNNNNNNGTTTAAGAGCTAA

Note: NNNNNNNNNNN refers sgRNA sequence

33. dNTPs mix.
34. Custom PCR and sequencing primers (Table 2).
35. SPRI select beads.
36. Magnetic stand.
37. 1-Kb plus DNA ladder.
38. 50-bp DNA ladder.
39. Agilent DNA 1000 reagents.
40. p24 ELISA kit.

3 Methods

3.1 Annealing of Single-Stranded Complementary Oligonucleotides (See Note 1*)*

1. Dissolve each oligonucleotide in a volume of annealing buffer to create a 100 μM stock solution.
2. Mix equal volumes of the dissolved complementary oligonucleotides in a PCR tube.
3. Heat to 95 °C and maintain the temperature for 2–5 min, cool to 25 °C over 45 min, and then cool to 4 °C for temporary storage.
4. Use a 1:20 dilution of annealed Oligos in ddH2O for the subsequent steps.

3.2 Cloning of Annealed Oligonucleotides (sgRNAs)

1. Add 2–10 μg of pU6-sgRNA-Ef1α-puro-T2A-BFP plasmid to 1 μl of BstXI, 1 μl of BlpI, 1 μl of 10× CutSmart buffer and water to 10 μl total volume, incubate the digestion mixture at 37 °C for at least 1 h.
2. Confirm successful plasmid digestion by agarose gel electrophoresis, and proceed to plasmid purification procedures, by using a commercially available plasmid purification kit (e.g., ZymoPure plasmid purification kit), according to the manufacturer's instructions.
3. Verify the concentration of the plasmid by spectrophotometric analysis.

Table 2
PCR primers for Illumina Hiseq 4000 sequencing

Primers	Sequence
Truseq 1-Forward	AATGATACGGCGACCACCGAGATCTACACTCTTTCCCTACACGACGCTCTTCCGATCTt**AAGTAGAG**gcacaaaaggaaactcaccct
Reverse	CAAGCAGAAGACGGCATACGAGAT**AAGTAGAG**GTGACTGGAGTTCAGACGTGTGCTCTTCCGATCTtcgactcggtgccactttttc
Truseq 2-Forward	AATGATACGGCGACCACCGAGATCTACACTCTTTCCCTACACGACGCTCTTCCGATCTat**ACACGATC**gcacaaaaggaaactcaccct
Reverse	CAAGCAGAAGACGGCATACGAGAT**ACACGATC**GTGACTGGAGTTCAGACGTGTGCTCTTCCGATCTatcgactcggtgccactttttc
Truseq 3-Forward	AATGATACGGCGACCACCGAGATCTACACTCTTTCCCTACACGACGCTCTTCCGATCTgat**CGCGCGGT**gcacaaaaggaaactcaccct
Reverse	CAAGCAGAAGACGGCATACGAGAT**CGCGCGGT**GTGACTGGAGTTCAGACGTGTGCTCTTCCGATCTgatcgactcggtgccactttttc
Truseq 4-Forward	AATGATACGGCGACCACCGAGATCTACACTCTTTCCCTACACGACGCTCTTCCGATCTcgat**CATGATCG**gcacaaaaggaaactcaccct
Reverse	CAAGCAGAAGACGGCATACGAGAT**CATGATCG**GTGACTGGAGTTCAGACGTGTGCTCTTCCGATCTcgatcgactcggtgccactttttc
Truseq 5-Forward	AATGATACGGCGACCACCGAGATCTACACTCTTTCCCTACACGACGCTCTTCCGATCTtcgat**CGTTACCA**gcacaaaaggaaactcaccct
Reverse	CAAGCAGAAGACGGCATACGAGAT**CGTTACCA**GTGACTGGAGTTCAGACGTGTGCTCTTCCGATCTtcgatcgactcggtgccactttttc
Truseq 6-Forward	AATGATACGGCGACCACCGAGATCTACACTCTTTCCCTACACGACGCTCTTCCGATCTatcgat**TCCTTGGT**gcacaaaaggaaactcaccct
Reverse	CAAGCAGAAGACGGCATACGAGAT**TCCTTGGT**GTGACTGGAGTTCAGACGTGTGCTCTTCCGATCTatcgatcgactcggtgccactttttc
Truseq 7-Forward	AATGATACGGCGACCACCGAGATCTACACTCTTTCCCTACACGACGCTCTTCCGATCTgatcgat**AACGCATT**gcacaaaaggaaactcaccct
Reverse	CAAGCAGAAGACGGCATACGAGAT**AACGCATT**GTGACTGGAGTTCAGACGTGTGCTCTTCCGATCTgatcgatcgactcggtgccactttttc
Truseq 8-Forward	AATGATACGGCGACCACCGAGATCTACACTCTTTCCCTACACGACGCTCTTCCGATCTcgatcgat**ACAGGTAT**gcacaaaaggaaactcaccct
Reverse	CAAGCAGAAGACGGCATACGAGAT**ACAGGTAT**GTGACTGGAGTTCAGACGTGTGCTCTTCCGATCTcgatcgatcgactcggtgccactttttc
Truseq 9-Forward	AATGATACGGCGACCACCGAGATCTACACTCTTTCCCTACACGACGCTCTTCCGATCTacgatcgat**AGGTAAGG**gcacaaaaggaaactcaccct
Reverse	CAAGCAGAAGACGGCATACGAGAT**AGGTAAGG**GTGACTGGAGTTCAGACGTGTGCTCTTCCGATCTacgatcgatcgactcggtgccactttttc
Truseq 10-Forward	AATGATACGGCGACCACCGAGATCTACACTCTTTCCCTACACGACGCTCTTCCGATCTt**AACAATGG**gcacaaaaggaaactcaccct
Reverse	CAAGCAGAAGACGGCATACGAGAT**AACAATGG**GTGACTGGAGTTCAGACGTGTGCTCTTCCGATCTtcgactcggtgccactttttc

(continued)

Table 2 (continued)

Primers	Sequence
Truseq 11-Forward	AATGATACGGCGACCACCGAGATCTACACTCTTTCCCTACACG ACGCTCTTCCGATCTat**ACTGTATC**gcacaaaaggaaactcaccct
Reverse	CAAGCAGAAGACGGCATACGAGAT**ACTGTATC**GTGACTGGA GTTCAGACGTGTGCTCTTCCGATCTatcgactcggtgccactttttc
Truseq 12-Forward	AATGATACGGCGACCACCGAGATCTACACTCTTTCCCTACACG ACGCTCTTCCGATCTgat**AGGTCGCA**gcacaaaaggaaactcaccct
Reverse	CAAGCAGAAGACGGCATACGAGAT**AGGTCGCA**GTGACTGGAG TTCAGACGTGTGCTCTTCCGATCTgatcgactcggtgccactttttc

3.3 Ligation of Annealed Oligonucleotides with Digested Plasmid Backbone

1. Add 100 ng of digested pU6-sgRNA-Ef1α-puro-T2A-BFP plasmid (*see* Subheading 3.2) to 2 μl of annealed oligonucleotides (1:20 dilution) (*see* Subheading 3.1), 2 μl of 10× T4 DNA ligase buffer, 1 μl of T4 DNA ligase and water to 20 μl total reaction volume.
2. Incubate at room temperature for 1–4 h or at 16 °C over night.

3.4 Transformation into DH5α Bacteria

1. Add 1 μl of individual ligation reaction (*see* Subheading 3.3) to 20 μl of thawed *DH5α* cells and incubate on ice for 30 min, followed by 42 °C for 45 s, and then immediately on ice for at least 2 min.
2. Add 1 ml of prewarmed sterile LB medium to the transformed *DH5α* cells, incubate them at 37 °C for at least 1 h for recuperation in a temperature-regulated shaker, set at 200 rpm.
3. Spread 100 μl of recuperated transformed *DH5α* cells on prewarmed LB-agar plates, supplemented with ampicillin (100 μg/ml) to achieve single colonies.
4. Incubate the plated LB-agar plates in the incubator at 37 °C overnight.
5. Pick several single colonies per plasmid-oligonucleotide construct and amplify colonies in LB medium, supplemented with ampicillin (100 μg/ml) at 37 °C for at least 10–16 h in a temperature-regulated shaker, set at 200 rpm.
6. Subject the amplified bacterial culture to plasmid extraction procedures, by using an appropriate plasmid purification kit (e.g., ZymoPure Plasmid purification kit or QIAprep Spin Miniprep kit), according to the manufacturer's instructions.
7. For individually extracted plasmid-oligonucleotide constructs, determine the concentration by spectrophotometric analysis

(e.g., Nanodrop) and verify the integrity and purity by agarose gel electrophoresis.

8. Verify the incorporation and sequence of individually ligated oligonucleotides inside the pU6-sgRNA-Ef1α-puro-T2A-BFP plasmid by sequencing, with the following sequencing primer: MP177-5′-gagatccagtttggttagtaccggg-3′.
9. To avoid unnecessary time- and cost-consuming vector preparation and sequencing for a larger sgRNA pool (more than 100 sgRNAs), the use of a robust Gibson Assembly PCR strategy [16] or purchase from a commercially available resource (e.g., Addgene) is recommended.

3.5 Amplification of Individual sgRNA or a Mixed sgRNA Pool

1. Mix a predefined number of sgRNA expression vectors, based on the set of genes to target, in equimolar amounts.
2. Dilute the mixed sgRNA sample to 100 ng/μl in 1× TE buffer.
3. Prewarm the recovery medium and plates to room temperature.
4. For each tube, add 1 μl of the mixed sgRNAs pool (100 ng/μl) to 50 μl of MegaX DH10B electrocompetent cells in an electroporation cuvette and electroporate at 1.8 kV and 180 Ω (*see* **Note 2**).
5. Add 1 ml of recovery medium into the cuvette, transfer the electroporated cells to a sterile cell culture falcon tube, and add an additional 3 ml of recovery medium.
6. Place the cell culture tubes in the shaking incubator for 1 h at 37 °C, set at 250 rpm.
7. For each of the transformed sgRNA pools, individually transfer 5 μl of the culture to an individual sterile 1.8-ml eppendorf tube, containing 995 μl of recovery medium and mix well by pipetting (**do not vortex!**).
8. From each tube (**step 7**), transfer 40 μl to an individual sterile 1.8-ml eppendorf tube, containing 200 μl of recovery medium, mix well by pipetting (**do not vortex!**), and replate on individual prewarmed LB-agar plates, supplemented with ampicillin (100 μg/ml).
9. Incubate the LB-agar plates at 37 °C for 14–16 h and count the number of colonies on each plate. To obtain the real colony number for each of the electroporations, multiply each counted colony number by the dilution factor of 4,000,000 (*see* **Note 3**).
10. For each electroporation, divide the remaining culture (3.995 ml) into two sterile 500 ml Erlenmeyer flasks, containing 250-ml prewarmed LB medium and supplemented with ampicillin (100 μg/ml). Incubate the eight Erlenmeyer flasks in the shaking incubator overnight at 37 °C, set to 250 rpm.

11. Collect and centrifuge each of the bacterial cultures in sterile falcon tubes (50 ml) by centrifugation at 4000 × *g* for 30 min.
12. Proceed with isolation and purification of the amplified sgRNA-containing plasmids by utilizing the EndoFree plasmid maxi kit, according to the manufacturer's instructions (*see* **Note 4**).
13. Determine the concentration by Nano-drop ND1000, and verify the integrity and purity of amplified pools of plasmids with sgRNA by electrophoresis (*see* **Note 5**).

3.6 Production of Lentiviral Particles (Fig. 2)

1. The day prior to transfection, plate 7.5 × 10^6 HEK293T cells in a total of 25 ml of complete medium in a 15-cm sterile culture dish.
2. Proceed only when the cell confluence has reached between 30% and 40%.
3. Prewarm 1× PBS, Opti-MEM, and corresponding culture dishes to 37 °C.
4. Remove complete medium and wash cells with sterile prewarmed 1× PBS to remove any residual FBS.
5. Mix 1500 μl of prewarmed Opti-MEM with 45-μl X-tremeGENE HP DNA transfection reagent in 1.8-ml tubes and incubate the mixtures at room temperature for 5 min.
6. To the 1.8-ml tube, add the desired lentiviral plasmid (9 μg), with the corresponding packaging vectors (8 μg of pCMV-dR8.91 and 1 μg of pMD2-G) (*see* **Note 6**).
7. Carefully close the cap of 1.8-ml microcentrifuge tube, mix well (**do not vortex!**), and incubate at room temperature for 15–30 min.
8. Pipette the mixtures slowly by drop-wise addition onto the cultured HEK293T cells in 10 ml of complete medium with 10% FBS.
9. The next day, add 8 ml of complete medium with 10% FBS and then allow viral particle production to continue for 48–72 h before harvesting (*see* **Note 7**).
10. Carefully collect all the culture supernatant (expected volume: ~18 ml), avoiding any detached cells.
11. Filter collected supernatant through a 0.45-μm filter and/or apply the collected supernatant to a Lenti-X concentrator and use it according to the manufacturer's instructions (*see* **Note 8**).
12. Incubate obtained mixture overnight at 4 °C, and centrifuge at 1500 × *g* for 45 min at 4 °C, resulting in a visible off-white pellet. Carefully remove the supernatant and gently resuspend the pellet in 1/10 or 1/100 of the original volume of complete DMEM, 1× PBS, or other appropriate media (*see* **Note 9**).

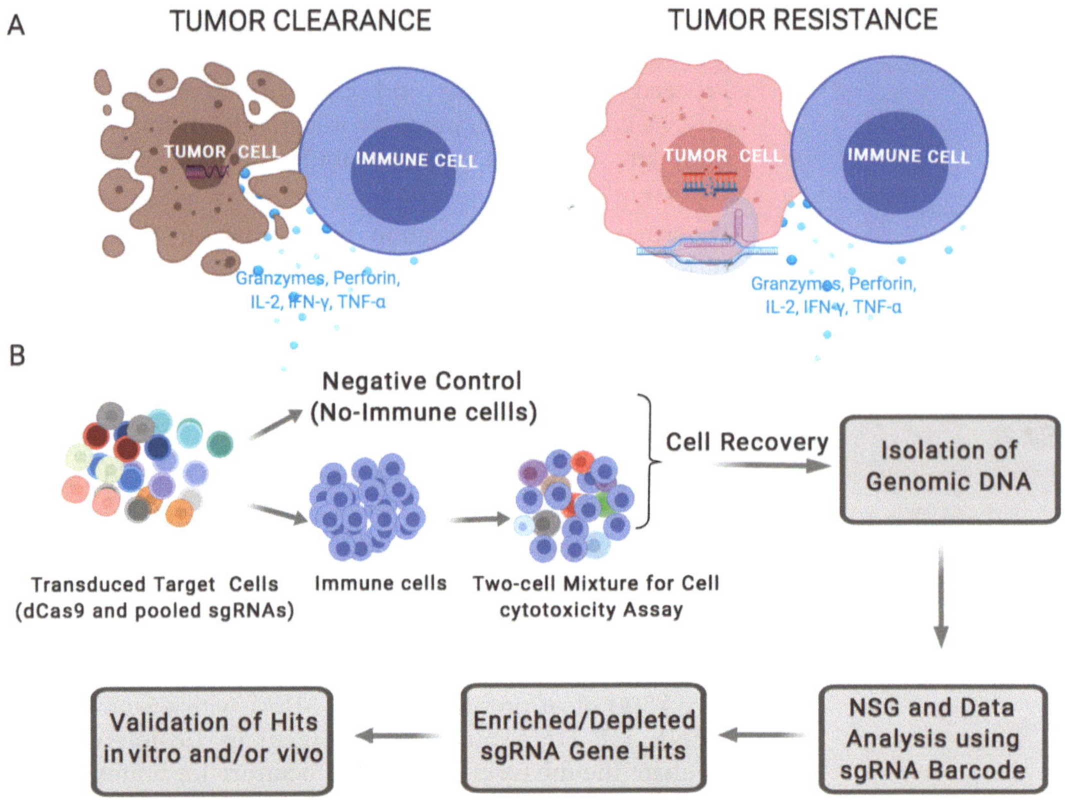

Fig. 2 General outline for genome-wide sgRNA gene-editing screen in two-cell type co-culture system. (**a**) On the left side: An immune cell kills a target tumor cell using granzymes and perforin, in addition to producing various cytokines. On the right side: CRISPR/dCas9 targets a gene in the tumor cell that makes it resistant to immune cell killing. (**b**) Screen steps are depicted starting with target cells with different sgRNAs, which are then added to immune cells. After a period of time for cell recovery, genomic DNA is isolated, sequenced, analyzed for enrichment, and validated as a target

13. The lentiviral titer estimation can be optimally achieved through the commercially available p24 ELISA kit, according to the manufacturer's instructions. The expected yield of lentiviral particles should be at least 1×10^6 TU/ml (*see* **Note 10**).
14. Lentiviral particles should be immediately aliquoted at desired concentrations and frozen at −80 °C. Lentiviral particles can be stored at 4 °C for about 3–7 days and at −80 °C for up to 1 year. Multiple freeze-thaw cycles are not recommended and result in low transfection efficiencies (*see* **Note 11**).

3.7 Generation of a Stable Cell Line with the Tet-Inducible dCas9 and sgRNA Pools

1. Prepare a batch of prewarmed complete medium (*see* **Note 12**), supplemented with polybrene (final concentration: 8 μg/ml) or DEAE (final concentration: 6 μg/ml).

2. Rapidly thaw the lentiviral particles containing TRE3G-KRAB-dCas9-mCherry or TRE3G-dCas9-VPR-mCherry at room temperature.
3. Seed 1–5 × 10^4/ml of either adherent or suspension target cells with complete medium, supplemented with polybrene (8 μg/ml) and the corresponding lentiviral particle dilution, into one well of a 6-well culture plate, with the total volume not exceeding 1 ml/well (*see* **Notes 13–15**).
4. Three days after adding the virus, carefully collect the cells from the well into a 15-ml falcon tube and pellet them by centrifugation at 450 × *g* for 5 min.
5. Carefully remove the culture medium, without disrupting the cell pellet and resuspend the cell pellet in a falcon tube with 3 ml of complete medium, transfer the cell suspension into a T25 flask, and add an additional 7 ml of complete medium.
6. Culture the cells for 3 days in 1 μg/ml tetracycline (*see* **Note 16**) and then FACS purify mCherry positive cells.
7. Expand the whole population of FACS purified mCherry positive cells for a frozen stock and further introduction of the sgRNA library.
8. Rapidly thaw the lentiviral particles containing a pool of individual pU6-sgRNA-Ef1α-puro-T2A-BFP plasmid aliquots, harboring the corresponding sgRNAs.
9. Seed 1–5 × 10^4/ml of cells from **step 7** with complete medium, supplemented with polybrene (8 μg/ml) and the lentiviral particle dilution (*see* **Notes 13–15**). For 50,000 gRNAs, divide lentivirus evenly into the wells of one 6-well plate.
10. Three days after adding the virus, carefully collect the cells from each well into 50-ml falcon tubes and pellet them by centrifugation at 450 × *g* for 5 min.
11. Carefully remove the culture medium, without disrupting the cell pellet, and resuspend the cell pellet in complete medium, transfer the cell suspension into T75 flasks, and add additional 30 ml complete medium, supplemented with puromycin (1–2 μg/ml) (*see* **Note 17**).
12. Culture the T75 flasks for an additional 3 days, monitoring each flask for the presence of dead cells and/or medium color change, on a daily basis (*see* **Note 18**).
13. Once transduced and polyclonal cell populations are growing well and have been sufficiently expanded (typically between days 10 and 14), cell stocks can be prepared, according to standard freezing protocols, or be used for downstream applications (*see* **Note 19**).

3.8 General Pooled sgRNA Library Screening Plan for a Two-Cell Type Coculture Assay (Fig. 3)

1. Three days before co-culture, set up three parallel cultures of the target cells, each with enough cells for complete library representation (500 cells/sgRNA) (*see* **Subheading 3.7**), in effector cell media (*see* **Notes 20** and **21**).
2. Two days before co-culture, add 1 μg/ml of tetracycline to activate dCas9 expression in two of the three cultures (*see* **Note 22**).
3. Co-culture the effector cells with the target cells at a predetermined effector (E): target (T) ratio (e.g., E:T, 0.25 to 50:1) and time span (e.g., 2–48 h) for one of the two cultures that received tetracycline. The remaining two groups of target cells (one that received tetracycline, and one that did not) are cultured without effector cells at the same conditions as the third group and serve as controls (*see* **Note 23**).
4. Mix cells in 50-ml falcon tubes and pellet cells, using centrifugation at 180 × *g* for 5 min.
5. Maintain cell pellets in 50-ml falcon tubes with a loosened lid and culture them in the CO_2 incubator for the predetermined period of time.
6. To improve the efficiency of gRNA library amplification from the target cells, the majority of the effector cells should be removed from the culture. This can be accomplished either by allowing the effector cells to die without supplying an essential growth factor for several days (e.g., T cells without IL-2) or based on any unique antigens on either cell line. Target cells are maintained for additional 48 h to 2 weeks in the cell culture incubator for recovery and stored by standard cryo-freezing procedures and/or processed for later downstream steps (e.g., sgRNA enrichment analysis) (*see* **Note 24**).

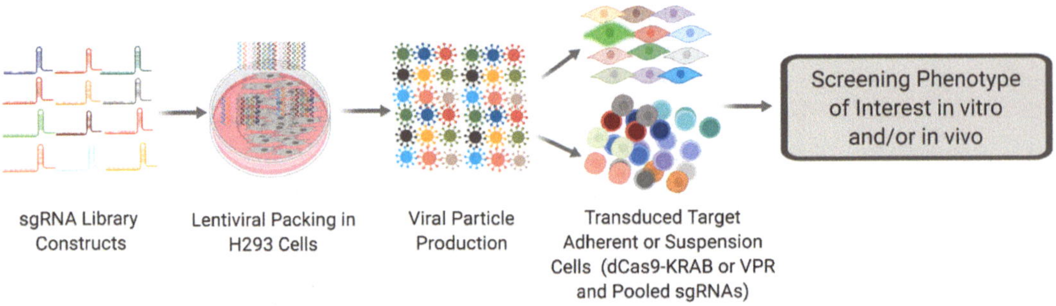

Fig. 3 General outline for the transduction of target cells lines with CRISPRi/a-dCas9 and pooled sgRNAs. The sgRNA library is transfected into 293 cells for lentiviral packaging and viral particle production. The viral particles are then used to transduce the target cells

3.9 Genomic DNA Isolation, PCR Amplification, and Next-Generation Sequencing for sgRNA Enrichment Analysis

1. Isolate genomic DNA (gDNA) from fresh or frozen cell pellets by using the NucleoSpin Blood Kits or Zymo research quick-gDNA MidiPrep Kit, according to the manufacturer's protocol (*see* **Notes 25** and **26**).
2. Digest 10 μg of the extracted gDNA with 5 μl SbfI-HF (400 U/mg), 5.0 μl of 10× CutSmart buffer, and water to a total volume of 50 μl at 37 °C overnight (*see* **Note 26**).
3. Prior to large agarose gel preparation, clean the corresponding gel tray and large well combs, by rinsing them well with ultrapure (RNAse and DNAse free) water.
4. Dry the gel tray and place previously cleaned combs.
5. Pour hand warm dissolved agarose (final percentage 0.8%) in 1× TBE, supplemented with SYBR safe into the preset gel tray.
6. Allow the agarose gel to solidify in the cold room for around 30 min.
7. Fill the gel tank with 1× TBE, aspirate any TBE that may have entered the wells.
8. Load digested gDNA sample per well with 1× loading dye and DNA ladder (50–2500 bp) (*see* **Note 27**).
9. Run samples at 150 V, for 1–1.5 h or until the purple dye front has travelled half way down the gel.
10. Excise gel pieces from each individual lane using a razor blade under long wave UV light, with the band size range corresponding between 350 and 700 bp. Make sure to change blades between different sample lanes.
11. Weigh and place each excised gel piece in individual 50-ml falcon tubes (*see* **Note 28**).
12. Extract digested gDNA fragments from individually excised gel pieces, by appropriate gel purification kit (e.g., Macherey Nagel gel extraction kit or Zymoclean gel DNA Recovery kit), according to the manufacturer's instructions.
13. Measure the concentration of extracted digested gDNA fragments.
14. Store size selected (350–700 bp) and purified gDNA fragments at −20 °C or continue directly with the next step.
15. Set up a PCR reaction with 10 μl of 10× buffer, 0.4 μl of 3′ common primer (stock concentration 100 μM), 0.4 μl of 5′ primer with index (stock concentration 100 μM), 2.0-μl dNTP (stock concentration 10 mM), 3.0-μl DMSO (stock concentration 100%), 1.0-μl Pfu Ultra II, 500-ng gRNA DNA, and water to a total reaction volume of 100 μl (*see* **Notes 29–32**).
16. Use the following PCR conditions: (step 1) 98 °C for 30 s, (step 2) 98 °C for 30 s, 56 °C for 15 s, 72 °C for 15 s, for 23 total cycles, (step 3) 72 °C for 10 min, (step 4) hold at 12 °C.

17. After the PCR completion, pool all PCR reactions for a given sample in a 5-ml or 15-ml tube and mix well. All of the sgRNAs (gDNA fragments) in the pool should be represented at much greater abundance, so you may proceed with only a fraction of the sample for PCR sample for subsequent purification. You can directly test the PCR product by running a small aliquot on a 20% TBE agarose gel with 50-bp DNA ladder, with expected enriched PCR product being at ~274 bp and primer dimers at ~150 bp (*see* **Note 33**).
18. Purify the PCR product, derived from individually excised gel pieces by an appropriate gel purification kit, according to the manufacturer's instructions.
19. Measure PCR cleanup yield by Nanodrop NT-100 and/or Qubit, and dilute a small portion of your sample to 400 pg/μl, based on the measurement.
20. Run purified samples on a Bioanalyzer (Agilent) with a High Sensitivity DNA chip. The enriched PCR product is expected to be at 274 bp, but the Bioanalyzer peak should appear around 276–280 bp.
21. Submit samples for sequencing using one pair of the twelve suggested primer pairs (*see* **Note 33** and Table 2).

4 Notes

1. The oligonucleotides do not require phosphorylation. When annealing, slow cooling to room temperature by putting the tubes on the lab bench rather than the default cooling ramp rate on a conventional PCR machine is critical. Annealed oligos can be stored at −20 °C and are stable through at least 2–3 freeze-thaw cycles. If many sgRNAs are to be cloned at once, it is recommended to buy the corresponding oligos in presuspended form at 100 μM each, in a plate or an arrayed format, which allows easy combining of complimentary oligos by multichannel handling. For higher throughput, it is advised to keep everything in PCR plates throughout the protocol until bacteria cultures have been plated.
2. Enough DNA (up to 400 ng) and 50–100 μl of competent bacteria have to be used, to ensure the representation of all sgRNAs of the corresponding hCRISPRi/a libraries.
3. The ideal efficiency should be equivalent to or greater than 200 colonies/per sgRNA, to ensure equal representation of each sgRNA in the library. Therefore, proceed only if the total number of colonies is at least 1.4×10^7.
4. Depending on the pellet size and weight, the use of multiple Endofree-Maxiprep columns is required, in order to avoid any column binding oversaturation during extraction.

5. Expected yield should be around 2 mg. It is recommended to aliquot the amplified libraries at 1000 ng/μl with 20 μl/tube. The amplified hCRISPRi/a libraries may be stored at −80 °C, and multiple freeze-thaw cycles should be avoided. Deep sequencing of the amplified and purified hCRISPRi/a libraries is strongly recommended, prior to be used in subsequent experimental steps or procedures.
6. The plasmids used in this protocol (e.g., pHR-TRE3G-dCas9) contain LTR repeat sequences, LTR-mediated recombination can happen in *E. coli* after transformation, resulting in a plasmid of reduced size. Hence, it is recommended to use Stbl3/MegaX competent cells for plasmid amplification, which will minimize any recombination events.
7. The cells start releasing viral particles into the culture supernatant 4–48 h post transfection, and usually after 24 h post transfection the supernatant containing the lentiviruses can be collected.
8. The use of the Lenti-X concentrator significantly increases the lentiviral titer. It is recommended to only use cellulose acetate or polyethersulfone (PES) (low protein binding) filters. Do not use nitrocellulose filters, as nitrocellulose binds to lentiviral envelope surface proteins, subsequently resulting in virus destruction.
9. All experimental waste has to be disposed, according to local hazardous waste disposal guidelines. According to Biohazard level 2+ waste disposal regulations, all equipment and reagents, which have come into contact with lentiviral particles, are required to undergo decontamination procedures with 10% bleach solution for at least 30 min, before waste disposal.
10. Viral titers are represented by functional infectious titer, measured in transduction units (TU/ml), or physical titer, measured in viral particles (VP/ml). The functional titer is a measurement of how much virus can infect the cell and is typically 100- to 1000-fold lower than the physical titer. Direct functional titer is a more accurate measurement for calculating MOI (Multiplicity of Infection), but it is more time-consuming and sometimes not feasible. The physical titer is a measurement of how much virus is present and is usually calculated based on the level of protein, such as p24 or viral nucleic acid. The physical titer is sufficient for most lentiviral experiments, and the functional titer can be calculated from physical titer.

    The yield of lentiviral particles and final lentiviral titer highly depends on the size of the lentiviral construct. Please note that large lentiviral constructs (e.g., dCas9 fusions) yield five- to tenfold lower titers compared to smaller lentiviral constructs (e.g., sgRNAs). The final lentiviral titer is also highly dependent on HEK293T cell health and transfection efficiency.

Virus producing HEK293T cells have a distinct round shape and phenotype. Do not allow the cultured HEK293T cells to become overconfluent and avoid cell clumps by ensuring that the cells are well trypsinized before plating, as clumps of cells can significantly reduce the final lentiviral titers. As viral production can vary between batches, it is recommended to purify and titer each lentiviral batch accordingly.

11. For experimental consistency, it is highly recommended to purify and titer the samples or store as single-use aliquots at −80 °C. Avoid multiple freeze-thaw cycles for the lentiviral supernatant.
12. Medium should be warmed to room temperature to minimize cell stress.
13. The health of the target cell line is critical for obtaining accurate results. Regularly check the cell culture for the absence of mycoplasma. Maintain the cells at ideal growing conditions for the specific cell line that you are working with. Thaw a new cell vial after 20–30 passages and restart the culture process. Do not add penicillin/streptomycin to the media during the lentiviral transduction.
14. Prior to the actual experiment, it is also advised to test a wide range of lentiviral particle dilutions (1:10–1:100), in order to determine the optimal dilution for the desired downstream applications. Please note that transducing too many cells, relative to the number of lentiviral particles, significantly reduces transduction efficiency, resulting in large amounts of cell death upon drug selection.
15. The lentiviral transduction of a final spin-infection step is optional for most protocols, as it generally results in an increased infection rate, but it is not absolutely required. In case of introducing this step during experimental procedures, the recommended spin should be carried out at 180–450 × g for 5 min and/or up to 2 h. It is advised to test several protocols, prior to the final experiment, as the setup will highly depend on the cell type to be transduced.
16. In our hands, this concentration of tetracycline works well, but different cell lines might require optimization.
17. The concentration of puromycin needs to be determined for each target cell as the minimal concentration that yields 100% death in 1 week of incubation.
18. Depending on the transduction efficiency, you will see different degrees of cell death upon antibiotic selection. In general, the infection efficiency and cell viability will vary between different cells. It is advised to test several transduced conditions, prior to the final experiment. With a transduction efficiency of 50%, expect approximately half of the cells to die within one

week of incubation. Media color should not over acidify and is typically changed every 3 days.

19. To achieve a stable cell pool, the antibiotic selection should last at least as long, as it takes the control (untransduced) cells to completely die. After that, the cells may undergo additional growth in the presence of the antibiotic. At this time, some researchers reduce the concentration of the antibiotic in culture or entirely remove the antibiotic entirely. If the antibiotic is reduced or removed from the culture, check the cells regularly to confirm transgene expression (either through mCherry and/or BFP expression by FACS or other gene and/or protein expression validation).
20. Target cells typically survive in effector cell medium over the course of 3 days. However, if the target cells loose viability, the media will need to be optimized such that the target cells survive and the effector cells still kill.
21. Most sgRNA-pooled libraries contain 3–6 sgRNAs per target gene, and maintaining the distribution of each sgRNA within the cell population is crucial. Hence, at the beginning of the screen, the working cell number of treated stably transformed cells should be kept at 300–500 cells for representing each individual sgRNA. The cell number is dependent on the screening library pool size.
22. Target cells should be validated for incorporation of the library by BFP and mCherry expression. Take approximately 100 μl of the corresponding cell suspension and fix it with 200–300 μl of 1% PFA for 20 min at room temperature, protecting the cell suspensions from light (e.g., wrapping the corresponding tubes with aluminum foil). Measure BFP and mCherry fluorescence of the cells by standard flow cytometry, looking for the percentage of cells expressing both fluorescence markers.
23. This screen approach can also be used with additional variable drug treatments, although the corresponding drug treatments and their individual concentrations have to be thoroughly tested (e.g., IC50) and optimized, in order to achieve the best drug effect with the least cell toxicity.
24. Cryo-freeze target cells in cryo-freezing tubes in a total volume of 1 ml freezing solution (90% FCS and 10% DMSO) and store them immediately at −80 °C or liquid nitrogen. Freezing can occur at multiple points of the screen to capture, as the cell population may change over time.
25. Frozen cell pellets or isolated gDNA can be stored at −20 °C for several months. Make sure to tighten the connection between the reservoir and the column during the extraction procedure. Centrifuge for sufficient time and speed to remove

any residual buffer in the column. An additional final dry spin is recommended for residual washing buffer removal. For maximum recovery of gDNA, if more than 150 million cells are harvested, two elution columns should be used with an expected yield of 1.0–1.5 mg of gDNA per column.

26. The sgRNA content of the sample up to the PCR step is at extremely low abundance, compared to plasmid or post-gDNA PCR. This makes the samples very susceptible to contamination, which can be falsely amplified during subsequent PCR steps. It is highly recommended to do all of the genomic extractions, digests, PCR setup on a separate bench, to avoid sample cross-contamination.
27. The SbfI digest and subsequent gel isolation can theoretically enrich for sgRNAs > 600-fold, but some material may be lost. Therefore, this step is optional, but highly recommended for samples, where a large number of cells (generally greater than 15 million) have been processed. For sgRNA pooled samples, where the amount of input DNA (assuming 6.6 pg of gDNA per cell), under these circumstances, the 500-fold sgRNA representation is 3.3 ng/per sgRNA.
28. The recommended input of 500 ng gDNA/per 100 μl reaction can be adjusted, although higher inputs (>1 μg) can lead to inhibition of PCR reactions or higher observed background. If very high number of PCRs, based on the 500 ng input, is set up, it is recommended that an unimportant stock sample, containing sgRNA pooled library-containing cells, is initially used, in order to optimize this step.
29. The PCR is designed to amplify the extracted gDNA fragments (containing sgRNAs), and for adding Illumina sequencing adapters for subsequent steps. For detailed primer information (Table 2), please refer to: https://weissmanlab.edu/CRISPR/illuminaSeqencingSamplePrep.
30. To minimize associated errors in sgRNA amplification, it is important to use a high-fidelity polymerase such as Pfu Ultra II (e.g., Agilent) or Kapa HiFi (Kapa Biosystems).
31. The corresponding sequencing primers are listed in Table 2.
32. Multiple samples can be pooled, prior to sequencing analysis. Alternatively, each sample can be analyzed in a separate channel, in order to obtain a more accurate concentration of the sequencing sample, prior to being pooled. Before proceeding, plan out how you want to pool/run your samples on the flow cell. Each sample should be PCR amplified, in order for them to have a unique index.
33. This protocol is optimized for the Illumina HiSeq4000 sequencer. If you are using other sequencers, consult your core facility or test the requirements for sequence diversity. Consult

with your sequencing facility to find out how many reads to expect per lane, and plan for 250–500 reads per sgRNA per sample. For example, a typical genome-scale 200,000 gRNA library screen should have at least 50 million reads per sample (e.g., control and experimental group).

Acknowledgments

We thank Dr. Jonathan Weissman at UCSF for providing research materials and expert advice. This work was supported by the Yale Cancer Center, NIH R21CA198561, NIH R21AI121993, and the Alliance for Cancer Gene Therapy.

Illumina Hiseq4000 compatible Primers (for PCR and sequencing)

Set A

Sample	TruSeq ID	Set A 5 Primer (orange indicates sample index for demultiplexing after sequencing)
1	12	aatgatacggcgaccaccgagatctacacgatcggaagagcacacgtctgaactccagtcacCTTGTAgcacaaaaggaaact caccct
3	14	aatgatacggcgaccaccgagatctacacgatcggaagagcacacgtctgaactccagtcacAGTTCCgcacaaaaggaaact caccct
5	3	aatgatacggcgaccaccgagatctacacgatcggaagagcacacgtctgaactccagtcacTTAGGCgcacaaaaggaaact caccct

Set B

Sample	Truseq ID	Set B 3 Primer
2	6	aatgatacggcgaccaccgagatctacacgatcggaagagcacacgtctgaactccagtcacGCCAATcgactcggtgccacttt ttc
4	10	aatgatacggcgaccaccgagatctacacgatcggaagagcacacgtctgaactccagtcacTAGCTTcgactcggtgccactttt tc
6	1	aatgatacggcgaccaccgagatctacacgatcggaagagcacacgtctgaactccagtcacATCACGcgactcggtgccacttt ttc

Common 3' Primer CAAGCAGAAGACGGCATACGAGATCGACTCGGTGCCACTTTTTC
Common 5' Primer CAAGCAGAAGACGGCATACGAGATGCACAAAAGGAAACTCACCCT

5' Sequencing Primer GTGTGTTTTGAGACTATAAGTATCCCTTGGAGAACCACCTTGTTG
3' Sequencing Primer CCACTTTTTCAAGTTGATAACGGACTAGCCTTATTTAAACTTGCTATGCTGT

References

1. Kampmann M (2018) CRISPRi and CRISPRa screens in mammalian cells for precision biology and medicine. ACS Chem Biol 13(2):406–416
2. Cyranoski D (2016) CRISPR gene-editing tested in a person for the first time. Nature 539(7630):479
3. Adli M (2018) The CRISPR tool kit for genome editing and beyond. Nat Commun 9(1):1911
4. Yang X, Boehm JS, Salehi-Ashtiani K, Hao T, Shen Y, Lubonja R, Thomas SR, Alkan O, Bhimdi T, Green TM et al (2011) A public genome-scale lentiviral expression library of human ORFs. Nat Methods 8(8):659–661
5. Carette JE, Guimaraes CP, Varadarajan M, Park AS, Wuethrich I, Godarova A, Kotecki M, Cochran BH, Spooner E, Ploegh HL et al (2009) Haploid genetic screens in human cells identify host factors used by pathogens. Science 326(5957):1231–1235
6. Koike-Yusa H, Li Y, Tan EP, Velasco-Herrera Mdel C, Yusa K (2014) Genome-wide recessive genetic screening in mammalian cells with a lentiviral CRISPR-guide RNA library. Nat Biotechnol 32(3):267–273

7. Gilbert LA, Larson MH, Morsut L, Liu Z, Brar GA, Torres SE, Stern-Ginossar N, Brandman O, Whitehead EH, Doudna JA et al (2013) CRISPR-mediated modular RNA-guided regulation of transcription in eukaryotes. Cell 154(2):442–451
8. Gilbert LA, Horlbeck MA, Adamson B, Villalta JE, Chen Y, Whitehead EH, Guimaraes C, Panning B, Ploegh HL, Bassik MC et al (2014) Genome-Scale CRISPR-mediated control of gene repression and activation. Cell 159(3):647–661
9. Chavez A, Scheiman J, Vora S, Pruitt BW, Tuttle M, E PRI, Lin S, Kiani S, Guzman CD, Wiegand DJ et al (2015) Highly efficient Cas9-mediated transcriptional programming. Nat Methods 12(4):326–328
10. Konermann S, Brigham MD, Trevino AE, Joung J, Abudayyeh OO, Barcena C, Hsu PD, Habib N, Gootenberg JS, Nishimasu H et al (2015) Genome-scale transcriptional activation by an engineered CRISPR-Cas9 complex. Nature 517(7536):583–588
11. Li YH, Lin TP, Chang CL, Ch SK, Tsai SM, Mai TM (1961) Bilateral familial pheochromocytoma. Zhonghua Wai Ke Za Zhi 9:349–352
12. Abdulkadir SA, Casolaro V, Tai AK, Thanos D, Ono SJ (1998) High mobility group I/Y protein functions as a specific cofactor for Oct-2A: mapping of interaction domains. J Leukoc Biol 64(5):681–691
13. Smith RE, Strieter RM, Phan SH, Lukacs N, Kunkel SL (1998) TNF and IL-6 mediate MIP-1alpha expression in bleomycin-induced lung injury. J Leukoc Biol 64(4):528–536
14. Lanza AM, Kim DS, Alper HS (2013) Evaluating the influence of selection markers on obtaining selected pools and stable cell lines in human cells. Biotechnol J 8(7):811–821
15. Mali P, Aach J, Stranges PB, Esvelt KM, Moosburner M, Kosuri S, Yang L, Church GM (2013) CAS9 transcriptional activators for target specificity screening and paired nickases for cooperative genome engineering. Nat Biotechnol 31(9):833–838
16. Gibson DG, Young L, Chuang RY, Venter JC, Hutchison CA III, Smith HO (2009 May) Enzymatic assembly of DNA molecules up to several hundred kilobases. Nat Methods 6(5):343–345

Chapter 16

The Propagation and Quantification of Two Emerging Oncolytic Viruses: Vesicular Stomatitis (VSV) and Zika (ZIKV)

Robert E. Means, Sounak Ghosh Roy, and Samuel G. Katz

Abstract

Developments in genetic engineering have allowed researchers and clinicians to begin harnessing viruses to target and kill cancer cells, either through direct lysis or through recruitment of antiviral immune responses. Two powerful viruses in the fight against cancer are the single-stranded RNA viruses vesicular stomatitis virus and Zika virus. Here, we describe methods to propagate and titer these two viruses. We also describe a simple cell-killing assay to begin testing modified viruses for increased potential killing of glioblastoma cells.

Key words Oncolytic virus therapy, Vesicular stomatitis virus, Zika virus, Glioblastoma, Immunotherapeutics, Viral titration, Plaque assay, Cell killing

1 Introduction

Oncolytic virus therapy (OVT) is one of the latest additions to the arsenal of immuno-oncology [1]. OVT takes advantage of the ability of viruses to specifically kill tumor cells while leaving the normal cells unharmed. While the idea of using a virus to kill tumor cells or enhance antitumor immunity is not entirely new, most of the path-breaking research has been conducted only in the past two decades owing to significant development in gene transfer technology. Multiple viruses spanning all of the Baltimore classification groups—type I (dsDNA—Herpes simplex virus, adenovirus, vaccinia virus), type II (ssDNA—parvovirus, chicken anemia virus), type III (dsRNA—reovirus), types IV/V (ssRNA—vesicular stomatitis, coxsackie, measles, Newcastle disease, Zika, Seneca Valley)—have been tried, with or without genetic modifications, for stalling the progression of different cancers (melanoma, breast, head, neck, prostate, urothelial, and glioblastoma) [2–10].

Samuel G. Katz and Peter M. Rabinovich (eds.), *Cell Reprogramming for Immunotherapy: Methods and Protocols*, Methods in Molecular Biology, vol. 2097, https://doi.org/10.1007/978-1-0716-0203-4_16,

One major breakthrough in the clinical use of OVT occurred in 2006 when the State Food and Drug Administration of the People's Republic of China approved the commercial use of *Oncorine*™/ *H101* (a modified adenovirus developed by Shanghai Sunway Biotech) for treatment of head and neck cancers [11, 12]. Almost a decade later, herpes simplex virus (*Talimogene Laherparepvec* (*T-VEC*)/*IMLYGIC*™) engineered to upregulate the production of granulocyte-macrophage colony stimulating factor was approved by the FDA for melanoma treatment in the United States. In a follow-up study, *T-VEC* increased the efficacy of an immune checkpoint (CTLA4) inhibitor, *Ipilimumab*, when used in combination in patients with advanced, unresectable melanomas [13, 14].

In this review, we shall focus on two single-strand RNA viruses, a (−) sense, Baltimore group V virus, vesicular stomatitis virus (VSV), and a (+) sense, group IV virus, Zika virus (ZIKV), where (−) and (+) refer to the RNA strand encoded by the virus. The genome of a (+) ssRNA virus can be loaded directly onto the ribosome, whereas a (−) ssRNA virus first needs its RNA to be converted into the (+) strand before translation. Both have been used for the treatment of multiple types of cancer, including glioblastoma.

The enveloped virus VSV (*Rhabdoviridae* family/*Vesiculovirus* genus) has emerged as one of the most favorable candidates for OVT primarily due to its low pathogenicity in humans, its natural hosts being horses, cattle, and pigs [15]. VSV was identified as an oncolytic virus in the early 2000s when it was found to selectively kill interferon nonresponsive human tumor cells [16]. Researchers demonstrated that this virus was cytotoxic against human melanoma xenografts implanted in mice. Since then, replication-competent strains of VSV have been tested for their efficiency in stopping the progression of numerous tumors—one of its early successes was against Multifocal Glioma and Metastatic Carcinoma in the brain [6, 7]. Studies on the mechanisms behind the antitumor properties of VSV have found that the key processes are the induction of apoptosis (mostly by the viral matrix protein) and a significant reduction in the amount of blood flow (hypoxia) to the tumor cells [17–20].

In addition to the native virus, a wide range of recombinant VSVs that express genes encoding cytokines to stimulate the immune system or encoding proteins that are cytotoxic to the tumor, like thymidine kinase or TP53, are also being tested for potential therapeutic value [6]. A common approach is to engineer a virus overexpressing a proinflammatory cytokine. A recent study using a recombinant VSV expressing interferon-γ shows high potential in the 4T1 mammary adenocarcinoma model [21]. This virus slowed tumor growth in an immune system-dependent manner. Another group reported that VSV engineered to express interferon-β and the sodium iodide symporter (NIS) was nonpathogenic and nontransmissible in a natural VSV host (pig) [8]. This work allays concerns about the potential transmission of VSV OVT from patients to the natural host and raises the possibility that it might be safe within humans as well.

Although identified in the middle of the last century, the neurotropic Zika virus (ZIKV) (*Flaviviridae* family) garnered widespread attention only in 2007, during an epidemic in the Yap Islands (Micronesia) [22, 23]. Almost a decade later, it was linked to debilitating diseases like Guillain–Barré syndrome and acute disseminated encephalomyelitis (ADEM) in adults. One of the very few flaviviruses capable of vertical transmission (mother-to-child), it is also responsible for microcephaly in infants [24, 25]. Ironically, the ability of ZIKV to trigger apoptotic cell death in neural progenitor cells (NPC)—a possible mechanism of microcephaly—is being used to harness this virus for OV therapies in glioblastoma (GBM) [3–5, 9, 26, 27]. Unlike other flaviviruses such as West Nile virus (WNV), ZIKV triggered apoptosis specifically in glioblastoma stem cells (GSCs) but had minimal effect on differentiated glioma cells (DGC) both in vitro and in brain organoids. Furthermore, intratumoral injection with a ZIKV-Dakar, mouse-adapted strain halted the progression of implanted GBMs in mice [4, 5, 27]. In an attempt to advance this finding, a live attenuated ZIKV vaccine candidate (ZIKV-LAV) was tested for its efficacy against GBMs in mice [9]. While exhibiting a marked decrease in neurotoxicity when compared to a licensed vaccine for another flavivirus, Japanese Encephalitis virus, ZIKV-LAV still effectively halted growth of GBMs. Encouraging findings for ZIKV have been reported for other brain tumors as well, including a recent study showing effectiveness against neuroblastoma [3]. This publication presented data that ZIKV is more likely to bind to neuroblastoma cells expressing the cell surface glycoprotein CD24, as poorly permissive cells lacked CD24 and were less prone to ZIKV-mediated cytopathic effects. In summary, these two viruses hold immense potential for OVT and more studies need to be dedicated to understanding the inherent mechanism.

2 Materials

2.1 VSV Propagation

1. BHK-21 (ATCC# CCL-10) (*see* **Note 1**).
2. BHK-21 growth medium: Eagle's Minimum Essential Medium (EMEM) supplemented with 10% fetal bovine serum (FBS) and 1% penicillin–streptomycin solution.
3. 10-cm tissue culture plates.
4. 15-ml conical Falcon tubes.

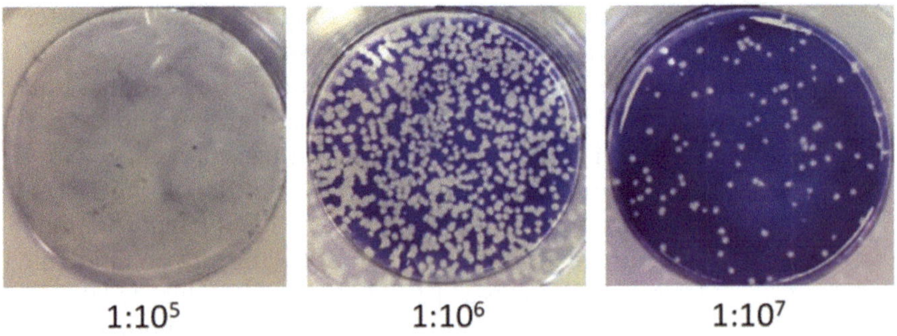

Fig. 1 Plaque assay of VSV-infected BHK-21. BHK-21 cells were plated in a 6-well plate and left to attach overnight. The following day, serial dilutions of virus were prepared (10^{-5} through 10^{-10}) and 100 μl added to the monolayer of BHK cells. After 30′ adsorption, the medium was swapped out for methylcellulose and let to solidify. The plates were then incubated at 37 °C for about 2 days. The overlay was then removed, and the cells were stained with 1% crystal violet for 20 min. The number of plaques was then counted after gently washing off the excess stain. The total number of plaque-forming units (*PFU*) was determined by dividing the average number of plaques by the dilution factor times the volume added

2.2 VSV Titration by Plaque Assay (Fig. 1)

1. 6-Well plates.
2. 2× DMEM.
3. FBS.
4. Methylcellulose.
5. Test tubes.
6. 1× PBS.
7. Crystal violet solution.

2.3 ZIKV Propagation

1. C6/36 (ATCC# CRL1660) (*see* **Note 2**).
2. VERO (ATCC# CCL81) (*see* **Note 3**).
3. T-75 flask.
4. 100× calcium–magnesium mix: dissolve 1.327 g $CaCl_2$ $2H_2O$ and 2.133 g $MgCl_2$ $6H_20$ in 100 ml of water and autoclave.
5. Flavivirus dilution media. In a sterile tissue culture hood, mix 437-ml autoclaved dH_2O, 50-ml autoclaved 1× PBS, 3 ml of filter-sterilized 35% BSA, 5 ml of autoclaved 1× calcium–magnesium mix, and 5 ml of 100× penicillin–streptomycin solution.
6. Viral propagation growth media. EMEM with 2-mM L-glutamine, 1-mM sodium pyruvate, 1.5-g/L sodium bicarbonate, 2% FBS, and penicillin–streptomycin.
7. 15- and 50-ml Falcon tubes.
8. Cryovials.
9. Dry ice.
10. Ethanol.
11. 100-kDA cutoff Amicon filter tube.

2.4 ZIKV Titration by Plaque Assay

1. VERO (ATCC #CCL81).
2. 12-Well plate.
3. RPMI 1640.
4. Agarose overlay: 50% low-melting point agarose, 45% 2× MEM, 5% FBS.
5. Ethanol.
6. Crystal violet solution.

2.5 Cell Killing Assay for Oncolytic Viruses

1. U-87 MG (ATCC HTB-14) (*see* **Note 4**).
2. Flat-well bottom 96-well plate.
3. DMEM.
4. 1× PBS.
5. FBS.
6. XTT assay kit.
7. Absorbance plate reader.

3 Methods

3.1 VSV Propagation

1. The day before infection split confluent BHK cells 1:10 onto sufficient numbers of 10-cm plates. Cells should be at 30–40% confluency at the time of infection.
2. On the day of infection inoculate cells at a multiplicity of infection (MOI) of 0.005 (*see* **Note 5**).
3. Incubate until 80–90% cell lysis (~48 h). Harvest supernatant, clarify by low-speed centrifugation (~400 × *g*), and store at −80 °C.

3.2 VSV Titration by Plaque Assay

1. The day prior to performing a titration, prepare one 6-well plate per virus preparation to be titrated. Split confluent BHK cells 1:10 to yield wells at 30–40% confluency the next day.
2. Make 2% methylcellulose by heating ~200 ml of dH_2O to 80 °C and then adding 10-g methylcellulose slowly while stirring. Bring the volume up to 500 ml with ice-cold dH_2O, continuing to stir. Chill entire mixture to 4 °C with stirring and continue to incubate for about 30 min until all reagent has dissolved. Filter sterilize through a 0.45-μm filter and then place into the freezer to eliminate any bubbles. Store long term at 4 °C.
3. Mix 12 ml per plate of equal parts 2% methylcellulose with 2× DMEM + 10% FBS.
4. On the day of the experiment, prewarm the 1% methylcellulose + DMEM and 5% FBS to 37 °C.

5. For each virus to be tittered, set out nine test tubes in the hood and label them 10^{-2} through 10^{-10}. Add 0.9 ml of DMEM without serum to each.
6. Add 100 μl of virus stock to the first tube labeled 10^{-2}. Vortex briefly and transfer 100 μl to the second tube with a fresh tip. Repeat with the remainder of the tubes.
7. Wash cells in 6-well plates with 1× PBS and add 0.5 ml of serum-free DMEM to each well.
8. Add 100 μl of the 10^{-5} dilution to the first well. The second well receives 100 μl of the 10^{-6} dilution, and so on through the 10^{-10} dilution.
9. Incubate the plates at 37 °C for 30 min, rocking every 10 min.
10. Aspirate the media carefully from each well and then add back 2 ml of the prewarmed 1% methylcellulose solution (from **step 4**).
11. Incubate plates at 37 °C for 48 h (*see* **Note 6**).
12. Aspirate media from all wells and carefully add 1% crystal violet stain and incubate at room temperature for 20 min.
13. Wash away excess stain by carefully immersing plate multiple times in a beaker containing room temperature dH_2O.
14. The resulting titer can be calculated from the well containing between 10 and 100 plaques using the formula:

    Plaque-forming units (pfu)/ml = number of plaques/(dilution factor ∗ volume of virus added in ml),

    e.g., 18 plaques in the 10^{-6} well → 18/(0.000001 ∗ 0.1) = 18 0,000,000 pfu/ml.

3.3 ZIKV Propagation

1. A day before the scheduled ZIKV infection, plate approximately 10^6 C6/36 or Vero cells in a T-75 flask (*see* **Note 7**).
2. After overnight incubation, remove growth medium and wash cells with flavivirus dilution media (*see* **Note 8**).
3. Dilute stock virus in 2 ml (*see* **Note 9**) of Flavivirus Dilution Media to an appropriate MOI and add this solution to the cells. While MOIs greater than 1 are preferred, we have used MOIs as low as 0.01 successfully.
4. Incubate the cells with the virus dilution for roughly 2 h at 28 °C (C6/36) or 37 °C (VERO) with constant shaking. If a shaker is not available, swirl the flasks every 15 min (*see* **Note 10**).
5. After the adsorption phase, add 10 ml viral propagation growth medium on top of the initial inoculum and keep the cells in the incubator for 2–4 days, until signs of cytopathic effects are visible (*see* **Note 11**).

6. Collect the supernatant from the flasks and transfer to a Falcon tube. For subsequent steps, solutions should be chilled or kept on ice.
7. Wash the attached cells with the growth medium in order to collect any viral particles still loosely bound to the surface of the host cells.
8. Combine the solutions from **steps 6** and 7 and spin at 200 × *g* for 5 min (at 4 °C).
9. OPTIONAL: Spin this solution at 1200 × *g* for 10 min (4 °C). After the spin, transfer the supernatant to a fresh Falcon tube and store on ice. Transfer 15 ml from this solution to a 100-kDA cutoff Amicon filter tube. Spin these tubes at 1200 × *g* for 45 min at 4 °C. After the spin, the supernatant left on top of the filter contains the concentrated virus. Use a P-200 pipette to transfer the concentrate to a fresh cryovial/microcentrifuge. Discard the flow-through. Repeat these steps until all supernatant has been concentrated—each filter can be used for two spins (*see* **Note 12**).
10. After the spin, discard the pellet of dead cells and other debris and aliquot the cleared supernatant into sterile cryovial tubes.
11. Quick freeze the aliquots of virus stock in dry ice/ethanol (95% ethanol chilled on dry ice).
12. After the quick freeze step, store these aliquots in −80 °C freezer.

3.4 ZIKV Titration by Plaque Assay

1. Plate 3 × 10^5 Vero cells into each well of a 12-well plate and allow attachment overnight (*see* **Note 13**).
2. Prepare tenfold serial dilutions (usually 10^2–10^6) of the viral stock in RPMI 1640 without supplements.
3. Add ~250 μl of each dilution, as well as the undiluted stock, to the corresponding wells.
4. Incubate the plates at 37 °C for about 2 h. Make sure to swirl the plates every 15 min.
5. After the adsorption, cover the cells with agarose overlay (*see* **Note 14**).
6. Let the overlay solidify inside the biosafety cabinet and incubate them at 37 °C for 3–4 days.
7. After 3–4 days, take out the plates and gently remove the overlay by adding a few drops of warm tap water or 1× PBS.
8. Add 1% crystal violet solution (in 30% ethanol) to the cells and leave them for about 20 min at room temperature.
9. Gently wash the wells with tap water and count plaque formation (as shown in Fig. 2).

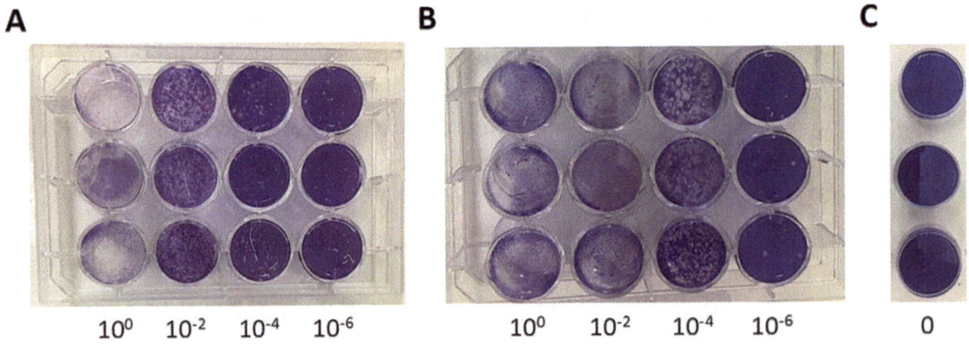

Fig. 2 Plaque assay of ZIKV-infected Vero cells. Vero cells were plated in a 12-well plate and left to attach overnight. The following day, serial dilutions of virus were prepared (10^{-2}, 10^{-4}, 10^{-6}) and 250 μl added to the monolayer of Vero cells (**a**, **b**). The increased number of plaques seen in **b** compared to **a**, indicates a higher viral titer. As a control, media without virus was added to Vero (**c**). After 2-h adsorption, agarose overlay medium was added and let to solidify. The plates were then incubated at 37 °C for about 3–4 days. The overlay was then removed, and the cells were stained with 1% crystal violet for about 20 min. The number of plaques was then counted after gently washing off the excess stain. The total number of plaque-forming units (*PFU*) was determined by dividing the average number of plaques by the dilution factor times the volume added. For example, if there were an average of 20 plaques in the 10^{-6} wells, the virus titer would be: $20/10^{-6} * 0.25$ ml = $8 * 1\ 0^{7}$ PFU/ml

3.5 Cell Killing Assay for Oncolytic Viruses

1. The day prior to infection, U-87 MG cells are plated at 3×10^4 cells per well of a flat-well bottom 96-well plate (cells should be ~80% confluent at the time of infection) and allowed to adhere overnight. Prepare a total of 12 wells per virus to be tested (six concentrations in duplicate).
2. The day of infection, prepare a master plate of virus dilutions. Add 100 μl of serum-free DMEM to each well of a 96-well plate, 12 wells per virus to be measured. The typical setup has four different viruses along the short axis diluted in duplicate. Wells in column 1 get no virus and act as the negative control. Wells in column 2 get 0.1 pfu/ml, column 3 get 1 pfu/ml, column 4 get 2.5 pfu/ml, column 5 get 5 pfu/ml, and column 6 get 10 pfu/ml. This can be repeated for the other half of the plate with additional viruses depending on the number needed.
3. Wash the plate containing the cells with 1× PBS and then add 100 μl of the virus dilution to the corresponding wells and incubate for 1 h.
4. At the end of the incubation, remove the virus and add 100 μl of DMEM + 2% FBS and then incubate another 72 h.
5. To determine cell viability, add XTT at 0.5 mg/ml according to the manufacturer's directions and incubate for 5 h.
6. Measure optical density at 450 nm. Uninfected cells are set to 100% viability.

4 Notes

1. These fibroblasts derived from the kidney of a Syrian golden hamster (*Mesocricetus auratus*) are used for both growth and tittering of VSV. Cells are incubated at 37 °C in 5% CO_2. Cultures are typically split two times per week at 1:8.
2. As the mosquito is the secondary host/vector for arbovirus, these *Aedes albopictus* clones are widely used for the propagation of ZIKV. Cells are maintained in RPMI 1640 media containing 10% FBS, 1% penicillin–streptomycin solution, 1% nonessential amino acids, 1% sodium pyruvate, 1% L-glutamine, and 3% NaHCO3. As an alternative to RPMI 1640, Eagle's Minimum Essential Medium (EMEM) can be used as the base media. Cells are incubated at 28 °C in 5% CO_2. Cells are typically split twice per week at 1:8.
3. These epithelial cells isolated from the kidney of an adult African green monkey (*Cercopithecus aethiops*) are used for the titration of ZIKV. Cells are maintained in EMEM supplemented with 10% fetal bovine serum (FBS) and 1% penicillin–streptomycin solution. Cells are incubated at 37 °C in 5% CO_2. Cultures are typically split two times per week at 1:8. In most cases, VERO has also substituted C6/36 as the preferred host cell for growing ZIKV thus circumventing the issue of slower growth rate which is usually observed in C6/36.
4. These epithelial cells were isolated from a human brain tumor, most likely a glioblastoma. They provide a useful target for determination of the killing potential of both VSV and ZIKA. Cells are maintained in EMEM supplemented with 10% fetal bovine serum (FBS) and 1% penicillin–streptomycin solution. Cells are incubated at 37 °C in 5% CO_2. Cultures are typically split three times per week at 1:3.
5. The MOI of infection is calculated by dividing the pfu/ml by the number of cells in the culture. For a typical 10-cm plate of BHK at time of infection, there are approximately $3.5 * 10^6$ cells. So, 0.005 MOI = X pfu/ml/$3.5 * 10^6$. X = 17,500 pfu/ml. For 30 ml of medium, add a total of $5.25 * 10^5$ pfu from your stock.
6. Be careful not to disturb plates during this time or the plaques will become fuzzy and difficult to count.
7. This is to ensure that the host cell is 70–80% confluent at the time of infection. While C6/36 cells give a higher viral yield, they grow very slowly compared to Vero cells.
8. This solution facilitates the binding of the viral particles to the host cells.
9. The volume of flavivirus dilution media should be kept as minimal as possible, in order to avoid dilution of the viral inoculum. The surface area of a T-75 flask is just covered by 2 ml.

10. Shaking during the adsorption step ensures that the virus inoculum contacts as many cells as possible.
11. The reduced serum ensures a slow rate of cell proliferation and a more efficient binding of the virus. The former also keeps the virus from being stressed due to overconfluence of host cells. It is advisable to stop the incubation and collect the virus when ~30% host cells are dead.
12. For selected strains where the titer is less than 10^5, additional purification and concentration of the viral stocks might be necessary. This is especially relevant when VERO is used instead of C6/36, to grow the virus.
13. Dispense the cells at the center of the well and gently rock the plate to ensure that the cells form a dispersed monolayer.
14. As an alternative, 1% carboxymethylcellulose (CMC)/2.5% α-MEM may be used instead of the agarose medium. Both of these have been used equivalently in the literature.

References

1. Kamta J, Chaar M, Ande A, Altomare DA, Ait-Oudhia S (2017) Advancing cancer therapy with present and emerging immuno-oncology approaches. Front Oncol 7:64. https://doi.org/10.3389/fonc.2017.00064
2. Fountzilas C, Patel S, Mahalingam D (2017) Review: Oncolytic virotherapy, updates and future directions. Oncotarget 8(60):102617–102639. https://doi.org/10.18632/oncotarget.18309
3. Mazar J, Li Y, Rosado A, Phelan P, Kedarinath K, Parks GD, Alexander KA, Westmoreland TJ (2018) Zika virus as an oncolytic treatment of human neuroblastoma cells requires CD24. PLoS One 13(7):e0200358. https://doi.org/10.1371/journal.pone.0200358
4. Wood H (2017) Neuro-oncology: a new role for Zika virus in glioblastoma therapy? Nat Rev Neurol 13(11):640–641. https://doi.org/10.1038/nrneurol.2017.138
5. Zhu Z, Gorman MJ, McKenzie LD, Chai JN, Hubert CG, Prager BC, Fernandez E, Richner JM, Zhang R, Shan C, Tycksen E, Wang X, Shi PY, Diamond MS, Rich JN, Chheda MG (2017) Zika virus has oncolytic activity against glioblastoma stem cells. J Exp Med 214(10):2843–2857. https://doi.org/10.1084/jem.20171093
6. Bishnoi S, Tiwari R, Gupta S, Byrareddy SN, Nayak D (2018) Oncotargeting by vesicular stomatitis virus (VSV): advances in cancer therapy. Viruses 10(2). https://doi.org/10.3390/v10020090
7. Ozduman K, Wollmann G, Piepmeier JM, van den Pol AN (2008) Systemic vesicular stomatitis virus selectively destroys multifocal glioma and metastatic carcinoma in brain. J Neurosci 28(8):1882–1893. https://doi.org/10.1523/JNEUROSCI.4905-07.2008
8. Velazquez-Salinas L, Naik S, Pauszek SJ, Peng KW, Russell SJ, Rodriguez LL (2017) Oncolytic recombinant vesicular stomatitis virus (VSV) is nonpathogenic and nontransmissible in pigs, a natural host of VSV. Hum Gene Ther Clin Dev 28(2):108–115. https://doi.org/10.1089/humc.2017.015
9. Chen Q, Wu J, Ye Q, Ma F, Zhu Q, Wu Y, Shan C, Xie X, Li D, Zhan X, Li C, Li XF, Qin X, Zhao T, Wu H, Shi PY, Man J, Qin CF (2018) Treatment of human glioblastoma with a live attenuated Zika virus vaccine candidate. MBio 9(5):e01683–e01618. https://doi.org/10.1128/mBio.01683-18
10. Baltimore D (1971) Expression of animal virus genomes. Bacteriol Rev 35(3):235–241
11. Garber K (2006) China approves world's first oncolytic virus therapy for cancer treatment. J Natl Cancer Inst 98(5):298–300. https://doi.org/10.1093/jnci/djj111
12. Liang M (2018) Oncorine, the world first oncolytic virus medicine and its update in China. Curr Cancer Drug Targets 18(2):171–176. https://doi.org/10.2174/1568009618666171129221503
13. Andtbacka RH, Kaufman HL, Collichio F, Amatruda T, Senzer N, Chesney J, Delman KA, Spitler LE, Puzanov I, Agarwala SS, Milhem M, Cranmer L, Curti B, Lewis K, Ross M, Guthrie T, Linette GP, Daniels GA, Harrington K, Middleton MR, Miller WH Jr, Zager JS,

Ye Y, Yao B, Li A, Doleman S, VanderWalde A, Gansert J, Coffin RS (2015) Talimogene Laherparepvec improves durable response rate in patients with advanced melanoma. J Clin Oncol 33(25):2780–2788. https://doi.org/10.1200/JCO.2014.58.3377

14. Chesney J, Puzanov I, Collichio F, Singh P, Milhem MM, Glaspy J, Hamid O, Ross M, Friedlander P, Garbe C, Logan TF, Hauschild A, Lebbe C, Chen L, Kim JJ, Gansert J, Andtbacka RHI, Kaufman HL (2018) Randomized, open-label phase II study evaluating the efficacy and safety of Talimogene Laherparepvec in combination with ipilimumab versus ipilimumab alone in patients with advanced, unresectable melanoma. J Clin Oncol 36(17):1658–1667. https://doi.org/10.1200/JCO.2017.73.7379
15. Betancourt D, Ramos JC, Barber GN (2015) Retargeting oncolytic vesicular stomatitis virus to human T-cell lymphotropic virus type 1-associated adult T-cell leukemia. J Virol 89(23):11786–11800. https://doi.org/10.1128/JVI.01356-15
16. Stojdl DF, Lichty B, Knowles S, Marius R, Atkins H, Sonenberg N, Bell JC (2000) Exploiting tumor-specific defects in the interferon pathway with a previously unknown oncolytic virus. Nat Med 6(7):821–825. https://doi.org/10.1038/77558
17. Koyama AH (1995) Induction of apoptotic DNA fragmentation by the infection of vesicular stomatitis virus. Virus Res 37(3):285–290
18. Gaddy DF, Lyles DS (2005) Vesicular stomatitis viruses expressing wild-type or mutant M proteins activate apoptosis through distinct pathways. J Virol 79(7):4170–4179. https://doi.org/10.1128/JVI.79.7.4170-4179.2005
19. Schache P, Gurlevik E, Struver N, Woller N, Malek N, Zender L, Manns M, Wirth T, Kuhnel F, Kubicka S (2009) VSV virotherapy improves chemotherapy by triggering apoptosis due to proteasomal degradation of Mcl-1. Gene Ther 16(7):849–861. https://doi.org/10.1038/gt.2009.39
20. Qi X, Du L, Chen X, Chen L, Yi T, Chen X, Wen Y, Wei Y, Zhao X (2016) VEGF-D-enhanced lymph node metastasis of ovarian cancer is reversed by vesicular stomatitis virus matrix protein. Int J Oncol 49(1):123–132. https://doi.org/10.3892/ijo.2016.3527
21. Bourgeois-Daigneault MC, Roy DG, Falls T, Twumasi-Boateng K, St-Germain LE, Marguerie M, Garcia V, Selman M, Jennings VA, Pettigrew J, Amos S, Diallo JS, Nelson B, Bell JC (2016) Oncolytic vesicular stomatitis virus expressing interferon-gamma has enhanced therapeutic activity. Mol Ther Oncolytics 3:16001. https://doi.org/10.1038/mto.2016.1
22. Dick GW (1952) Zika virus. II. Pathogenicity and physical properties. Trans R Soc Trop Med Hyg 46(5):521–534
23. Duffy MR, Chen TH, Hancock WT, Powers AM, Kool JL, Lanciotti RS, Pretrick M, Marfel M, Holzbauer S, Dubray C, Guillaumot L, Griggs A, Bel M, Lambert AJ, Laven J, Kosoy O, Panella A, Biggerstaff BJ, Fischer M, Hayes EB (2009) Zika virus outbreak on Yap Island, Federated States of Micronesia. N Engl J Med 360(24):2536–2543. https://doi.org/10.1056/NEJMoa0805715
24. Cauchemez S, Besnard M, Bompard P, Dub T, Guillemette-Artur P, Eyrolle-Guignot D, Salje H, Van Kerkhove MD, Abadie V, Garel C, Fontanet A, Mallet HP (2016) Association between Zika virus and microcephaly in French Polynesia, 2013-15: a retrospective study. Lancet 387(10033):2125–2132. https://doi.org/10.1016/S0140-6736(16)00651-6
25. Rasmussen SA, Jamieson DJ, Honein MA, Petersen LR (2016) Zika virus and birth defects—reviewing the evidence for causality. N Engl J Med 374(20):1981–1987. https://doi.org/10.1056/NEJMsr1604338
26. Souza BS, Sampaio GL, Pereira CS, Campos GS, Sardi SI, Freitas LA, Figueira CP, Paredes BD, Nonaka CK, Azevedo CM, Rocha VP, Bandeira AC, Mendez-Otero R, Dos Santos RR, Soares MB (2016) Zika virus infection induces mitosis abnormalities and apoptotic cell death of human neural progenitor cells. Sci Rep 6:39775. https://doi.org/10.1038/srep39775
27. Lubin JA, Zhang RR, Kuo JS (2018) Zika virus has oncolytic activity against glioblastoma stem cells. Neurosurgery 82(5):E113–E114. https://doi.org/10.1093/neuros/nyy047

Part III

Monitoring Reprogrammed Immune Cells In Vivo

Chapter 17

Radiolabeling and Imaging of Adoptively Transferred Immune Cells by Positron Emission Tomography

Amer M. Najjar

Abstract

Positron emission tomography (PET) using ^{89}Zr is a clinically relevant imaging modality that enables long–term monitoring of adoptively transferred immune cells. This article describes a two-step radiometal labeling procedure utilizing the bifunctional siderophore p-isothiocyanatobenzyl-desferrioxamine (DFO-Bz-NCS) that chelates ^{89}Zr with high affinity and binds covalently to primary amines of cell-surface proteins via its isothiocyanate moiety. Cells labeled with ^{89}Zr-DFO-Bz-NCS remain viable and retain the radiolabel, enabling repetitive PET imaging of adoptively transferred immune cells with high sensitivity and specificity for up to 2 weeks.

Key words Positron emission tomography, PET, ^{89}Zr, Zirconium, p-Isothiocyanatobenzyl-desferrioxamine, Immune cells, Imaging

1 Introduction

The expanding application of cell-based therapies utilizing chimeric antigen receptor (CAR) T cells and natural killer (NK) cells to target cancer cells has created a need for imaging techniques to assess the effectiveness of the adoptively transferred immune cells by monitoring their in vivo persistence and trafficking to their intended tumor targets. A number of cell labeling and imaging approaches have been developed over the last two decades for single-photon emission computerized tomography (SPECT) and positron emission tomography (PET). These methods entail transfer of ^{111}In, ^{64}Cu, ^{99m}Tc, or ^{89}Zr into cells mediated by antibodies or chelators such as oxine and pyruvaldehyde-bis(N4-methylthiosemicarbazone) (PTSM) [1–4].

The feasibility of these methods, however, has been limited by a number of factors. The sensitivity of SPECT imaging using ^{111}In or ^{99m}Tc is relatively low and is exacerbated by the systemic distribu-

Samuel G. Katz and Peter M. Rabinovich (eds.), *Cell Reprogramming for Immunotherapy: Methods and Protocols*, Methods in Molecular Biology, vol. 2097, https://doi.org/10.1007/978-1-0716-0203-4_17,

tion of infused immune cells. Furthermore, efflux of the radiometals within hours of oxine- or PTSM-mediated labeling compromises the specificity and sensitivity of imaging [2, 3, 5]. ^{111}I labeling has also been shown to have profound negative effects on cellular metabolism, proliferation, and viability [3, 6].

The chemical and physical properties of ^{89}Zr are highly compatible with the biological function of adoptively transferred cells. The long half-life of ^{89}Zr (78.4 h) enables reliable long–term, repetitive PET imaging for up to 2 weeks, a suitable time frame for monitoring the persistence and distribution of the immune cells. ^{89}Zr also enables high image resolution due to the low translational energy (395.5 keV) of decay to ^{89}Y through positron emission and electron capture [7, 8].

The method described here has been used extensively for the labeling of antibodies with ^{89}Zr [9, 10] and has more recently been adapted to labeling cell surface proteins and imaging of live cells [11]. The cellular labeling process entails two steps as illustrated in Fig. 1. In the first step, ^{89}Zr is chelated to p-isothiocyanatobenzyl-desferrioxamine (DFO-Bz-NCS), a bifunctional siderophore that binds radiometals with high affinity via three hydroxamate groups. This ^{89}Zr-DFO-Bz-NCS complex exhibits high stability with minimal loss of chelated ^{89}Zr in serum [12]. Cell labeling by conjugation of the ^{89}Zr-DFO-Bz-NCS complex with primary amines on cell surface proteins is carried out in the second step. Conjugation is facilitated by the terminal isothiocyanate group of DFO-Bz-NCS.

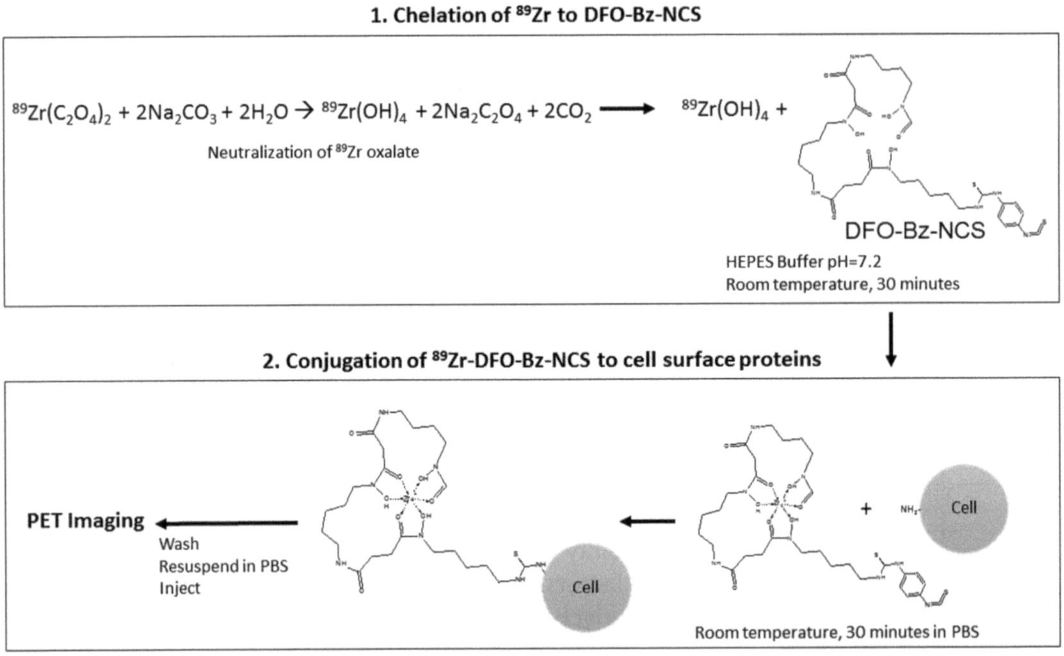

Fig. 1 Two-step chelation and cell labeling scheme. ^{89}Zr oxalate is neutralized prior to chelation with DFO-Bz-NCS. The ^{89}Zr-DFO-Bz-NCS complex is then conjugated with primary amines of cell surface proteins. Cells are washed prior to infusion and imaging by PET

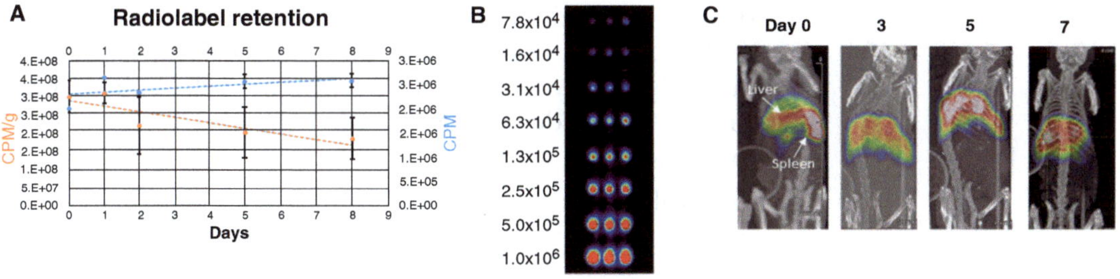

Fig. 2 Viability and PET imaging of ^{89}Zr-DFO-Bz-labeled T cells. (**a**) ^{89}Zr-DFO-Bz-labeled Jurkat cells were incubated in RPMI medium with 10% fetal bovine serum. Radioactivity (counts per minute, CPM) was measured at indicated days from a sampling of cells. Absolute radioactivity levels remained steady for up to 8 days in each sampling. When radioactivity is normalized to cellular mass (CPM/g), the levels decrease, indicating dilution of the radiolabel as the cells divide in culture. (**b**) Serial dilution of ^{89}Zr-DFO-Bz-NCS-labeled Jurkat cells illustrates detection of less than 10^4 cells in a 50-μL volume. Labeling was achieved at a specific activity of 0.16 MBq (4.3 μCi)/10^6 cells diluted two-fold in a series of six transfers. (**c**) ^{89}Zr-DFO-Bz-CAR T cells (10^6/100 μL PBS) were imaged for up to 7 days postintravenous infusion. The T cells can be detected in the liver and spleen throughout the experimental duration

This method of cell labeling with ^{89}Zr offers numerous advantages. The cells can be labeled in less than 1 h immediately prior to infusion. The covalently bound ^{89}Zr-DFO complexes result in rigid cellular retention of the radiolabel without compromising cellular viability or function (Fig. 2). Exclusive retention of the radioisotope by the labeled cells yields enhanced signal-to-background ratios that translate into high sensitivity and specificity. Furthermore, the long half-life of ^{89}Zr offers a logistical advantage enabling shipping of the radioisotope to remote locations eliminating the need for adjacent cyclotron facilities [13].

2 Materials

Deionized ultra-pure water must be used for the preparation of stock solutions. Chelation and cell labeling reactions are performed at room temperature.

2.1 ^{89}Zr-Labeled p-Isothiocyanatobenzyl-Desferrioxamine (DFO-Bz-NCS)

1. ^{89}Zr oxalate solution: ^{89}Zr is typically provided by cyclotron facilities in 1 M oxalic acid at an activity concentration of 740 megabequerels (MBq)/mL (20 mCi/mL) (*see* **Note 1**).
2. 1 M Sodium carbonate (Na_2CO_3) solution: Dissolve 106.0 mg in 1 mL of water. Solution may be stored at room temperature.
3. 0.5 M HEPES buffer, pH 7.2: Dissolve 11.9 g HEPES in 80 mL of water and adjust the pH to 7.2 using 1 M NaOH. Add water to a final volume of 100 mL. Solution may be stored at room temperature.

4. 10 mM p-isothiocyanatobenzyl-desferrioxamine (DFO-Bz-NCS) solution: Dissolve 7.5 mg of DFO-Bz-NCS (molecular weight = 752.9) in 1 mL of cell culture-grade dimethylsulfoxide (DMSO). Mix thoroughly to dissolve completely and store at 4 °C.
5. Dose calibrator: Programmed to measure ^{89}Zr activity within experimental range of 0–74 MBq (2 mCi).

2.2 Cell Suspension

1. Phosphate-buffer saline (PBS).
2. Cell suspension (10^7–10^8 cells/4 mL PBS): Count cells and transfer desired number into a 15-mL conical tube. Centrifuge at 500 × *g* for 5 min to pellet the cells. Wash twice with 10 mL PBS and resuspend the cell pellet in 4 mL PBS. Keep on ice until the labeling reaction is initiated (*see* **Note 2**).

3 Methods

3.1 Preparation of ^{89}Zr-DFO-Bz-NCS Complex for Cell Labeling

1. To minimize radiation exposure, all radiolabeling procedures must be performed in hot cells or lead-shielded ventilated hoods. Prepare and place all needed reagents and radioactive disposal containers within the work area in advance for seamless and efficient workflow.
2. Transfer 50 μL of ^{89}Zr oxalate solution (equivalent to ~37 MBq) into a 1.5-mL Eppendorf tube.
3. Neutralize ^{89}Zr oxalate to $^{89}Zr(OH)_4$ by adding an equivalent volume (50 μL) of 1 M Na_2CO_3. Neutralization is necessary for effective chelation with DFO-Bz-NCS and subsequent labeling of live cells at physiological pH.
4. Mix by gently pipetting two to three times and keep at room temperature for 5 min (*see* **Note 3**). The pH should equate to about 7.2. To verify, measure by placing a 5 μL drop on a pH strip.
5. Add 400 μL 0.5 M HEPES buffer, pH 7.2 and mix thoroughly.
6. Add 10 μL DFO-Bz-NCS solution and mix thoroughly to initiate the chelation process with ^{89}Zr (*see* **Note 4**). Incubate at room temperature for 30 min.

3.2 Cell Labeling

1. Add the entire ^{89}Zr/DFO-Bz-NCS complex solution (~500 μL) to the cell suspension in the 15-mL conical tube and incubate at room temperature for 30 minutes with continuous and gentle rotation. During this step, ^{89}Zr-DFO-Bz-NCS will covalently conjugate with primary amines on cell surface proteins.

2. Centrifuge at 500 × *g* for 5 min to pellet the cells, remove the supernatant, and resuspend the cell pellet in 4 mL PBS. Repeat this wash step two more times to thoroughly remove unconjugated ^{89}Zr-DFO-Bz-NCS complexes and free ^{89}Zr ions.
3. Resuspend cells in PBS to achieve the desired cell concentration for infusions. Measure total activity using a dose calibrator.
4. Calculate the specific activity by dividing total radioactivity by the cell number. Typical yields are in the range of 1.1–7.4 MBq (30–200 μCi)/10^7 cells.
5. The cells may be administered via intravenous, intraperitoneal, or intracardiac routes depending on the experimental design.
6. Infused animals may be repetitively imaged in a PET scanner for up to 14 days. Ample senstivity of detection can be achieved with infusion doses of 0.19-0.74 MBq (5–20 μCi) in a mouse model (*see* **Note 5**).

4 Notes

1. Variations in the activity concentration of the stock ^{89}Zr oxalate solution are well tolerated by this labeling method. Reduced specific activity of cell labeling may result from lower initial activity concentrations, but may be still adequate to achieve required sensitivity for PET imaging.
2. Cells must be washed thoroughly to remove any extraneous serum proteins. Binding of ^{89}Zr-DFO-Bz-NCS to soluble proteins will lower the efficiency of cell labeling and decrease specific activity. To maintain optimal cellular viability, prepare cells during the chelation step and immediately prior to labeling with ^{89}Zr-DFO-Bz-NCS.
3. The neutralization process will generate CO_2 gas. As a precaution, leave the tube uncapped for a few seconds to allow the gas to escape before recapping.
4. It is essential to keep the percentage of DMSO in the chelation mixture to a minimum to prevent precipitate formation. Addition of 10 μL DFO-Bz-NCS/DMSO solution represents 2% of the total volume and does not alter the solubility of any components in the mixture.
5. With isotopic decay of ^{89}Zr, it may be necessary to prolong PET scan times to increase signal gain and maximize the signal-to-noise ratio for optimal detection. PET imaging may be potentially extended to 3 weeks depending on the initial dose of infusion.

References

1. Sato N, Wu H, Asiedu KO, Szajek LP, Griffiths GL, Choyke PL (2015) (89)Zr-oxine complex PET cell imaging in monitoring cell-based therapies. Radiology 275(2):490–500
2. Adonai N, Adonai N, Nguyen KN, Walsh J, Iyer M, Toyokuni T, Phelps ME, McCarthy T, McCarthy DW, Gambhir SS (2002) Ex vivo cell labeling with 64Cu-pyruvaldehyde-bis(N4-methylthiosemicarbazone) for imaging cell trafficking in mice with positron-emission tomography. Proc Natl Acad Sci U S A 99(5):3030–3035
3. Kuyama J, McCormack A, George AJ, Heelan BT, Osman S, Batchelor JR, Peters AM (1997) Indium-111 labelled lymphocytes: isotope distribution and cell division. Eur J Nucl Med 24(5):488–496
4. Hughes DK (2003) Nuclear medicine and infection detection: the relative effectiveness of imaging with 111In-oxine-, 99mTc-HMPAO-, and 99mTc-stannous fluoride colloid-labeled leukocytes and with 67Ga-citrate. J Nucl Med Technol 31(4):196–201; quiz 203–4.
5. Brenner W, Aicher A, Eckey T, Massoudi S, Zuhayra M, Koehl U, Heeschen C, Kampen WU, Zeiher AM, Dimmeler S, Henze E (2004) 111In-labeled CD34+ hematopoietic progenitor cells in a rat myocardial infarction model. J Nucl Med 45(3):512–518
6. Gildehaus FJ, Haasters F, Drosse I, Wagner E, Zach C, Mutschler W, Cumming P, Bartenstein P, Schieker M (2011) Impact of indium-111 oxine labelling on viability of human mesenchymal stem cells in vitro, and 3D cell-tracking using SPECT/CT in vivo. Mol Imaging Biol 13(6):1204–1214
7. Holland JP, Sheh Y, Lewis JS (2009) Standardized methods for the production of high specific-activity zirconium-89. Nucl Med Biol 36(7):729–739
8. Deri MA, Zeglis BM, Francesconi LC, Lewis JS (2013) PET imaging with (8)(9)Zr: from radiochemistry to the clinic. Nucl Med Biol 40(1):3–14
9. Vosjan MJ, Perk LR, Visser GW, Budde M, Jurek P, Kiefer GE, van Dongen GA (2010) Conjugation and radiolabeling of monoclonal antibodies with zirconium-89 for PET imaging using the bifunctional chelate p-isothiocyanatobenzyl-desferrioxamine. Nat Protoc 5(4):739–743
10. Perk LR, Vosjan MJ, Visser GW, Budde M, Jurek P, Kiefer GE, van Dongen GA (2010) p-Isothiocyanatobenzyl-desferrioxamine: a new bifunctional chelate for facile radiolabeling of monoclonal antibodies with zirconium-89 for immuno-PET imaging. Eur J Nucl Med Mol Imaging 37(2):250–259
11. Bansal A, Pandey MK, Demirhan YE, Nesbitt JJ, Crespo-Diaz RJ, Terzic A, Behfar A, DeGrado TR (2015) Novel (89)Zr cell labeling approach for PET-based cell trafficking studies. EJNMMI Res 5:19
12. Meijs WE, Herscheid JD, Haisma HJ, Pinedo HM (1992) Evaluation of desferal as a bifunctional chelating agent for labeling antibodies with Zr-89. Int J Rad Appl Instrum A 43(12):1443–1447
13. Zhang Y, Hong H, Cai W (2011) PET tracers based on Zirconium-89. Curr Radiopharm 4(2):131–139

Chapter 18

Functional Analysis of Human Hematopoietic Stem Cells In Vivo in Humanized Mice

Yuanbin Song, Rana Gbyli, Xiaoying Fu, and Stephanie Halene

Abstract

Ex vivo generation and expansion of functional hematopoietic stem cells represents the holy grail of reprogramming and would constitute a major advance in stem cell therapies and generation of blood cellular products. In vivo testing is critical to assure proper cell intrinsic function in an organismal context. Here we describe methods for the generation of human hematopoiesis chimeric mice and evaluation of hematopoietic stem cell function. The choice of mouse model, stem cell source, and transplantation route can be adjusted to suit the desired application.

Key words Humanized mice, Cytokine humanization, Xenotransplantation, Immunodeficient mouse models, Hematopoietic stem cell, Myeloid differentiation

1 Introduction

The murine host has remained a readily available and ethically acceptable model for the study of human diseases and therapeutic testing. Immunodeficient mouse models support engraftment of human hematopoietic stem cells but with limitation in efficiency and mature lineage representation. Systematic genetic alterations of the immunodeficient host have generated progressively more permissive and supportive murine hosts (*see* Table 1). The most frequently used immunodeficient mouse strains lack T-, B-, and NK-cells and as a result also have defective dendritic cell function. *Nod Scid gamma* (NSG) mice in addition carry a polymorphism in the *Sirpa* gene, expressed in murine macrophages, that allows enhanced binding to the human CD47 ligand (providing a "don't-eat-me" signal) [12]. Human SIRPA (S), introduced as transgene or knockin, e.g., into the Rag−/− γ−/− (RG) background, replicates this effect and significantly improves engraftment of human hematopoietic cells (SRG) [8]. Mutation of the murine stem cell factor receptor c-kit compromises murine HSC retention in the

Samuel G. Katz and Peter M. Rabinovich (eds.), *Cell Reprogramming for Immunotherapy: Methods and Protocols*, Methods in Molecular Biology, vol. 2097, https://doi.org/10.1007/978-1-0716-0203-4_18,

Table 1
Various immunodeficient mouse models

Common name	NSG™, NOD *scid* gamma [1, 2]	NSGS, NOD *scid* gamma Il3- GM-SF (NSG-SGM3) [3–5]	NSGW41 [6, 7]	*SIRPA Rag gamma* [8]	MISTRG [9–11]
Mature B cells	Absent	Absent	Absent	Absent	Absent
Mature T cells	Absent	Absent	Absent	Absent	Absent
Natural killer cells	Absent	Absent	Absent	Absent	Absent
Dendritic cells Macrophages	Defective, due to T-cell defects				Defective, due to T-cell defects and lack of murine *M-Csf*
Complement	Absent	Absent	Absent	Present	Present
Benefits	Supports engraftment of human hematopoietic stem cells. Study of human tissues and tumors	Enhances human myelopoiesis and terminal differentiation. Increased efficiency of engrafting human acute myeloid leukemia (AML) Increased $CD4^{+}$ $FoxP3^{+}$ regulatory T cell population	Long-term engraftment of human hematopoietic stem cells without irradiation. Improved human erythro- and megakaryopoiesis from cord blood	Supports engraftment of human hematopoietic stem cells. Study of human tissues and tumors	Support efficient engraftment of adult human HSC, functional innate immune system, and secondary transplantation. Support engraftment of MDS cell stem cells and erythro- and megakaryopoiesis from adult and diseased stem cells
Considerations	Sensitive to irradiation	Compromised human stem cell regeneration and reduction of human erythro- and B- lymphopoiesis	Compromise of murine stem cells and erythropoiesis due to defect in murine c-kit		Mice are severely immuno-compromised. Competent human macrophages may phagocytose mouse cells

HSC niche, allowing murine and human stem cell engraftment without prior irradiation [6, 7, 13]. The relevance of "humanization" of the HSC niche is evident in models where human mesenchymal stem cells are transplanted and stimulated to form "ossicles" [14]. Cytokine humanization improves myeloid cell maturation, albeit at the expense of hematopoietic stem cell maintenance and with lineage skewing when introduced as transgenes [3–5]; knockin of several non-cross reactive human cytokines into the corresponding murine loci in the SRG strain on the other hand has enhanced not only engraftment and maintenance of human HSCs but has also resulted in improved reconstitution of the adaptive and innate immune systems and generation of mature myeloid lineage cells [15–17]. Combination of these cytokine humanizations for *M-CSF*, *IL3/GM-CSF*, and *Thrombopoietin* in the *SRG* background, termed "MISTRG," has resulted in a highly permissive and supportive niche for normal and diseased human hematopoietic stem cells [9–11, 18, 19].

In this chapter we specifically focus on protocols in MISTRG mice. Various transplantation methods through diverse injection routes have been established to direct the human HSCs to the supportive niche. We here focus on two transplantation routes, specifically intrahepatic injection into newborn 1–3 day old [20] and intrafemoral injection into adult mice. Human HSCs (enriched in the $CD34^+$ fraction) can be obtained from several sources and in various conditions, including cryopreserved or fresh cells from human fetal liver (FL), umbilical cord blood (UCB), G-CSF mobilized adult peripheral blood stem cells (PBSC), or bone marrow (BM). For PBSC and BM xenotransplantations, either alone or in addition to $CD34^+$ enrichment, we recommend the use of the anti-human CD3 antibody OKT3 to efficiently deplete human T-cells to prevent graft-versus-host disease [21].

In summary, in this unit, we provide various approaches for generating human xenograft mice starting with the collection and isolation of human $CD34^+$ hematopoietic stem and progenitor cells (HSPCs), detailed description of intra-femoral and intrahepatic transplantation routes, and basic analysis protocols. The unique aspects and advantages of each method and important factors to be taken into consideration are highlighted throughout this unit.

2 Materials

2.1 Purification of Human $CD34^+$ HSPC

1. Human $CD34^+$ HSPC can be obtained from human bone marrow (BM), G-CSF mobilized human peripheral blood mononuclear cells (PBMCs), umbilical cord blood (UCB), and fetal liver (FL) from consented donors at the respective institution or from commercial sources. Isolation of HSPCs can also be

achieved from other sources, such as differentiated induced pluripotent stem or embryonic stem cells (*see* **Note 1**).

2. 5, 10 ml serological pipettes.
3. 100 mm petri dish.
4. Razor blades.
5. 50 ml Falcon tubes.
6. 1.5 ml Eppendorf tubes.
7. Hemocytometer.
8. Sterile phosphate buffered saline (PBS).
9. Collagenase D stock solution: 1% Collagenase D stock solution, DMEM (Dulbecco's Modified Eagle Medium). Dissolve 2.5 g collagenase in 250 ml DMEM. Store desired aliquots at −20 °C.
10. Ficoll-Paque density gradient media.
11. Trypan Blue.
12. Human $CD34^+$ enrichment kit (e.g. CD34-Microbead-Kit, Miltenyi-Biotech, or equivalent products from other vendors).
13. Red Blood Cell (RBC) lysis buffer.
14. Freezing solution: 90% FBS (Fetal Bovine Serum), 10% DMSO (Dimethyl sulfoxide). Chill on ice prior to use (*see* **Note 2**).
15. Cell freezing container (Mr. Frosty, ThermoFisher). Change the isopropanol in Mr. Frosty after every fifth use. In-between uses keep Mr. Frosty at 4 °C or −20 °C and remove just prior to use.

2.2 Xeno-transplantation

1. Immunodeficient mice (Table 1): Various mouse models with variable permissiveness for human HSC engraftment exist. A table detailing the various models including their genetic backgrounds, cellular components, and distinguishing features is provided (*see* **Notes 3** and **4**).
2. X-RAD 320 irradiator (Rad Source, Buford, GA).
3. $CD34^+$ selected cells.
4. 70% ethanol, PBS.
5. Anti-human CD3 antibody (clone Okt3).
6. Laminar airflow hood.
7. Hamilton syringe (50 μl with 22-G removable needle).
8. Hamilton syringe (50 μl with 27-G needle on PTFE Luer Lock).
9. Plexi-glass irradiator pie (Braintree Scientific).

10. Electric razor.
11. 70% ethanol.
12. Sterile gloves.
13. 1 ml syringe with 27½-G needle.
14. Isoflurane anesthetic: 30% Isoflurane, 70% 2-propylene glycol.
15. Heating lamp.
16. Gauze sponges.
17. Carprofen: 50 mg/ml in sterile saline. Administer 5 mg/kg intraperitoneally or subcutaneously every 24 h for up to 3 days.

2.3 Tissue Harvest

1. Scissors and various sized tweezers.
2. Sterilized mortar and pestle.
3. 27½-G needle.
4. 1 ml syringes.
5. EDTA coated tubes.
6. PBS.
7. FACS buffer: PBS containing 0.5% BSA and 2 mM EDTA. Weigh 2.5 mg Bovine Serum Albumin and suspend in 500 ml PBS. Add 2 ml 0.5 M EDTA (Ethylenediaminetetraacetate). Filter sterilize and keep at 4 °C.
8. 50 ml conical centrifuge tubes.
9. 40-μm sterile cell strainers (BD, Franklin Lakes, NJ).
10. Table top centrifuge for 5, 15, 50 ml tubes.
11. Petri dishes.

2.4 Analysis of Engraftment and Hematopoietic Lineage Representation

1. 5 Laser flow cytometer.
2. Antibodies for flow cytometry, *see* Table 2.
3. RBC lysis buffer. Dilute 10× RBC lysis buffer 1:10 in sterile deionized water. Store at 4 °C.
4. 5 ml Falcon Round-Bottom Polystyrene Tubes with cell strainer cap.

3 Methods

3.1 Purification of Human CD34$^+$ Hematopoietic Stem and Progenitor Cells (*See* Note 5)

1. For isolation of FL CD34$^+$ HSCPC cut FL into small fragments with sterile razor blades, treat with 1% Collagenase D solution for 45 min at 37 °C and gently pipette up and down with serologic pipette to prepare cell suspension. Filter using 40-micron sterile cell strainers before next step.

Table 2
Flow cytometry antibody panels

Panel	Antibody name	Clone	Maker	Dilution
	Human BD Fc block		BD	1:100
HSPC panel				
	muCD45 APC/Cy7	30-F11	Biolegend	1:300
	muTer119 APC/Cy7	Ter-119	Biolegend	1:300
	hCD45 BV510	HI30	Biolegend	1:100
	huCD123 PE/Dazzle594	6H6	Biolegend	1:100
	huCD38 BV421	HIT2	Biolegend	1:100
	huCD34 PE	561	Biolegend	1:100
	huCD90 PerCP/Cy5.5	5E10	Biolegend	1:100
	huCD135 APC	BV10A4H2	Biolegend	1:100
	huCD10 PE/Cy7	HI10a	Biolegend	1:200
	huCD45RA AF488	HI100	Biolegend	1:300
Lymphoid cell panel				
	muCD45 APCcy7	30-F11	Biolegend	1:300
	muTer119 APCcy7	Ter119	Biolegend	1:300
	hCD45 BV510 amCyan	HI30	Biolegend	1:100
	huCD33 APC	WM53	Biolegend	1:100
	huCD34 PE	561	Biolegend	1:100
	huCD3 FITC	OKT3	Biolegend	1:200
	huCD19 PeCy7	HIB19	Biolegend	1:200
	huCD335 NKp46 BV421	9E2	Biolegend	1:100
	huCD56 Percpcy5.5	MEM-188	Biolegend	1:100
	huCD16 PE/Dazzle594 PE TxRed	3G8	Biolegend	1:100
Monocyte panel				
	muCD45 APCcy7	30-F11	Biolegend	1:300
	muTer119 APC cy7	Ter119	Biolegend	1:300
	hCD45 BV510 amCyan	HI30	Biolegend	1:100
	huCD64 FITC	10.1	Biolegend	1:100
	huCD11b APC	ICRF44	Biolegend	1:100
	huCD10 PeCy7	HI10a	Biolegend	1:200

(continued)

Table 2
(continued)

Panel	Antibody name	Clone	Maker	Dilution
	huCD15 PE	W6D3	Biolegend	1:100
	huCD14 PerCP Cy5.5	63D3	Biolegend	1:100
	huCD16 PE/Dazzle594 PE TxRed	3G8	Biolegend	1:100
	huCD13 PB	WM15	Biolegend	1:100

2. Dilute fetal liver suspension from step 1, G-CSF mobilized PBMCs, BM or UCB 1:1 with PBS (*see* **Note 6**). Pipet up to 25 ml Ficoll-Paque into 50 ml Falcon tube with the final ratio of 1:1 Ficoll-Paque to sample. Slowly layer sample on top of Ficoll using a serological pipette; avoid mixing Ficoll and sample.
3. Centrifuge in a tabletop centrifuge at 800 × *g* for 20 min at room temperature with brakes OFF (*see* **Note 7**).
4. Aspirate the upper layer, which contains plasma and platelets and discard unless desired for other purposes. Do not disturb the mononuclear cell (MNC) layer at the sample/Ficoll interface.
5. Using a sterile 10 ml pipette, collect MNCs from the interface minimizing the aspiration of Ficoll and transfer the cells into a new sterile 50 ml conical tube. Interface cells from a maximum of two 50 ml tubes can be combined into one wash tube.
6. Add 1× PBS to bring the volume up to 50 ml. Gently resuspend cells with a serological pipette to wash cells.
7. Centrifuge at 400 × *g* for 10 min at 4 °C with brakes ON. Discard the supernatant.
8. Repeat **steps 6** and **7** for a second wash.
9. If the cell pellet is contaminated with a significant number of RBC add a RBC lysis step (**step 10**), otherwise proceed to **step 11**.
10. RBC lysis: Resuspend cell pellet in 5 ml of 1× RBC lysis buffer and incubate for a maximum of 5 min at RT. Repeat **steps 6** and **7** to wash the cells and proceed to **step 11**.
11. Add 1× PBS to bring the volume up to 45 ml. Gently pipette up and down to mix.
12. Count MNCs using a Hemocytometer. Exclude dead cells by trypan blue exclusion.

13. If cells are to be cryopreserved, spin cells at 400 × *g* and discard supernatant. Resuspend cells at 1–5 × 10^7/ml in 1 ml cold freezing medium (90% FBS, 10% DMSO) and place in freezing container in −80 °C freezer overnight and transfer to liquid nitrogen the following day.
14. If cells are to be used immediately, proceed to next steps for $CD34^+$ enrichment with the CD34-Microbead-Kit according to vendor's protocol.

3.2 Xeno-transplantation (See Note 8)

3.2.1 Intrahepatic Injection into Newborn Mice (1–3 Days Old)

The rationale for intrahepatic injection into newborn mice is several-fold. Liver and bone marrow contribute to perinatal hematopoiesis, a time of significant expansion of the hemato-lymphoid system. HSCs and colony-forming progenitor cells (CFCs) migrate from liver to BM in the perinatal period indicative of a receptive stem cell niche [22]. Technically, the newborn liver is readily accessible via transcutaneous injection either via a suprasternal or a direct transcutaneous route requiring minimal time and with high survival rates. Xenotransplantation during the neonatal stage (ideally D2–3) in addition takes advantage of the first 4–6 weeks of life to allow grafts to take and expand, significantly reducing the number of days required to maintain animals and mouse costs. A disadvantage lies in the need to plan experiments ahead and time delivery with desired experiments. Description of intrahepatic injection procedure is shown in Fig. 1.

1. Gently remove newborn mice (1–3 days old) from cage and place in 10 cm^2 petri dish covered with sterile gauze. Sublethally irradiate newborn mice (X-ray irradiation with X-RAD 320 irradiator, PXi) twice 4 h apart at 150 cGy (*see* **Note 9**). Place newborns back in cage in-between irradiation doses.
2. Resuspend freshly selected or thawed $CD34^+$ cells at desired concentration in sterile PBS in an Eppendorf tube. Approximately 20 min prior to injection add OKT3 antibody at 5 μg/100 μl to cell suspension and maintain cells on ice. Maximum injectable cell suspension volume/newborn mouse is 20 μl.
3. After the second irradiation place cage in laminar airflow hood. Remove pups from cage and place on clean tissue paper.
4. Draw 20 μl of the human $CD34^+$ cell suspension (prepared beforehand) into the Hamilton syringe with 22-G Hamilton needle, maximum volume is 20 μl/pup.
5. Hold newborn mouse between thumb and index finger. Insert the needle with bevel up subcutaneously above the sternal notch and advance *under* the skin and *above* the sternum toward the left upper quadrant of the abdomen where the liver is visible. Advance the needle approximately 3 mm beyond the

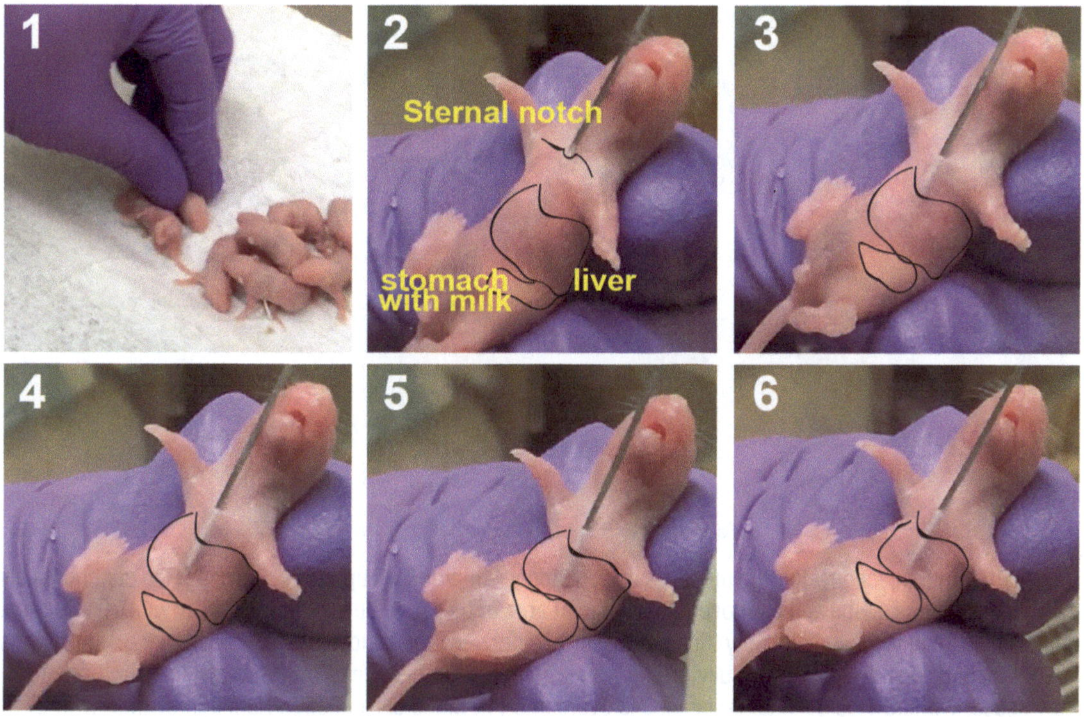

Fig. 1 For injection of human CD34+ cells, newborn mice should be held between thumb and index finger (1, 2). Insert the needle with bevel up subcutaneously above the sternal notch (2) and advance ***under*** the skin and ***above*** the sternum toward the left upper quadrant of the abdomen (3) where the liver is visible. Advance the needle approximately 3 mm beyond the rib cage (4) and insert the needle tip until the entire bevel is covered by liver tissue (5) and gently discharge content of the needle (6). Withdraw the needle, gently wipe blood off pup and place the pup gently in the cage with its mother. Aspirate 70% ethanol with the syringe and discard and rinse syringe and needle several times with sterile PBS. Proceed with the next pup. When all pups have been injected, assure that mom is actively gathering pups and feeding. Monitor pups 1–2 h after injection and daily thereafter

rib cage, and insert the needle tip until the entire bevel is covered by liver tissue and gently discharge content of the needle. Withdraw the needle, gently wipe blood off pup, and place the pup in the cage with its mother.

6. Aspirate 70% ethanol (in Eppendorf) with the syringe, and discard several times. Then rinse syringe and needle several times with sterile PBS. Proceed with the next pup.
7. Provide postoperative care as described below.

3.2.2 Intrafemoral Injection in Adult Mice (Minimum Age 6 Weeks, Ideal 8–10 Weeks)

If strict timing is needed the intra-venous route for larger cell numbers and highly engraftable cells and the intra-femoral route for rarer and more difficult to engraft cell sources should be chosen. Intrafemoral engraftment carries the advantage of directly "depositing" cells into the bone marrow niche. Disadvantages lie in the somewhat more difficult surgical procedure, the possible injury to

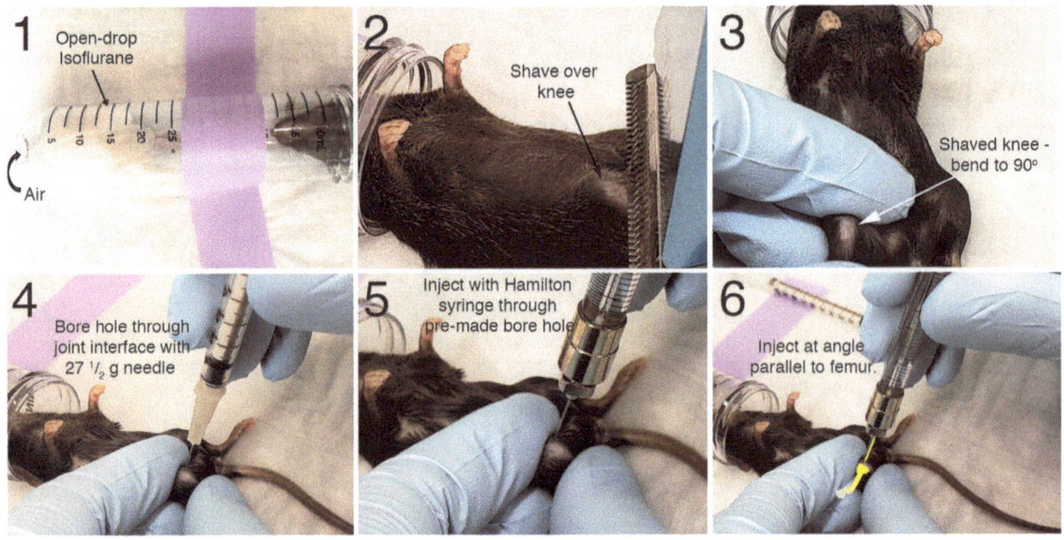

Fig. 2 For femur injection, after the mouse is asleep, apply isoflurane via open-drop devise (1) to continuuously maintain anesthesia. Shave the fur over the knee joint area of the leg to be injected (2), spray with 70% ethanol and flex the knee to 90° and identify the femoral surface between the condyles (3). By twisting the 1 ml syringe with attached 27½-G needle carefully create a hole through the distal end of the femur into the bone marrow space; withdraw as soon as resistance is overcome (4). Make sure to align Hamilton syringe with 27-G injection needle along the femur long axis (5, 6), and insert the needle tip through the tunnel into the top of the bone marrow cavity and inject (6)

the mouse bone, and the need for mice of at least 6–10 weeks of age (femur or tibial bone have to have sufficient diameter to accommodate the injection needle). It is particularly useful for small cell numbers and single cell assays. It is also described in detail elsewhere [23]. Description of intrafemoral injection procedure is shown in Fig. 2.

1. Cells should be prepared as for intrahepatic injection under point 2, but with a maximum injectable volume of 5 μl for each mouse.
2. Sublethally irradiate adult mice (6–8 weeks or older) twice, 4 h apart at 250 cGy (X-ray irradiation with X-RAD 320 irradiator, PXi) in a plexi-glass irradiator pie.
3. After the second irradiation, transfer cage to laminar airflow hood equipped with inhalational isoflurane device.
4. Place the mouse into the inhalational anesthesia chamber supplying 1% (v/v) inhaled Isoflurane in 1 L/min of oxygen. Once the mouse is asleep (check via paw pinch) apply isoflurane via open-drop method to maintain anesthesia. For the latter place a gauze soaked with 0.5 ml isoflurane/propylene glycol (30/70 v/v) mix in 50 ml conical tube that has been

cut open at the tip. Place the mouse's nose into the opening of the tube.

5. Apply heat via heating lamp to maintain body temperature.
6. Draw 20 µl of the human $CD34^{+}$ cell suspension (prepared beforehand) into the Hamilton syringe with 27-G Hamilton needle. Maximum volume is 5 µl.
7. With the razor, shave the fur over the knee joint area of the leg to be injected.
8. Disinfect the area with 70% ethanol, and change to sterile gloves.
9. Flex the knee to 90°, and identify the femoral surface between the condyles.
10. With a 1 ml syringe with attached 27.5-G needle carefully create an intra-femoral tunnel into the bone marrow cavity by gently twisting the needle. As soon as no resistance is felt, stop and remove the 27.5-G needle.
11. Insert the 27-G Hamilton needle tip through the tunnel into the top of the bone marrow cavity by aligning the Hamilton syringe needle along the femur long axis. Change direction carefully if resistance is felt during the injection. After injection, hold the syringe for 5–10 s, then pull out the needle.
12. Provide postoperative care as described below.

3.2.3 Postoperative Care

1. For intrahepatic injection of newborns, gently place pups in cage with their mother. Assure that the mother gathers and nurses pups. Check presence of milk filled stomach in pups and adequate growth.
2. For intra-femoral injection of adults, remove the mouse from the inhalational anesthesia mouthpiece or open-drop tube. Apply Carprofen s.c. after the procedure for post-procedure analgesia. Place the mouse in a cage under a heat lamp for approximately 20 min until recovered from anesthesia and fully mobilized. In the next 24 h monitor the mouse for discomfort or infection. Assure normal grooming and feeding. Repeat the Carprofen dose every 24 h as needed during the first 3 post-procedural days.

3.3 Analysis of Engraftment and Hematopoietic Lineage Representation

After xenotransplantation mice should be monitored regularly as per local regulations. Engraftment levels can be assessed at different time points by performing interim analysis, which includes peripheral blood analysis (cell counts (CBC) and immunophenotyping) as well as bone marrow aspiration. Final analysis can be completed by quantitative procedures and histologic bone marrow analysis.

We here provide our basic analysis protocol for human hematopoiesis reconstituted mice including flow cytometry antibody panels for the evaluation of peripheral blood, bone marrow, and other hematologic and non-hematologic organs (Table 2). Peripheral blood from the mouse can be obtained via several routes, including from the retro-orbital plexus, the facial vein or mandibular veins, and others. Institutional guidelines and training should be followed. A detailed protocol and video for bone marrow aspiration is published elsewhere [24]. For terminal analysis the mouse should be anesthetized and euthanized according to institutional guidelines. Peripheral blood, bone marrow, spleen, liver, and other organs should be harvested depending on desired experimental parameters. For some studies, such as of the hematopoietic stem cell or to expand disease models for therapeutic purposes secondary transplantation may be necessary. The entire donor mouse marrow with mixed human and mouse cells or enriched human cell fractions may be transplanted into secondary recipients are performed for primary recipients.

3.3.1 Peripheral Blood

1. Collect 50–500 μl of whole blood into EDTA coated tubes.
2. Aliquot each blood sample into one 1.5 ml microcentrifuge tubes with 100 μl FACS buffer for each desired panel. Save unstained cells as negative control.
3. Save the remaining sample for other potential assays as needed (e.g. CBC, gDNA, etc.).
4. Apply human FC block at 1:100 and incubate for 15 min on ice.
5. Stain each sample with the desired antibody cocktail at final concentrations as indicated. Examples are given in Table 2. Stain cells for 20–30 min on ice.
6. For the first wash step apply to each sample 1 ml 1× RBC lysis buffer and incubate at RT for 5 min. Spin at 500 × *g*, discard supernatant, and wash two more times with 1 ml FACS buffer.
7. Resuspend cells in 200 μl of FACS buffer, and apply cells to 5 ml Round-Bottom Tubes with Cell Strainer Cap.
8. Set up the acquisition panel on a flow-cytometer equipped with the appropriate lasers. For all multi-color flow cytometry experiments it is essential to include compensation controls and "Fluorescence Minus One" (FMO) controls to correctly identify gating boundaries. We acquire our data with FACSDiva on a LSR Fortessa (BD Biosciences) equipped with five lasers. Acquire at least the minimum number of events to result in sufficient events in the gate of interest and analyze data with available analysis software. We acquire at least 20,000 events

from biological replicates and analyze our data using FlowJo V10 software (FlowJo, LLC).

3.3.2 Tissues

1. Harvest bone marrow (from one femur, tibias and as needed from spine), spleen, liver, and other desired organs. Generate single cell suspensions.
   (a) For flushed bone marrow (femur, tibia) gently disrupt with a 1 ml syringe with a 27-G needle.
   (b) For crushed bone marrow (femur, tibia, pelvic bones, spine) filter cells through a 40 μm cell strainer.
   (c) For spleen, liver, and other tissues cut these into small pieces with razor blades, disrupt small fragments by gentle pipetting with a P1000 in FACS buffer and filter through a 40 μm cell strainer.
2. Save one femur for histology if desired.
3. Count cells and resuspend the necessary cell number at the desired concentration. In general ~$0.5–1 \times 10^6$ cells should be stained for each assay. This number can be adjusted based on the expected frequency of the population of interest.
4. Add FC block to the cell suspensions (only block the cells to be stained), and incubate on ice for 15 min.
5. Prepare antibody panel solutions at 2× concentration for a final volume of 100 μl/~1×10^6 cells for all tissues to be stained.
6. Prepare one Eppendorf tube for each sample and antibody panel.
7. Pipette 50 μl of 2× antibody cocktail into the respective Eppendorf tube.
8. Pipette 50 μl of each cell suspension into the Eppendorf tube with the desired antibody cocktail.
9. Stain cells for 30 min on ice.
10. Proceed with **steps 6–8** as for peripheral blood (Subheading 3.3.1).

4 Notes

1. The outcomes of hematopoietic stem cell transplants (HSCT) have established a clear relationship between the sources of the grafts and the number of huCD34$^+$ HSPC that can be obtained, their engraftment properties, and their differential homing. Bone marrow (BM), mobilized peripheral blood, umbilical cord blood, and fetal liver are among the preferred resources for CD34$^+$ HSPCs. The number of human CD34$^+$

cells obtained varies with each product. Fetal liver between 14 and 20 weeks gestation is richest for huCD34+ while UCB is a more accessible source. Efficiency of engraftment is dependent upon the proliferative capacity of the hematopoietic stem cell and decreases with age in the order of FL > UCB > BM ≥ G-CSF mobilized PBMCs [25, 26]. Grafts derived from FL or UCB CD34+ cells are likely to be more immune tolerant than from BM and PBMCs. Use of BM or G-CSF mobilized PBMCs enables the investigator to select donors with specific genetic polymorphisms or affected by diseases of interest and thus to generate disease-specific patient derived xenografts (PDXs). The quality of reconstitution varies based on donor age, disease, and other clinical factors. Other human tissues such as liver, bone, and thymus can be co-transplanted into the recipient hosts, as in the bone-liver-thymus (BLT) model to provide a human microenvironment for proper development and education of human cells in the murine host [27, 28].

2. Do not add individual components directly to cells as dilution of DMSO releases heat which can damage cells.
3. We here detail experiments using MISTRG mice. $MIS^{h/h}TRG$ mice in the $Rag2^{-/-}$ $Il2rg^{-/-}$ 129xBalb/c genetic background with homozygous knockin replacement of the endogenous murine *Csf1*, *Il3*, *Csf2*, *Tpo*, and *Sirpa* genes with their human counterparts are bred to $MIS^{m/m}TRG$ mice (that express murine *Sirpa*) to generate human cytokine homozygous and hSIRPA heterozygous $MIS^{h/m}TRG$ mice (labeled MISTRG throughout the manuscript) [10, 29].
4. Immunodeficient mice are particularly vulnerable to infections. Mice should be maintained in specific pathogen-free (SPF) rooms with restricted access and maximum barrier precautions. Investigators may choose to maintain xenografted immunodeficient mice on specific antibiotics according to their facilities' rules and regulations and veterinary care recommendations.
5. All cell processing occurs at RT unless otherwise mentioned. All processing should occur under sterile technique in a tissue culture dedicated laminar air flow hood. All work with human samples is considered BL2.
6. All source tissues for CD34+ HSPC should be processed as soon after harvest as possible to maximize cell viability. We recommend a maximum delay of up to 24 h.
7. Brakes are kept off to assure deceleration does not disrupt the density gradient.
8. All animal work and procedures must be approved by the local Institutional Animal Care and Use Committee.

9. Irradiation dose should be adjusted based on tolerance of the mouse strain and desired engraftment levels. Provided doses are for MISTRG newborn mice.

Acknowledgements

This study was supported by the Edward P. Evans and the Frederick Deluca Foundations (to S.H.). Y.S. is supported by the Young Scientists Fund of the National Natural Science Foundation of China (Grant No. 81800122). This study was supported by the Animal Modeling Core of the Yale Cooperative Center of Excellence in Hematology (NIDDK U54DK106857).

References

1. Ishikawa F, Yasukawa M, Lyons B, Yoshida S, Miyamoto T, Yoshimoto G, Watanabe T, Akashi K, Shultz LD, Harada M (2005) Development of functional human blood and immune systems in NOD/SCID/IL2 receptor {gamma} chain(null) mice. Blood 106(5):1565–1573. https://doi.org/10.1182/blood-2005-02-0516
2. Shultz LD, Lyons BL, Burzenski LM, Gott B, Chen X, Chaleff S, Kotb M, Gillies SD, King M, Mangada J, Greiner DL, Handgretinger R (2005) Human lymphoid and myeloid cell development in NOD/LtSz-scid IL2R gamma null mice engrafted with mobilized human hemopoietic stem cells. J Immunol 174(10):6477–6489
3. Wunderlich M, Chou FS, Link KA, Mizukawa B, Perry RL, Carroll M, Mulloy JC (2010) AML xenograft efficiency is significantly improved in NOD/SCID-IL2RG mice constitutively expressing human SCF, GM-CSF and IL-3. Leukemia 24(10):1785–1788. https://doi.org/10.1038/leu.2010.158
4. Nicolini FE, Cashman JD, Hogge DE, Humphries RK, Eaves CJ (2004) NOD/SCID mice engineered to express human IL-3, GM-CSF and Steel factor constitutively mobilize engrafted human progenitors and compromise human stem cell regeneration. Leukemia 18:341–347
5. Billerbeck E, Barry WT, Mu K, Dorner M, Rice CM, Ploss A (2011) Development of human CD4+FoxP3+ regulatory T cells in human stem cell factor-, granulocyte-macrophage colony-stimulating factor-, and interleukin-3-expressing NOD-SCID IL2Rgamma(null) humanized mice. Blood 117:3076–3086
6. Cosgun KN, Rahmig S, Mende N, Reinke S, Hauber I, Schafer C, Petzold A, Weisbach H, Heidkamp G, Purbojo A, Cesnjevar R, Platz A, Bornhauser M, Schmitz M, Dudziak D, Hauber J, Kirberg J, Waskow C (2014) Kit regulates HSC engraftment across the human-mouse species barrier. Cell Stem Cell 15(2):227–238. https://doi.org/10.1016/j.stem.2014.06.001
7. Rahmig S, Kronstein-Wiedemann R, Fohgrub J, Kronstein N, Nevmerzhitskaya A, Bornhauser M, Gassmann M, Platz A, Ordemann R, Tonn T, Waskow C (2016) Improved human erythropoiesis and platelet formation in humanized NSGW41 mice. Stem Cell Reports 7(4):591–601. https://doi.org/10.1016/j.stemcr.2016.08.005
8. Strowig T, Rongvaux A, Rathinam C, Takizawa H, Borsotti C, Philbrick W, Eynon EE, Manz MG, Flavell RA (2011) Transgenic expression of human signal regulatory protein alpha in Rag2−/−{gamma}c−/− mice improves engraftment of human hematopoietic cells in humanized mice. Proc Natl Acad Sci U S A 108(32):13218–13223. https://doi.org/10.1073/pnas.1109769108
9. Rongvaux A, Takizawa H, Strowig T, Willinger T, Eynon EE, Flavell RA, Manz MG (2013) Human hemato-lymphoid system mice: current use and future potential for medicine. Annu Rev Immunol 31:635–674. https://doi.org/10.1146/annurev-immunol-032712-095921
10. Deng K, Pertea M, Rongvaux A, Wang L, Durand CM, Ghiaur G, Lai J, McHugh HL, Hao H, Zhang H, Margolick JB, Gurer C, Murphy AJ, Valenzuela DM, Yancopoulos

GD, Deeks SG, Strowig T, Kumar P, Siliciano JD, Salzberg SL, Flavell RA, Shan L, Siliciano RF (2015) Broad CTL response is required to clear latent HIV-1 due to dominance of escape mutations. Nature 517(7534):381–385. https://doi.org/10.1038/nature14053

11. Song Y, Rongvaux A, Taylor A, Jiang T, Tebaldi T, Balasubramanian K, Bagale A, Terzi YK, Gbyli R, Wang X, Zhao J, Podoltsev N, Xu M, Neparidze N, Wong E, Torres R, Bruscia EM, Kluger Y, Manz MG, Flavell RA, Halene S (2019) A highly efficient and faithful MDS patient-derived xenotransplantation model for pre-clinical studies. Nat Commun 10(1):366
12. Takenaka K, Prasolava TK, Wang JC, Mortin-Toth SM, Khalouei S, Gan OI, Dick JE, Danska JS (2007) Polymorphism in Sirpa modulates engraftment of human hematopoietic stem cells. Nat Immunol 8(12):1313–1323. https://doi.org/10.1038/ni1527
13. Waskow C, Madan V, Bartels S, Costa C, Blasig R, Rodewald HR (2009) Hematopoietic stem cell transplantation without irradiation. Nat Methods 6(4):267–269. https://doi.org/10.1038/Nmeth.1309
14. Reinisch A, Thomas D, Corces MR, Zhang X, Gratzinger D, Hong WJ, Schallmoser K, Strunk D, Majeti R (2016) A humanized bone marrow ossicle xenotransplantation model enables improved engraftment of healthy and leukemic human hematopoietic cells. Nat Med 22(7):812–821. https://doi.org/10.1038/nm.4103
15. Rathinam C, Poueymirou WT, Rojas J, Murphy AJ, Valenzuela DM, Yancopoulos GD, Rongvaux A, Eynon EE, Manz MG, Flavell RA (2011) Efficient differentiation and function of human macrophages in humanized CSF-1 mice. Blood 118(11):3119–3128. https://doi.org/10.1182/blood-2010-12-326926
16. Rongvaux A, Willinger T, Takizawa H, Rathinam C, Auerbach W, Murphy AJ, Valenzuela DM, Yancopoulos GD, Eynon EE, Stevens S, Manz MG, Flavell RA (2011) Human thrombopoietin knockin mice efficiently support human hematopoiesis in vivo. Proc Natl Acad Sci U S A 108(6):2378–2383. https://doi.org/10.1073/pnas.1019524108
17. Willinger T, Rongvaux A, Takizawa H, Yancopoulos GD, Valenzuela DM, Murphy AJ, Auerbach W, Eynon EE, Stevens S, Manz MG, Flavell RA (2011) Human IL-3/GM-CSF knock-in mice support human alveolar macrophage development and human immune responses in the lung. Proc Natl Acad Sci U S A 108(6):2390–2395. https://doi.org/10.1073/pnas.1019682108
18. Saito Y, Ellegast JM, Rafiei A, Song Y, Kull D, Heikenwalder M, Rongvaux A, Halene S, Flavell RA, Manz MG (2016) Peripheral blood CD34+ cells efficiently engraft human cytokine knock-in mice. Blood 128(14):1829–1833. https://doi.org/10.1182/blood-2015-10-676452
19. Theocharides AP, Rongvaux A, Fritsch K, Flavell RA, Manz MG (2016) Humanized hemato-lymphoid system mice. Haematologica 101(1):5–19. https://doi.org/10.3324/haematol.2014.115212
20. Traggiai E, Chicha L, Mazzucchelli L, Bronz L, Piffaretti JC, Lanzavecchia A, Manz MG (2004) Development of a human adaptive immune system in cord blood cell-transplanted mice. Science 304(5667):104–107. https://doi.org/10.1126/science.1093933
21. Wunderlich M, Brooks RA, Panchal R, Rhyasen GW, Danet-Desnoyers G, Mulloy JC (2014) OKT3 prevents xenogeneic GVHD and allows reliable xenograft initiation from unfractionated human hematopoietic tissues. Blood 123(24):e134–e144. https://doi.org/10.1182/blood-2014-02-556340
22. Wolber FM, Leonard E, Michael S, Orschell-Traycoff CM, Yoder MC, Srour EF (2002) Roles of spleen and liver in development of the murine hematopoietic system. Exp Hematol 30(9):1010–1019
23. Zhan Y, Zhao Y (2008) Hematopoietic stem cell transplant in mice by intra-femoral injection. Methods Mol Biol 430:161–169. https://doi.org/10.1007/978-1-59745-182-6_11
24. Chung YR, Kim E, Abdel-Wahab O (2014) Femoral bone marrow aspiration in live mice. J Vis Exp (89):e51660. https://doi.org/10.3791/51660
25. Lepus CM, Gibson TF, Gerber SA, Kawikova I, Szczepanik M, Hossain J, Ablamunits V, Kirkiles-Smith N, Herold KC, Donis RO, Bothwell AL, Pober JS, Harding MJ (2009) Comparison of human fetal liver, umbilical cord blood, and adult blood hematopoietic stem cell engraftment in NOD-scid/gammac−/−, Balb/c-Rag1−/−gammac−/−, and C.B-17-scid/bg immunodeficient mice. Hum Immunol 70(10):790–802. https://doi.org/10.1016/j.humimm.2009.06.005
26. Holyoake TL, Nicolini FE, Eaves CJ (1999) Functional differences between transplantable human hematopoietic stem cells from fetal liver, cord blood, and adult marrow. Exp Hematol 27(9):1418–1427
27. Lan P, Tonomura N, Shimizu A, Wang S, Yang YG (2006) Reconstitution of a functional human immune system in immunodeficient mice through combined human fetal thymus/

liver and CD34+ cell transplantation. Blood 108(2):487–492. https://doi.org/10.1182/blood-2005-11-4388

28. Bonig H, Papayannopoulou T (2012) Mobilization of hematopoietic stem/progenitor cells: general principles and molecular mechanisms. Methods Mol Biol 904:1–14. https://doi.org/10.1007/978-1-61779-943-3_1

29. Rongvaux A, Willinger T, Martinek J, Strowig T, Gearty SV, Teichmann LL, Saito Y, Marches F, Halene S, Palucka AK, Manz MG, Flavell RA (2014) Development and function of human innate immune cells in a humanized mouse model. Nat Biotechnol 32(4):364–372. https://doi.org/10.1038/nbt.2858

Part IV

Clinical Design for Cell Therapy

Chapter 19

Monitoring Allogeneic CAR-T Cells Using Flow Cytometry

Agnieszka Jozwik, Alan Dunlop, Katy Sanchez, and Reuben Benjamin

Abstract

Chimeric antigen receptor T cell (CAR-T) therapies have now entered mainstream clinical practice with two approved autologous CAR-T products targeting CD19 and numerous other products in early and late phase clinical trials. This has led to a demand for highly sensitive, specific, and easily reproducible methods to monitor CAR-T cells in patients. Here we describe a flow cytometry based protocol for detection of allogeneic CAR-T cells and for monitoring their phenotype and numbers in blood and bone marrow of patients following CAR-T treatment.

Key words CAR-T cells, Allogeneic, Flow cytometry, Gene knockout, Immunotherapy, Cancer

1 Introduction

Recent years have seen the emergence of genetically engineered Chimeric Antigen Receptor T cells entering clinic with two autologous CAR-T cell products (made from patients own peripheral blood mononuclear cells), Tisagenlecleucel and Axicabtagene ciloleucel, getting FDA and EMA approval. These therapies are not conventional drugs, as instead of being biomolecules or chemical substances, they are actually live cells with a capacity to replicate and differentiate.

Current data confirms that CAR-T cell expansion and persistence correlates with clinical response and achieving remission in patients [1]. Therefore, establishing reliable methods of tracking CAR-T cell numbers is particularly important not only in the context of effectiveness of this treatment but also patient safety. This is because one of the most severe side effects of CAR-T therapy—Cytokine Release Syndrome (CRS) coincides with CAR-T cell expansion [2].

Generation of autologous CAR T-cells not only depends on availability of patient derived T cells but is also time-consuming and expensive. Therefore, numerous efforts have been concentrat-

Samuel G. Katz and Peter M. Rabinovich (eds.), *Cell Reprogramming for Immunotherapy: Methods and Protocols*, Methods in Molecular Biology, vol. 2097, https://doi.org/10.1007/978-1-0716-0203-4_19,

ing on developing universal, allogeneic CAR-T cells produced from healthy donors T cells [3]. These universal CAR-T cells share certain features relevant for their detection with the autologous products, like the chimeric antigen receptor itself, but at the same time some of them carry knockouts of genes that if expressed would potentially make them easily rejected or a cause of graft vs host disease (GVHD).

Numerous universal, allogeneic CAR-T cell products have a knockout of the T cell receptor alpha constant (TRAC) gene which causes disruption of the TCRαβ complex [4], leading to lack of its expression on the cell surface. This way allogeneic cells are disabled in terms of foreign antigen recognition via the T cell receptor (TCR), picking up antigen only via the CAR receptor, and so, do not pose a risk of GVHD as long as the product has been purified to a sufficient standard. Additionally, a number of groups have introduced knockouts of β2 microglobulin (β2M) [5] a protein that is an integral part of the MHC class I complex. This intervention is designed to prevent allogeneic CAR-T cells being recognized by the host's immune system, as HLA mismatch can lead to CAR-T cell rejection and lack of efficacy.

The basic CAR receptor in autologous as well as allogeneic products has its unique domains such as a binding moiety, i.e., single chain variable fragment, scFV, as well as features shared with native T cells like intracellular signaling CD3ζ or costimulatory domains, 41BB or CD28 which need to be taken into consideration when designing methods of detection.

Additional modifications, in the form of "safety switches," have been introduced into some of the CAR-T cell constructs due to the potential risks of insertional mutagenesis in cells transduced with retroviral or lentiviral vectors or uncontrolled CAR-T cell proliferation in response to target cells. These domains expressed within a CAR-T cell allow rapid elimination when exposed to an approved drug. One such example is a safety domain targetable by Rituximab, an anti-CD20 monoclonal antibody [6].

Other modifications include knocking out genes to render allogeneic CAR-T cells resistant to certain lymphodepleting drugs. Lymphodepletion pre CAR-T cell infusion is required to eliminate host T cells as well as to facilitate the release of homeostatic cytokines and may be even more important for the successful engraftment of HLA mismatched allogeneic CAR-T cells. One such modification involves knocking out the CD52 gene in the infused T cells, which allows the use of Alemtuzumab, an anti-CD52 antibody for lymphodepletion [7].

Detection of elements introduced into T cells, like the Chimeric Antigen Receptor as well as elements knocked out, like the TRAC gene, can be a part of strategies designed to monitor CAR-T cells.

One of the best established methods for tracking CAR-T cells ex vivo is qPCR, with numerous protocols published [8]. Amongst the different approaches, the one using primers against psi (Ψ) sequence (part of the lentiviral packaging signal) combined with primers against one of the housekeeping genes (GAPDH, albumin) would be a universal strategy for monitoring CAR-T cells where the receptor has been introduced via lentiviral transduction. Other molecular methods include next generation sequencing (NGS), especially ImmunoSEQ involving amplification of the hypervariable complementarity-determining region 3 (CDR3). Although these methods show high sensitivity, they do have some limitations. Vector copy number (VCN) generated by molecular methods gives a quantifiable measure of the number of CAR-T receptors per unit of DNA or per cell but does not say anything about the viability or phenotype of the CAR-T cells.

Furthermore, with current manufacturing processes the final allogeneic CAR-T cell product is not 100% pure for the CAR or for the knockout genes, with unmodified cells present in the infused population. This makes flow cytometry a particularly attractive and rapid method for monitoring numerous features of single cells including CAR-T receptor expression, presence or lack of other molecules, e.g., TCRαβ, CD52 expression as well as target tumor cells.

In this chapter we describe a flow cytometry based strategy for monitoring allogeneic CAR-T cells, where the CAR is derived from the scFv of an anti-human CD19 murine monoclonal antibody, and where the TRAC and CD52 genes have been knocked out.

We introduce one of the possible approaches and technical details to consider when designing and performing flow cytometric analysis intended to detect and quantify allogeneic CAR-T cells post infusion in patient blood and bone marrow samples.

2 Materials

1. Blood and bone marrow samples are collected in EDTA tubes. It is essential to get information about cellularity of blood and bone marrow (BM) samples based on white blood cell (WBC) measurement typically from a hematology analyzer (e.g., XP-300 Sysmex Haematology System). Samples may be kept at room temperature and stored at 2–8 °C overnight.
2. Prepare fresh PharmLyse Reagent (BD cat. no. 555899) by diluting it 1:10 with deionized water. Make sure PharmLyse is pre-warmed to room temperature before use.
3. Make sure a decontamination container with freshly prepared virkon disinfectant is available for all clinical waste disposal.
4. Pre-dilute lyophilized anti-mouse PE antibody (R-Phycoerythrin AffiniPure $F(ab')_2$ Fragment Goat Anti-

Mouse IgG, $F(ab')_2$ fragment specific—(Jacksons Immunoresearch cat. no. 115-115-164) with 1 ml deionized water and centrifuge if it does not look clear. Pre-diluted antibody is stable for 6 months from the point of resuspension.

5. Prepare mouse serum (Sigma I8765-10MG) as 5 μl single aliquots to be stored in −20 °C freezer.
6. Assure access to a clinical grade flow cytometer, with maintenance (e.g., regular cleaning, service etc.) and QC's (e.g., variability and validation of PMT settings (voltage and gain)) being done on regular basis.
7. Spherotech 8-peak Rainbow beads.

3 Methods

3.1 Cell Lysis

1. Pre-label the appropriate tube (*see* **step 4**) with patient's anonymized trial number, time point pre or post CAR-T cell infusion, and sample type, i.e., blood or BM.
2. Based on the WBC count of blood/BM, aliquot a volume equivalent to one million white blood cells (example: If sample WBC = $0.2 \times 10^3/\mu l$ then one needs $1 \times 10^6/0.2 \times 10^3 = 5000$ μl, i.e., lyse 5 ml of blood/BM) (*see* **Note 1**).
3. When working with BM make sure all clots of the sample are filtered out using a 100 μM cell strainer.
4. Depending on the volume of clinical material to be lysed, prepare an appropriate tube to allow the recommended ratio of blood/BM and PharmLyse (1:10). Using a Pasteur or serological pipette measure out an appropriate volume of PharmLyse and mix with clinical material by closing the tube and inverting it gently until evenly mixed (maximum five inversions).
5. Incubate blood/BM with PharmLyse for 10 min at room temperature.
6. Spin the tube at 570 × *g* for 5 min.
7. Discard the supernatant into a container with Virkon by single inversion of the tube or by aspiration using a Pasteur pipette being careful not to disrupt the cell pellet.
8. Break the cell pellet by vortexing the tube gently.
9. Using a serological pipette wash the cell pellet with 10 ml of Phosphate buffered saline (PBS) for larger volumes (1–10 ml) of lysed blood/BM and 3 ml for smaller volumes (<1 ml).
10. Spin the tube at 570 × *g* for 5 min.
11. Discard the supernatant into Virkon. Transfer the cells into a FACS tube if lysis was performed in a bigger volume Falcon and continue with staining procedure.

3.2 FACS Staining

1. Add 5 μl of Goat Anti-Mouse IgG $F(ab')_2$ fragment specific PE conjugated antibody to the residual volume of PBS (*see* **Note 2**).
2. Vortex and incubate for 15 min at RT in dark.
3. Using a Pasteur or serological pipette top up the tube with 3 ml of PBS and spin for 5 min at 570 × *g*.
4. Discard the supernatant and add 3 μl of mouse serum to residual volume of PBS (*see* **Note 3**).
5. Vortex and incubate for 5 min in the dark.
6. After 5 min continue with addition of remaining antibodies from the staining panel, vortex, and incubate in the dark for a further 20 min. Below we present an example of a staining panel (Table 1) that enables detection of two distinct features of allogeneic CAR-T cells (TRAC and CD52 knockout) (*see* **Note 4**).
7. Wash the cells with 3 ml of PBS and spin at 570 × *g* for 5 min.
8. Resuspend the cells in 300 μl PBS and add 1.5 μl 7AAD for the sample to be analyzed within the next half an hour.

3.3 Sample Acquisition

1. Samples are run on a validated, clinical grade Flow Cytometer (BD FACS Canto II is used in this protocol).
2. For acquisition purposes label the file with patient's anonymized trial number, time point (day pre/post infusion), and type of material being analyzed (blood/BM).

Table 1
Flow cytometry antibodies specification for the purpose of allogeneic CAR-T cell monitoring protocol

Antibody	Fluorochrome	Purpose	Clone	Vendor
7AAD	Readout PerCPCy5.5 channel	Gating for viable cells		BD
CD45	V500	Gating leukocytes	2D1	BD
CD4	V450	Gating host and CAR CD4+ T cells	RPA-T4	BD
CD8	PE Cy7	Gating host and CAR CD8+ T cells	RPA-T8	BD
TCRαβ	FITC	Monitoring CAR T cells (CAR+ TCRαβ−) as well as detecting host T cells (CAR− TCRαβ+)	WT31	BD
CD52	APC H7	Monitoring CD52 expression on CAR and host T cells as well as on leukemic cells	4C8	BD
CD19	APC	Monitoring target B-cells and leukemic cells	6D5	Biolegend

V Brilliant Violet, *FITC* fluorescein isothiocyanate, *PE* phycoerythrin, *PerCP* peridinin chlorophyll protein, *Cy* cyanin, *APC* allophycocyanin, *BD* Becton Dickinson (BD Biosciences, San Jose, CA), *BC* Beckman-Coulter (Fullerton, CA), *BD* Biolegend (San Diego, CA)

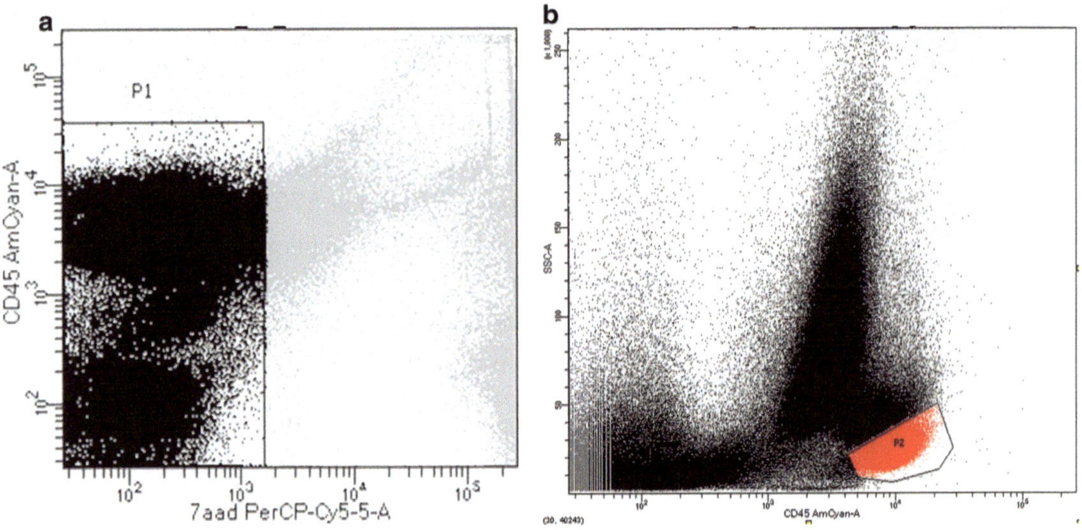

Fig. 1 As an initial approach all events are gated for viable cells (P1 parent gate) (**a**), with a lymphocyte gate (P2) set up based on CD45 expression against side scatter parameter (**b**). A minimum of 500,000 events should be recorded in the P2 leukocyte gate

3. Before running clinical samples for CAR-T cell detection, a series of optimization experiments is essential. Assay validation needs to be performed for each flow cytometry based CAR-T cell protocol variation to demonstrate its accuracy, specificity, reproducibility, and robustness over a designed range of measurements.
4. Setting up compensation controls is performed individually based on guidelines of certified Immunophenotyping laboratories (refer to Subheading 3.4).
5. The advised number of events to be acquired is 500,000 within the CD45+ WBC gate (Figs. 1a, b, and 2).
6. Recording will stop automatically if the pre-set number of events is achieved. In case of highly lymphopenic samples and low number of events, record as many events as possible and stop recording manually.
7. Export the FSC file to an appropriate destination to be accessed for the analysis.

3.4 Parameters Setup

Below we briefly describe an example of FACS Canto setup and provide a reference for compensation guidelines. To ensure accurate and reproducible results are generated a number of calibration and set up procedures are required to be run. These steps ensure that appropriate forward and side scatter (Fsc v Ssc) for each specimen is used, and by setting the PMT voltages on a quarterly basis based on target values for eight-peak rainbow beads we can be

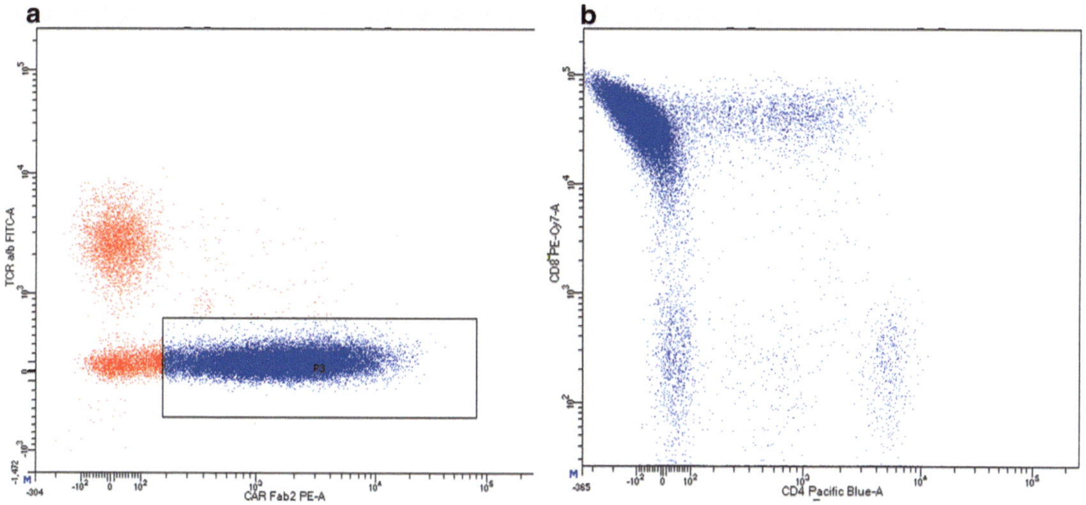

Fig. 2 P2 lymphocyte gate is a parent to subsequent gates. By plotting CAR Fab PE signal against TCRαβ the user can visualize CAR T cells with TRAC knockout (**a** P3 in blue). These can be further gated as CD4 and CD8 CAR T cells (**b** in blue)

confident that the results produced are reproducible. A customized compensation matrix is set up using generic labels for FITC, PE, PerCpCy5.5, APC, V450/421, and V500, the antigens tested on the APC Cy7/H7 and PeCy7 have label/fluorochrome specific compensations.

3.4.1 Setting Up Forward Scatter (FSC) and Side Scatter (SSC) Parameters

1. Pipette 50 μl of fresh blood into a FACS tube, and add 2 ml of freshly prepared PharmLyse and mix gently.
2. Incubate for 10 min at room temperature and then centrifuge at 570 × *g* for 5 min.
3. Discard the supernatant, and re-suspend the pellet in 2 ml of PBS.
4. Centrifuge at 570 × *g* for 5 min, discard the supernatant, and resuspend the cells by adding 250 μl of PBS and mixing.
5. Acquire and record 10,000 events in lymphocyte gate, then verify the mean FSC and SSC values for gated lymphocytes.
6. If required adjust FSC and SSC voltage to achieve mean target values for the gated lymphocyte population: FSC: 75,000/SSC: 22,000.

3.4.2 Setting Up PMT Voltages for Target Fluorescence Channels

1. Turn compensation off while setting fluorescence PMT voltages for Target MFI setting by unclicking the "Enable Compensation" option in the inspector window.
2. Import new experiment: "eight-peak Rainbow beads Target channels".

3. Prepare freshly a solution containing one drop of Spherotech eight-peak Rainbow beads in 1 ml of dH_2O.
4. Gently vortex to mix the beads in solution.
5. Measure and acquire, without recording, the eight-peak Rainbow beads solution at "LOW" flow rate.
6. Gate singlet beads in the FSC versus SSC bivariate dot plot according to the manufacturer's instructions.
7. Gate the eighth or seventh PEAK (the brightest peak or the next one downwards, respectively)—Population P2 in the FITC versus PE bivariate dot plot (*see* **Note 5**).
8. Continue the acquisition of the eight-peaks Rainbow bead suspension and adjust/fine-tune PMT voltages in all fluorescence channels to reach target MFI values according to the values supplied with the beads. For example, when setting up the BV421 channel a target MFI of 54,000 ± 15% (45,900–62,100) should be used.
9. Record information for about 2000 events and check the MFI for the eight or seventh Peak beads (*see* **Note 5**).
10. Correct the PMT voltages if necessary.
11. Repeat record/correct until you reach the target MFI value desired for all fluorescence channels.
12. Once Target MFI values for the eighth or seventh Peak are reached, record the Target MFI achieved and the corresponding final PMT values on the printout from the previous FSC/SSC setup.
13. For your records, print screen with Worksheet "Target MFI" and instrument settings, save as .jpg picture. This should be stored in D: EXPORT: TARGET MFI.
14. Compare PMT values to your notes, double check that "Target MFI" values and PMT values are correct (*see* **Note 6**).
15. Our clinical laboratory follows guidelines on fluorescence compensation from the EuroFlow consortium [9]; however, many other protocols are available.

3.5 Results Analysis and Interpretation

For consistency and robustness of results, we recommend using established flow cytometry software like Diva or Infinicyt. The protocol shown here is based on results and validation procedures performed using Infinicyt. When using Infinicyt software, before analyzing the files, a predefined profile with population tree and analysis strategy should ideally be pre-set and used consistently for every analysis. Infinicyt software allows auto-population of results into a customized report specifying abundance of all populations

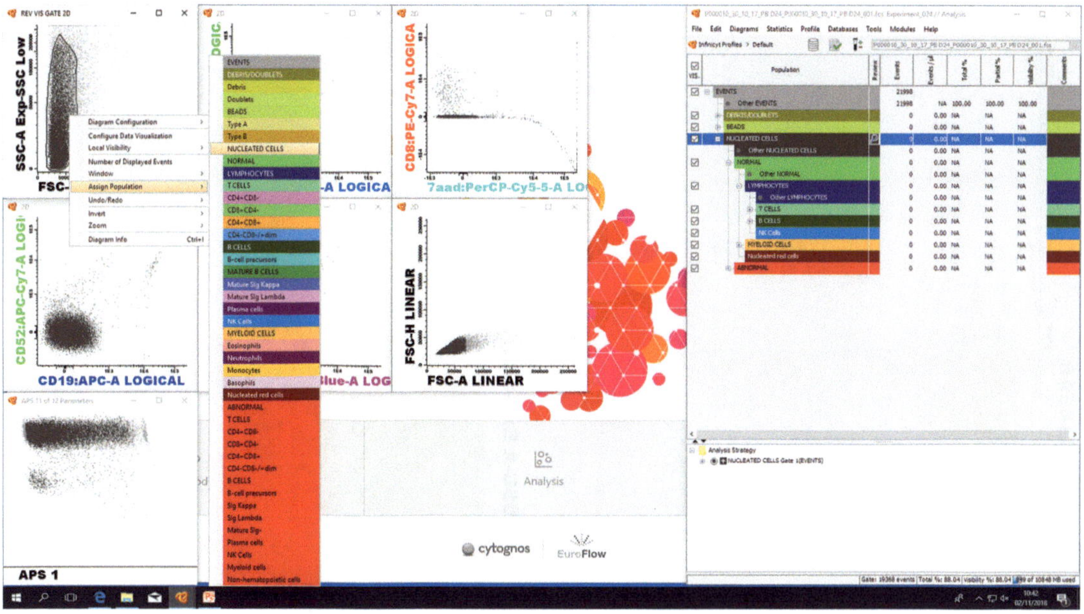

Fig. 3 Default template view of the analysis Infinicyt window when opening an FCS file

investigated as percentage or absolute number (Fig. 9). Generating a final report is the last step of the analysis. Files can be saved as analyzed ".cyt" files for further overview by a different operator. Below we describe some guidelines helpful for starting the analysis and creating tools for the analysis of multiple files together with a report of the final results.

1. Open software by double clicking on Infinicyt icon.
2. Select the "Manual Analysis" option.
3. Choose your FCS file from the destination where it has been exported, to be analyzed from a file selection window. When you open the file, you will see an example of a template with a set of default diagrams and a population tree (Fig. 3) (*see* **Note 7**).
4. To select a population to be included in analysis strategy, draw a region around the cells of interest. The region will be initially shown in black.
5. Go to the Population Tree and click on a label for the selected population so the selected events gate changes color to the one selected. The events included in the region delimited by the line you drew will take the same color as the label.
6. Create gates to discriminate for *debris* and give an appropriate name to the gate (Fig. 4a, b).

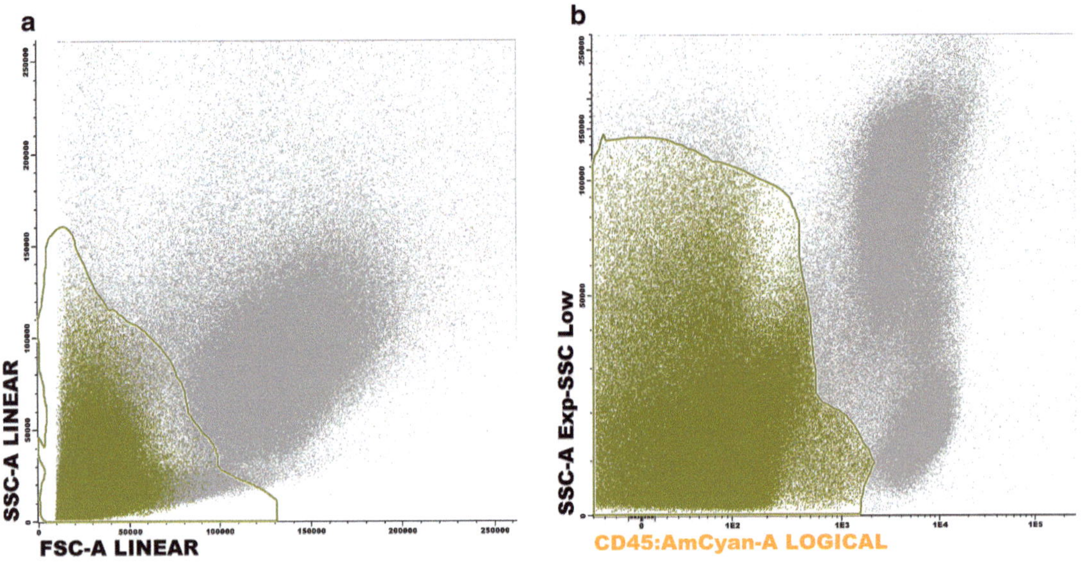

Fig. 4 Cell debris are gated out by highlighting a population of cells with low SSC and FSC (green) (**a**) as well as those of low SSC and negative for CD45 expression (**b**)

7. From the gate excluding the debris, create three gates defining *nucleated cells* (Fig. 5a–c).
8. Use the nucleated cells gate as a parent to set up the lymphocyte gate (Fig. 6).
9. Use the lymphocyte gate to select three major populations of cells: CD4+ (Fig7A), CD8+ (Fig. 7b), and CD4− CD8− (Fig. 7c).
10. Use the pre-set CD4, CD8, and double negative gates to select CAR-T cell population (Fig. 8a–c).
11. Once you have defined the diagrams, population tree, report, and the other internal configurations, go to "Profile" in the Manual Analysis Window Menu Bar, and choose Save Profile. Automatically the software will ask you where to save it and the extension applied will be ".inp" (Infinicyt™ profile).
12. Enter the numbers into a customized report. Total number of CAR-T cells can be represented as % of TNC (total nucleated cells) after adding up all three populations of CAR-T+ cells (CD8+, CD4+, and CD4−CD8−). Additionally, absolute number of CAR-T cells can be included in the report after providing WBC (while blood count) in the Statistics part of the software. A customized report is the final stage of generating results and reporting on abundance of the chosen populations (Fig. 9).

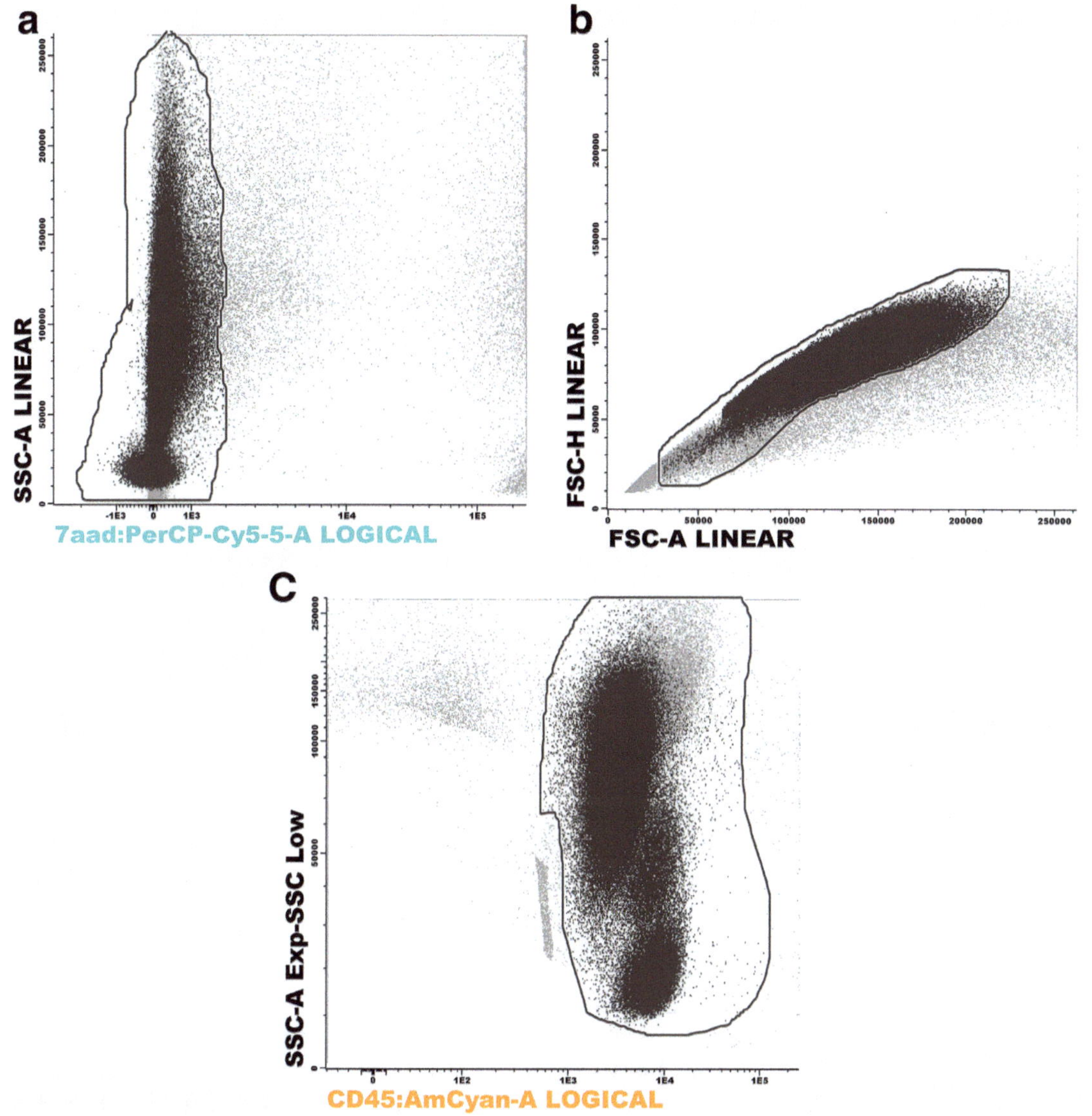

Fig. 5 Nucleated cells are selected based on a negative signal for the 7AAD cell death marker against SSC (**a**), singlets (**b**), and CD45 positive leukocytes (**c**). *LOGICAL* logarithmic

4 Notes

1. Patients can remain leukopenic post lymphodepletion and CAR-T cell infusion; therefore, it is essential to aim for one million white blood cells rather than a certain volume of blood/BM. This is feasible for adults but may not always be possible for pediatric patients. The sensitivity of the flow

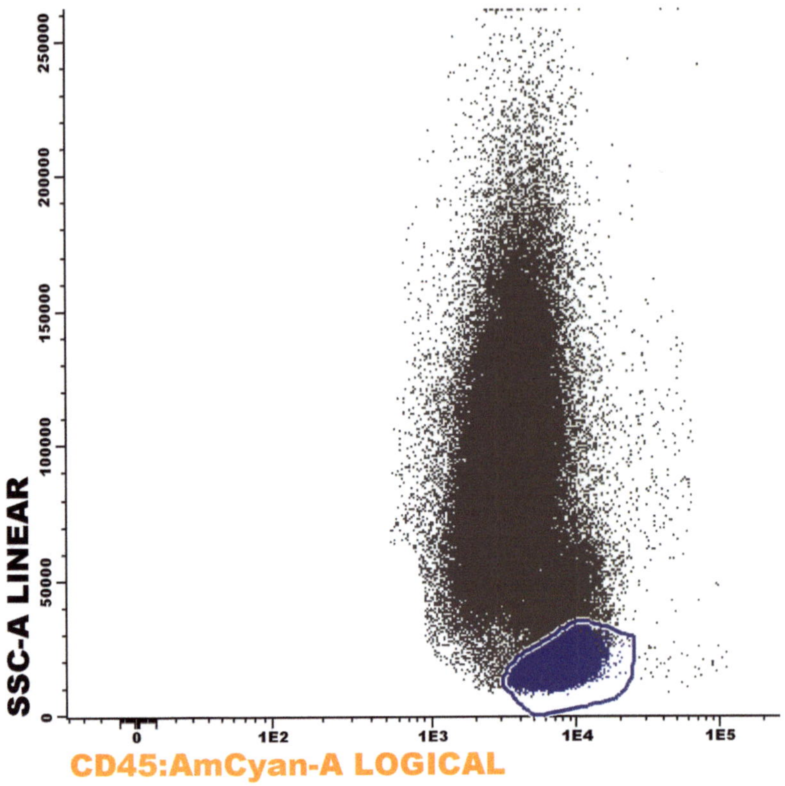

Fig. 6 Lymphocyte gate is selected on the diagram by plotting CD45 expression against SSC from the leukocytic gate. Use high magnification (diagram can be enlarged by dragging one of the corners to desired size) to clearly visualize the lymphocyte gate

cytometry method used for monitoring CAR-T cell numbers in this protocol has been established to be 0.1%, with background non-specific staining with anti-mouse antibody estimated to be 0.01–0.04%. Individual users carrying out this protocol are strongly recommended to establish themselves the limit of detection for their assay.

2. This method of CAR-T cell detection is applicable only for CAR constructs carrying a murine component as fully human CARs will not be detected by the anti-mouse antibody. Ideally custom made, highly specific, idiotypic anti-CAR T cell antibodies would be used for CAR detection although the method described here could potentially be used for CAR constructs with a murine domain. It is down to individual users to establish which anti murine antibody is able to detect the product they are investigating.

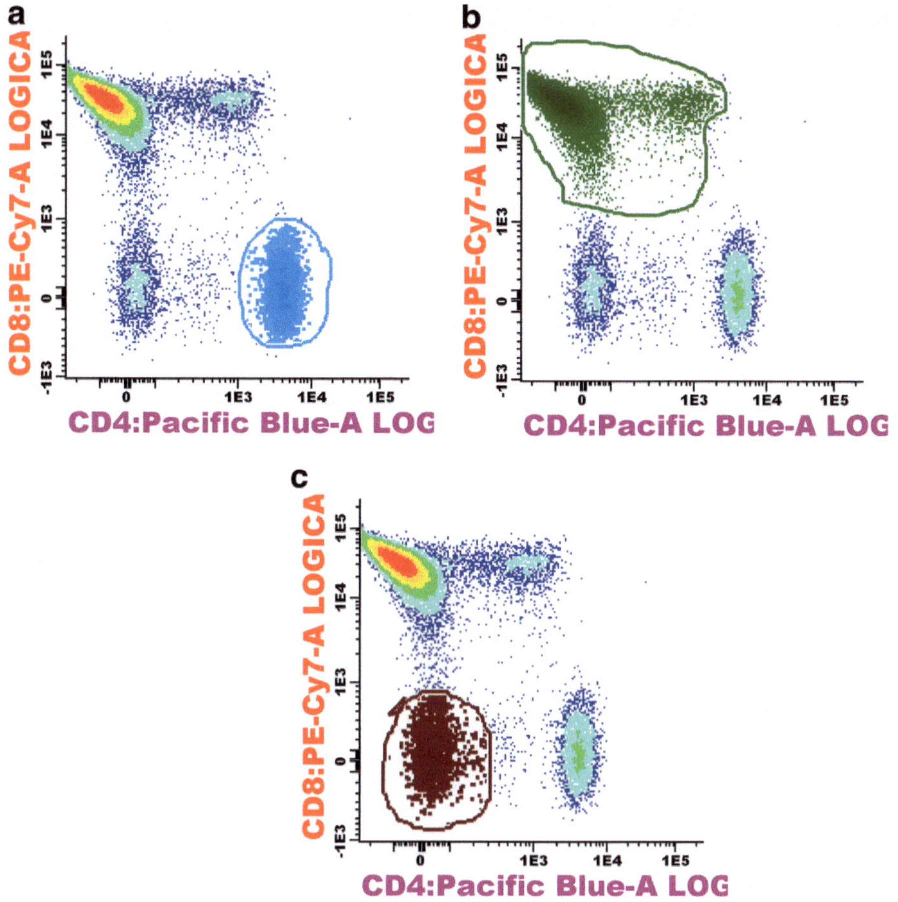

Fig. 7 Leukocyte gate is used for selecting CD4+ (**a**) and CD8+ (**b**) T lymphocytes as well as a double negative population (**c**)

3. This method uses an anti-mouse antibody for detection of the murine part of the CAR construct. The blocking step involving addition of mouse serum prevents high background and binding of mouse monoclonal antibodies to any remaining anti-mouse antibody.
4. The FACS staining panel should be designed based on the features of allogeneic CAR-T cells (such as the introduced knockouts). Antibodies from the panel should be titrated according to reliable protocols previously described [10].
5. Choosing the eighth or the seventh peak as the reference peak to set PMT voltages will depend on the fluorescence inten-

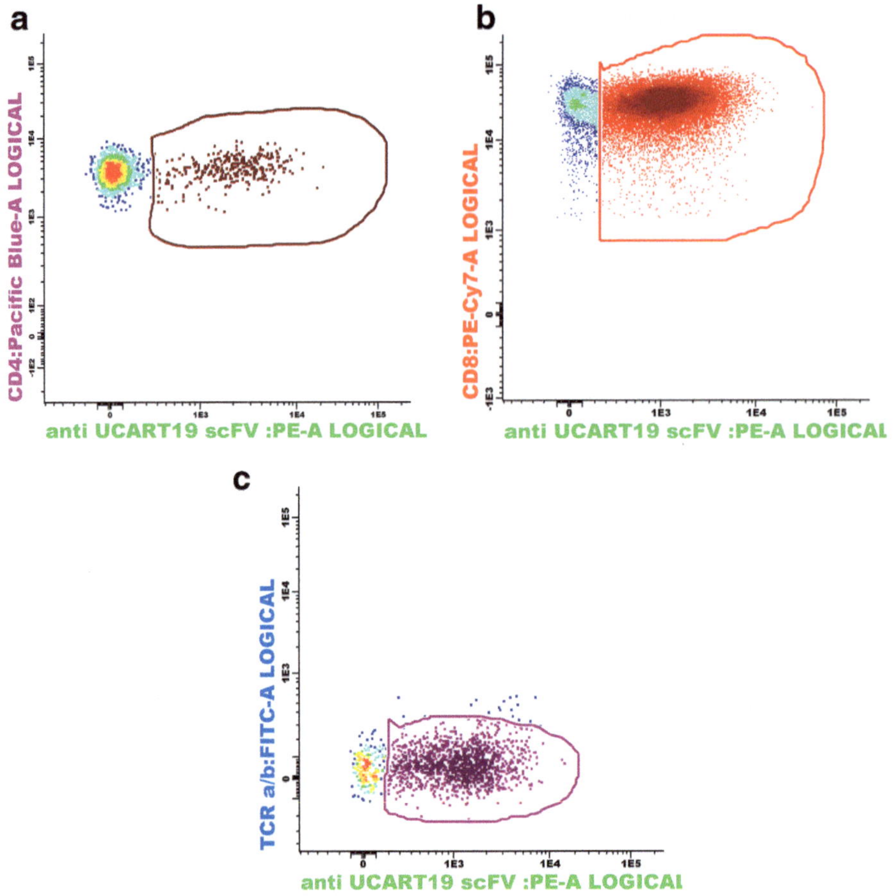

Fig. 8 CAR-T positive cells are gated within CD4+ T cells (**a**), CD8+ T cells (**b**) as well as within a population that is CD4−CD8− (**c**)

sity of the eight-peak rainbow beads lot. Target MFI values vary for different lots, and peaks obtained for the eight-peak Rainbow beads may differ from those listed in Table 1. Updated information on lot- and peak-specific target MFI values is available at the EuroFlow web site (www.euroflow.org).

6. Sometimes PMT values vary unexpectedly when a "New" tube is created. Double checking PMT is essential.
7. The population tree color coding and labels can be edited depending on the design of the experiment.

Doctor:	
Institution/Department:	
Date of colletion:	
Acquisition Date:	
Gender:	
Age:	
Diagnosis Observation:	
Type of Sample:	

RESULTS

Population	Events	Total %	%TNC	WBC/ul
NUCLEATED CELLS	3029	6.18	100.00	110.00
LYMPHOCYTES	2602	5.31	85.90	94.49
T CELLS	2560	5.22	84.51	92.97
f CD4+/CD8+ ratio: 0.10				
CD4+CD8-	175	0.36	5.77	6.36
CD4+ CAR Pos	127	0.26	4.19	4.61
CD8+CD4-	1821	3.71	60.11	66.13
CD8+ CAR Pos	1526	3.11	50.37	55.42
CD4-CD8-/+dim	509	1.04	16.81	18.48
4/8 NEG car pos	347	0.71	11.46	12.60

Fig. 9 Infinicyt software allows users to automatically generate a customized report with CAR-T cells percentages for particular populations as well as total numbers. Table above is an example of cell populations to be included in the report

References

1. Porter D, Hwang W, Frey N, Lacey S, Shaw P, Loren A, Bagg A, Marcucci K, Shen A, Gonzalez V, Ambrose D, Grupp S, Chew A, Zheng Z, Milone M, Levine B, Melenhorst J, June C (2015) Chimeric antigen receptor T cells persist and induce sustained remissions in relapsed refractory chronic lymphocytic leukemia. Sci Transl Med 7:303–139
2. Giavridis T, van der Stegen SJC, Eyquem J, Hamieh M, Piersigilli A (2018) Sadelain CAR T cell-induced cytokine release syndrome is mediated by macrophages and abated by IL-1 blockade. Nat Med 24(6):731–738
3. Graham C, Jozwik A, Pepper A, Benjamin R (2018) Allogeneic CAR-T cells: more than ease of access? Cells 7(10). https://doi.org/10.3390/cells7100155
4. Torikai H, Reik A, Liu PQ, Zhou Y, Zhang L, Maiti S, Huls H, Miller JC, Kebriaei P, Rabinovich B, Lee DA, Champlin RE, Bonini C, Naldini L, Rebar EJ, Gregory PD, Holmes MC, Cooper LJ (2012) A foundation for universal T-cell based immunotherapy: T cells engineered to express a CD19-specific chimeric-antigen-receptor and eliminate expression of endogenous TCR. Blood 119(24):5697–5705
5. Ren J, Liu X, Fang C, Jiang S, June CH, Zhao Y (2017) Multiplex genome editing to generate universal CAR T cells resistant to PD1 inhibition. Clin Cancer Res 23(9):2255–2266

6. Philip B, Kokalaki E, Mekkaoui L, Thomas S, Straathof K, Flutter B, Marin V, Marafioti T, Chakraverty R, Linch D, Quezada SA, Peggs KS, Pule M (2014) A highly compact epitope-based marker/suicide gene for easier and safer T-cell therapy. Blood 124(8):1277–1287
7. Poirot L, Philip B, Schiffer-Mannioui C, Le Clerre D, Chion-Sotinel I, Derniame S, Potrel P, Bas C, Lemaire L, Galetto R, Lebuhotel C, Eyquem J, Cheung GW, Duclert A, Gouble A, Arnould S, Peggs K, Pule M, Scharenberg AM, Smith J (2015) Multiplex genome-edited T-cell manufacturing platform for "off-the-shelf" adoptive T-cell immunotherapies. Cancer Res 75(18):3853–3864
8. Wang H, Du X, Chen WH, Lou J, Xiao HL, Pan YM, Chen H, An N, Zhang QX (2018) Establishment of a quantitative polymerase chain reaction assay for monitoring chimeric antigen receptor T cells in peripheral blood. Transplant Proc 50(1):104–109
9. Kalina T, JG Marvelde TE, van der Velden VHJ, Flores-Montero J, Thűrner D, Böttcher S, Cullen M, Lhermitte L, Sedek L, Mendonça A, Hrusak O, van Dongen JJM, Orfao A (2012) EuroFlow standardization of flow cytometer instrument settings and immunophenotyping protocols. Section 3. Design and evaluation of EuroFlow standard operating procedure for establishing optimal compensation settings. Leukemia 26:1986–2010
10. Stewart CC, Stewart SJ (1997) Titrating antibodies. In: Robinson JP et al (eds) Current protocols in cytometry. J. Wiley & Sons, New York, pp 4.1–4.4

Chapter 20

Key Factors in Clinical Protocols for Adoptive Cell Therapy in Melanoma

Bryden Considine and Michael E. Hurwitz

Abstract

Adoptive cell therapy (ACT) with autologous tumor infiltrating lymphocytes (TIL) has been studied for patients with advanced metastatic cancers (primarily melanoma) for decades and has changed significantly during that period. Treatment with TIL includes ex vivo cell activation and expansion followed by re-infusion of these cells into the patient. After cell infusion, patients receive Interleukin-2 (IL-2). Objective response rates up to 52% have been seen in patients with metastatic melanoma. Efforts to improve TIL therapy include better selection and expansion of tumor-reactive lymphocytes, optimization of IL-2 or other T cell activating cytokine dosing, and, potentially, genetic manipulation of the immune cell product. Here we describe methods involved in the collection, expansion, and treatment with TIL for patients with metastatic melanoma.

Key words Adoptive cell therapy, Tumor infiltrating lymphocytes, Melanoma, Interleukin-2, Lymphodepletion, T cells

1 Introduction

Despite advances in cancer chemotherapy and immunotherapy, not all patients with metastatic malignancies will respond to treatment, and few will have durable remissions. In patients with metastatic melanoma, the combination of the checkpoint inhibitors, nivolumab, and ipilimumab provides an overall survival at 3 years of 58% [1]. Despite this recent success, most patients either do not respond to checkpoint inhibitor therapy or initially respond but eventually develop progressive disease. In these patients, the disease is uniformly lethal.

Adoptive cell therapy (ACT) is the transfer of cells with anti-tumor reactivity into the tumor bearing host [2]. The use of tumor infiltrating lymphocytes (TIL) involves removing tumor from the patient, ex vivo activation, and expansion of immune cells from that tumor followed by the transfer of these cells back into the

Samuel G. Katz and Peter M. Rabinovich (eds.), *Cell Reprogramming for Immunotherapy: Methods and Protocols*, Methods in Molecular Biology, vol. 2097, https://doi.org/10.1007/978-1-0716-0203-4_20,

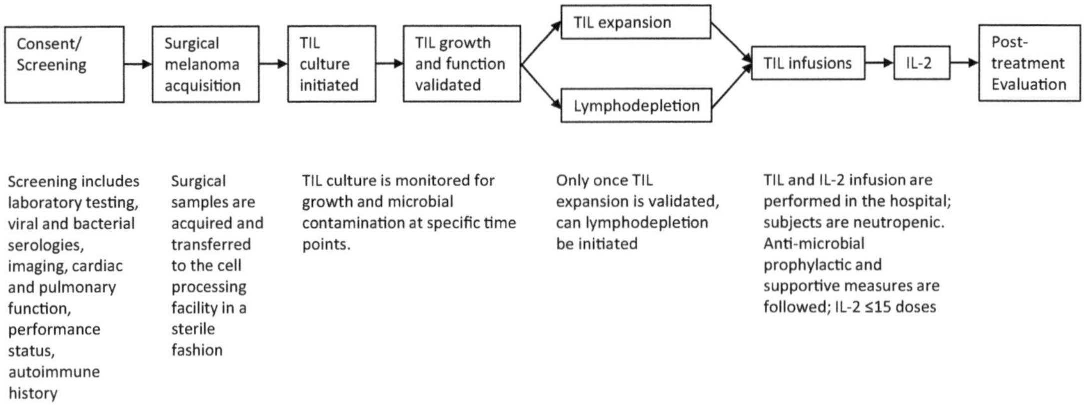

Fig. 1 Schema of adoptive cell therapy using TIL

same patient and then treatment of the patient with multiple doses of IL-2 to boost immune cell function (Fig. 1). By growing resected tumor in IL-2 containing medium, the only cells that remain after expansion are T-cells. ACT with TIL is presumed to work because T cells in the tumor microenvironment recognize the tumor but are suppressed and unable to perform their killing functions. Removing T cells from this environment, promoting their proliferation, and activating them with cytokines relieve the suppression, allowing them to attack tumors. Each of these steps has evolved over the last several decades, and multiple sites worldwide have ACT protocols open to continue to improve the techniques.

2 History of Adoptive Cell Therapy with Tumor Infiltrating Lymphocytes

The first successful use of TIL in patients with metastatic cancer was performed in phase I/II studies in the late 1980s by both Rosenberg and colleagues at the United States National Cancer Institute and Kradin and colleagues at Massachusetts General Hospital [3–6]. In these early TIL protocols, preparation began with resection of tumors that were then dissected into 5 mm^3 pieces and incubated with enzymes (e.g. hyaluronidase, DNase, collagenase) to disrupt tissues. Cells were then strained through filters to remove large tissue fragments, spun, washed, expanded in the presence of IL-2 to 2–4 × 10^{10} cells, and then infused into patients [3]. In one early study, of 12 patients treated with TIL (six melanoma, four renal cell (RCC), one breast, and one colon cancer), partial responses were achieved in one each of the melanoma (3 months), RCC (3 months), and breast cancer (<1 month) [4]. These early phase I/II studies showed response rates in the range of 17–55%, but only 1 out of the 70 subjects in these trials had a response of >1 year, and only 5 out of 16 responders had responses

>6 months [4–7]. The process of ACT with TIL has a number of elements, each of which has been further refined over the last 30 years. The improvements include the choice of cells to transfer, optimization of in vitro cell growth, pre-treatment, and post-therapy treatment for patients (Table 1). Within the next section, we discuss some of the improvements with the most progress.

3 Optimizing ACT Regimens

3.1 Lymphodepletion

Despite encouraging response rates in some of the early TIL trials, durable responses and persistence of infused cells were not observed. In murine models, pre-treatment therapy with a lymphodepleting regimen augmented the anti-tumor response. Murine studies have identified at least three defined mechanisms of action. First, lymphodepletion removes a large number of T regulatory cells capable of suppressing anti-cancer immune responses [8]. Second, lymphodepletion appears to open up a 'niche' in which infused cells can proliferate. Third, the level of endogenous anti-inflammatory cytokines is decreased as is the number of cells competing for activating cytokines [9–11]. It has been theorized that lymphodepleting chemotherapy causes translocation of gut microbiota into body tissues, which leads to systemic inflammation with concomitant improved T cell killing function though this has not been demonstrated experimentally in ACT models [12].

The role of lymphodepletion prior to TIL therapy in patients with metastatic melanoma was investigated by Dudley et al. Thirteen patients received a lymphodepleting chemotherapy regimen beginning on day −7 from infusion consisting of cyclophosphamide 30 or 60 mg/kg for 2 days with concomitant MESNA which is used in high dose cyclophosphamide regimens to protect against hemorrhagic cystitis. This was followed by fludarabine 25 mg/m^2 for 5 days leading up to TIL cell infusion on day 0. This study also assessed different doses of IL-2, which will be discussed below. Five patients demonstrated mixed responses, and HD-IL-2 was better tolerated following lymphodepletion [13]. This study established the safety and potential role of chemotherapeutic lymphodepletion prior to TIL therapy.

Non-myeloablative versus ablative lymphodepletion with total body irradiation (TBI) was investigated by the same group [14]. All three cohorts received non-myeloablative chemotherapy with cyclophosphamide 60 mg/m^2 and fludarabine 25 mg/kg as described above followed by TIL infusion. Two cohorts received TBI with either 2 Gy or 12 Gy following non-myeloablative chemotherapy. In these cohorts after the TIL infusion, patients were given autologous CD34$^+$ stem cell rescue. The third cohort did not receive TBI or stem cell rescue. Objective response rates were

Table 1
ACT clinical trials

Study	Disease	Design	Infused product	*n*	ORR	Response duration	Survival	Cell production success	Time of cell production
Rosenberg et al. [2]	• Melanoma (7) • CRC (9) • RCC (3) • Sarcoma (4) • Lung adeno (1)	Pilot	LAK	25	44%	• CR 10+ months • No information on PR responses			
Topalian et al. [4]	• Melanoma (6) • CRC (1) • RCC (4) • BrCA (1)	Phase I	TIL—used conditioned medium from LAK, not feeder cells	12	17%	3 months for both PRs			19–48 days
Kradin et al. [6]	Lung adenocarcinoma	Phase I	TIL—used radiated PBL and PHA	7	0				28–112 days
Rosenberg et al. [7]	Melanoma	Phase II	TIL	20	55%	2, 3, 3, 4, 4, 4, 6, 7, 7, 9, 13+ months		20/37: seven progression, eight not enough lymphocyte growth, two bacterial contam of lymphs	28–56 days
Kradin et al. [5]	• Melanoma (13) • RCC (7) • Lung (8)	Phase II	TIL—used radiated PBL and PHA	31	18%	3, 3, 5, 6, 11 months			
Rosenberg et al. [40]	Melanoma	Phase II	TIL—used conditioned medium from LAK, not feeder cells	86	33%	• CR: 20, 21+, 38+, 46+ • PR: 1, 1, 2, 4, 4, 4, 4, 5, 5, 6, 7, 7, 7, 9, 53+	• 2 year: 22% • 4 year: 14%		Days of growth: • 11–20: 1 (1%) • 21–30: 17 (20%) • 31–40: 33 (38%) • 41–50: 19 (22%) • 51–60: 11 (13%) • >60: 5 (6%)

Figlin et al. [38]	RCC after nephrectomy	Phase III	IL-2 + TIL	79	10%	
			IL-2 alone	81	11%	
Dudley et al. [41]	Melanoma	Phase I	• gp100 responsive TIL or PBL from patients immunized with gp100 • Grown with irradiated PBMCs, IL-2, and OKT-3	11	0	
Dudley et al. [13]	Melanoma	Phase I	Pts all received g209-2M peptide vaccination (from gp100), 5 TIL, 10 PBL	15	0	
Dudley et al. [42]	HLA-A2+ melanoma	Phase II	TIL—grown in irradiated PBMC, IL-2 (6000 IU/ml), OKT3 antibody	13	46%	2, 8, 9+, 10+, 15+, 24+
Dudley et al. [43]	Melanoma	Phase II	TIL—grown in irradiated PBMC, IL-2 (6000 IU/ml), OKT3 antibody	35	52%	Mean 11.5 ± 2.2

(continued)

Table 1
(continued)

Study	Disease	Design	Infused product	*n*	ORR	Response duration	Survival	Cell production success	Time of cell production
Kershaw et al. [44]	Ovarian cancer	Phase I	T cells genetically modified with anti-FR single-chain Ab linked to Fc-gamma	14	0				
Dudley et al. [14]	Melanoma	Phase II	TIL—grown in irradiated PBMC, IL-2 (6000 IU/ml), OKT3 antibody; for pts. who received TBI, they also got 2x10E6 CD34+ self-cells previously frozen down	93	49%	PR range 2–64+; CR all <47+ months			
					52%	PR 3–33+; CR 25+, 37+			
					72%	PR 3–14+; CR 8+–17+			
Besser et al. [25]	Melanoma	Pilot	Selected TIL: secrete >200 pg/ml IFNgamma on Ag stimulation; multiple cultures from tissue frags	33	16%	0.5–7.5 months and one not achieved	1.6–23.3 months and two still alive	21/33 excluded: • Cells failed IFNgamma selection: 8 • No autologous tumor target available: 7 • Patient deterioration: 4	Time to establishment of cx (after this, both undergo REP for ~14 days): • Arm 1: 27.1 ± 7.4 days • Arm 2: 15.5 ± 3.6 days
			Unselected TIL from multiple tissue fragments	14	38%	2.4, 3 months, and six not achieved	3.1 and seven still alive (4+–13+ months)	3/14 excluded	

Johnson et al. [45]	Melanoma	Pilot	PBL transduced with TCR recognizing gp100	36	19%	3, 4, 14+			
			PBL transduced with TCR recognizing MART-1		30%	3, 4, 9, 16+, 17+, 17+			
Besser et al. [46]	Melanoma	Phase II	Unselected TIL	27	50%	•PR: 3, 3+, 4, 6+, 9, 10+, 13+, 20+ • CR: 4+,20+	PFS 7.3 months; OS 9.3 months	Three did not generate enough cells, three deteriorated clinically, one withdrew	To establish xc: 9–28 days (14.2 ± 4.6 day mean) then REP × 14 days
Robbins et al. [47]	NY-ESO-1-expressing melanoma and synovial cell sarcoma	Phase I	NY-ESO-1 specific TCR transduced PBMC in melanoma	17	45%	PR: 3, 8, 9+CR: 20+, 22+			
			NY-ESO-1 specific TCR transduced PBMC in synovial cell sarcoma		67%	5, 8, 10, 18			
Itzhaki et al. [24]	Melanoma	Phase II	Unselected TIL	55	48%	• PR: 3, 4, 4, 4+, 5, 5, 6, 9, 21+, 23, 24+ • CR: 4, 7+, 15+, 31+	Med PFS ~4.5 monthsmed OS can't be calculated (16+ in responders, 7 months in non-responders	Three did not generate enough cells, one continued to have melanoma cells in culture after 4 weeks, ten deteriorated clinically, two withdrew	9–33 days (+14 day REP)

(continued)

Table 1
(continued)

Study	Disease	Design	Infused product	*n*	ORR	Response duration	Survival	Cell production success	Time of cell production
Radvanyi et al. [30]	Melanoma	Phase II	TIL—grown in irradiated PBMC, IL-2 (6000 IU/ml), OKT3 antibody	31	43%		med PFS 7.6 months; med OS not reached		This is not a fast protocol
Dudley et al. [29]	Melanoma	Phase II	Unselected TIL	101	35%	• Unselected: PR: 3, 5, 7, 10, 16, 19+, 26+, 30+, 31+, 32+, CR: 29+, 30+		17 no TIL growth, 15 rapid progression	35–35.5 days
			CD8+ selected TIL		22%	• CD8+ selected: PR: 4, 4, 12, 24+, CR: 7, 11, 22+, 33+			
Besser et al. [28]	Melanoma	Phase II	Unselected TIL	80	41%		med OS 15.2 months (for 57 pts);12 months OS 46%, 24 months OS 33%	8 no TIL growth, 3 withdrew, 11 clinical deterioration	Approximately 28 days

Andersen et al. [23]	Melanoma	Phase II	Unselected TIL	25	41%	8, 11, 12 months for three of the PRsAll other PR/CR ongoing (6–45 months)	21.8 months OS3.9 months PFS	96% grew	Median 21 days (range 13–36)
Chandran et al. [48]	Metastatic uveal melanoma	Phase II	Unselected TIL	21	35%	3, 4, 5, 9, 4+, 4+, 20+ months		21/27 grew	
Mullinax et al. [49]	Melanoma	Phase I	Unselected TIL	13	38%	All PR/CR ongoing: 34+, 38+, 46+, 53+ months	PFS 7.3 monthsOS not reached	100% grew	Median 51 days (range 39–77)

52% and 72% respectively compared to 49% without TBI though overall survival did not differ among the arms.

In a second randomized trial, 101 patients with metastatic melanoma were randomized to receive non-myeloablative chemotherapy with or without the addition of 12 Gy TBI before TIL cell infusion and high-dose IL-2. The patients who received 12 Gy TBI, also received $CD34^+$ stem cell rescue following TIL. The complete response (CR) rate was 24% in both groups, with overall survival of 38.2 versus 36.6 months in patients who received non-myeloablative versus non-myeloablative with TBI respectively. In the group who received TBI, 27% of patients developed thrombotic microangiopathy which was unique to this group [15]. Based on this study, ablative regimens with TBI did not add to survival and were associated with significant toxicities so they were not pursued further.

3.2 Role of IL-2

IL-2 is produced by the immune system, is a master regulator of T-cell function, and is required for T-cell activation, expansion, and anti-tumor responses in ACT [16, 17]. Thus, in most protocols, post lymphodepletion subjects receive IL-2 infusions after cells are re-infused to replace endogenous IL-2.In general, the doses used are high (600,000–720,000 IU/kg IV every 8 h for 6–15 doses) based on the experience with IL-2 as a monotherapy, which is used for metastatic melanoma and metastatic renal cell carcinoma. HD-IL-2 results in more durable responses than lower dose regimens (72,000 IU/kg IV every 8 h) [18, 19].

A theoretical reason behind the superiority of HD-IL-2 is that effector T cells, which kill tumors, express both a high affinity IL-2 receptor and a low affinity IL-2 receptor, whereas regulatory T cells, which suppress the anti-tumor immune response, only express the high affinity receptor. By increasing the dose of IL-2, both the high and low affinity receptors become saturated, and the relative activation of effector T cells to regulatory T cells increases [17]. In the setting of ACT, however, regulatory T cells have been depleted by lymphodepleting chemotherapy so it is unclear whether such high doses are required.

Treatment with HD-IL-2 is associated with a severe flu-like syndrome, hypotension requiring IV fluid, and sometimes pressor support, renal failure, cardiac toxicity, and central nervous system toxicity [20]. In fact, in early studies of HD-IL-2 alone (it is an approved therapy for metastatic melanoma), >50% of patients required vasopressor support, and 4% experienced treatment related death [21]. Despite toxicities associated with IL-2, most TIL studies have used HD-IL-2 following lymphodepletion.

Given the toxicities, lower dosing regimens of IL-2 are being tested. The role of IL-2 dose in ACT for metastatic melanoma was investigated in a phase 1 trial in 2002. All patients received a lymphodepleting chemotherapy regimen prior to TIL infusion followed by either no IL-2, LD IL-2 (72,000 IU/kg every 8 h for 5 days), or HD IL-2 (720,000 IU/kg every 8 h for 4 days). No objective partial responses were seen, defined as a 50% of greater decrease in measurable lesions for at least 1 month. However some patients in both the IL-2 cohorts experienced marked reduction at individual tumor sites [13]. In a pilot study by Ellebaek et al. the feasibility of subcutaneous LD IL-2 was investigated in six patients with metastatic melanoma following ACT [22]. Durable CRs were seen in 2/6 patients without any grade 3–4 (out of 4) IL-2 related toxicities. In a phase I/II study by Andersen et al., intravenous IL-2 was administered in a decrescendo fashion to patients with metastatic melanoma after lymphodepletion and TIL infusion. This decrescendo regimen consisted of a continuous infusion of IL-2 with a decreasing dose starting with 18 million IU/m^2 over 6, 12, then 24 h followed by 4.5 million IU/m^2 over 24 h for 3 days. Of 25 patients treated, the objective response rate was 42% with three patients achieving CRs and median overall survival 21.8 months [23]. At our center, we are currently performing a trial using 1/10 the standard dose of IL-2 (72,000 IU/kg IV for up to ten doses). Currently, the optimal dose is unknown, but using IL-2 is probably beneficial.

4 Tumor Infiltrating Lymphocyte Cell Product

4.1 Selection of TIL

Methods of cell expansion in vitro have evolved considerably. In most initial treatment protocols, cells were selected for tumor antigen reactivity based on in vitro cytokine release assays [13]. Specifically, tumors were mechanically separated into small pieces, and T cells were cultured from each piece separately. Then T cell reactivity for each culture was performed, usually by assessing γ-interferon production by T cells that were activated either by incubation with autologous tumor or with anti-CD3 antibodies. TIL cultures that produced a threshold amount of γ-interferon were then expanded (on average 1000-fold) to the large numbers required for therapy (10^9–10^{11} cells) [24]. Subsequent studies showed that these in vitro assays did not predict subsequent in vivo tumor response, and attempts to select cells by tumor antigen reactivity could result in longer ex vivo expansion times and fewer cells transferred as well as greater expense. Long expansion times were particularly problematic because patients waiting for therapy could deteriorate before the cells were fully expanded. Thus, cell expansion protocols that used all of the cells in single cultures were developed; these are often referred to as 'bulk TIL', 'young TIL' or 'unselected TIL' protocols. In a study by Besser et al. TIL cultures were established on average in 27.1 days with an 800-fold expansion for selected TIL versus 15.5 days and 1100-fold expansion in young unselected TIL [25]. Unselected expansion of TIL resulted in faster and better cell yields and a higher likelihood of successfully getting the cells into patients [25–29]. Large-scale production of young-TIL is rapid with successful cultures established in 93% of patients in one study [24]. Notably, young TIL protocols still use IL-2 and anti-CD3 antibody to help cells proliferate and become activated.

Another approach to improving the cell product was to enrich for $CD8^+$ T cells since those kill tumor, as opposed to $CD4^+$ cells. It has been observed that increased numbers of infused $CD8^+$ cells and a higher percentage of $CD8^+$ versus $CD4^+$ cells are significantly associated with improved responses [24, 30]. A study by Yao et al. revealed that the number of $CD4^+$ T-regulatory cells in the peripheral blood was higher in nonresponders than in responders [31]. A small trial (n = 34) randomized subjects to young TIL vs $CD8^+$ selected TIL. Response rates were 35% vs 20% (young vs selected), which was not significantly different [29].

5 ACT Therapy

5.1 Patient Selection and Preparation

ACT requires lymphodepleting regimens that are similar in intensity to those for autologous bone marrow transplant and treatment with HD IL-2, which requires hospitalization. An ECOG PS (Eastern Cooperative Oncology Group Performance Score) of 0–2 is required for most trials. The ECOG PS system is a standardized measurement tool used to describe a patient's level of functioning. ECOG PS of 0 representing a fully functioning patient and ECOG PS 2 capable of self-care and ambulatory >50% of waking hours. In addition to taking into account the patient's current performance, the fact that tumor harvest and cell expansion takes 3–6 weeks must be considered as patients with refractory disease may decompensate while awaiting cell product. For that reason, in our protocols, we allow other treatments during the time between TIL harvest and lymphodepletion. Alternatively, this problem can be circumvented by incorporating TIL harvest earlier in the disease process so that cells are available at time of progression, though this approach also would require a clinical protocol given the requirement of an excisional biopsy.

Prior to treatment, patients must undergo viral and some bacterial testing because of the intense immunosuppression caused by high dose chemotherapy. We also require pulmonary function testing and echocardiography because of potential toxicity from IL-2.

5.2 Tumor Harvest and TIL Expansion

Harvesting tumor from patients with metastatic melanoma can occur at any time in the disease process. Tumor specimen collection occurs in the operating room and all TIL processing must occur under sterile conditions. After removal, the tumor is placed into a transport container with sterile medium and brought to the cell processing facility. There, using sterile technique, the tumor is dissected and enzymatically dissociated. For unselected TIL production, initial TIL cell expansion occurs over 1–4 weeks at which point cells are transferred to specialized flasks for large scale production of expanded TIL [32]; the expansion step takes approximately 2 weeks. For expansion, TILs need to be exposed to a source of activating chemokines (e.g. IL-2) and ligands (e.g. anti-CD3 antibody). One way to accomplish this is to incubate the TILs with peripheral blood mononuclear cells (PBMCs), which are a source of both chemokines and ligands. The PBMCs, referred to as feeder cells, are first irradiated to prevent their proliferation. In addition, anti-CD3 antibodies and IL-2 are often added in the presence of feeder cells to activate T cells to proliferate. The expanded T cells can be cryopreserved at multiple steps for later expansion and use.

5.3 TIL Infusion

As noted above, prior to TIL infusion patients receive lymphodepleting chemotherapy. The most common regimen is cyclophosphamide/MESNA/fludarabine described in Subheading 3.1. Cell infusion, itself, usually takes between 30 min and 1 h and is benign with the exception of occasional infusional side effects like fever and chills. Following the cell infusion patients receive IL-2. IL-2 dosing is described in Subheading 3.2.

5.4 Coordination with TIL Production Laboratory

Since ACT requires growth of cells, which can take a variable amount of time (3–6 weeks), and patients need 7 days of lymphodepletion prior to receiving cells, there needs to be close coordination between the clinicians treating the patient and the cell processing facility. Only once the cell processing facility has ensured that growth appears adequate for cell infusion and that cells have not acquired any contamination can patients undergo lymphodepleting chemotherapy.

5.5 Toxicities and Supportive Care

During non-myeloablative lymphodepletion patients usually experience standard high-dose chemotherapy-related toxicities. Once cells are infused, neutropenia usually resolves in 1–2 weeks with growth factor support. To reduce the risk of infection during periods of cytopenias patients are started on antimicrobial prophylaxis often with acyclovir, fluoroquinolone, and anti-fungal therapy until counts recover.

Following TIL infusion and IL-2, fevers and hypotension are common. As noted earlier in Subheading 3.2, HD-IL-2 is associated with serious side effects. Management of HD IL-2 toxicities has been reviewed elsewhere [20].

6 ACT in the Current Treatment Landscape

Treatment with TIL remains investigational at select centers worldwide. Most ACT trials were performed in the era prior to the approval of immune checkpoint inhibitors for melanoma. One study reported ACT in metastatic melanoma patients, some of whom had received ipilimumab (anti-CTLA4) previously. The overall response rate (ORR) was 40% (23/57) for all subjects with an almost identical ORR of 38% (5/13) in ipilimumab pre-treated subjects [28]. In the era of immune-checkpoint inhibitors overall survival is significantly improved in patients with metastatic melanoma treated with combination nivolumab and ipilimumab. While anecdotally some subjects receiving ACT have achieved responses after receiving anti-PD1 alone or in combination with anti-CTLA4 therapies, response rates have not yet been reported from any studies.

7 Future Directions

This review has described a current standard for adoptive cell therapy. Multiple ways forward exist:

1. Cell expansion: cells are currently grown in the presence of IL-2 and anti-CD3 antibodies primarily but other combinations of cytokines and stimulators could be used. For example, a recently published paper demonstrated improved cell growth with the addition of anti-41BB/CD137 along with IL-2 and anti-CD3 [33]. Different cocktails of cytokines and antibodies could be used to optimize T cell features such as killing ability, proliferation potential, and/or persistence.
2. Selection of T cell subpopulations: the rationale behind ACT with TIL is that lymphocytes that have invaded into the tumor contain subpopulations that recognize tumor neoantigens. To increase the potency of the cell product, there are a number of methods to isolate different subpopulations that can be tested to determine which is optimal. Data exist that suggest that PD1+ TIL are the ones that recognize tumor neoantigens [34, 35]. PD1+ TIL can be isolated and expanded. Tran and colleagues used a very sophisticated method in which they sequenced tumor DNA, identified potential neoantigens, expressed those neoantigens on MHC, and screened pools of TIL to identify the neoantigen specific TIL and expand those. This method resulted in a long term durable response in a treatment-refractory cholangiocarcinoma [36].
3. T cell engineering: T cells can be engineered to express or suppress specific genes or gene combinations including chimeric

antigen receptors (CARs), metabolic enzymes to improve T cell function in the tumor microenvironment, anti-apoptotic proteins to keep cells alive, immune-function modulating proteins, or a host of other possible manipulations. This type of genetic engineering could be transient or permanent, depending on the particular needs of the cell product.

4. Combinations with other therapies such as immune checkpoint therapies.
5. Non-TIL cell products: NK cell products [37] and products from peripheral blood mononuclear cells (PBMCs) are also being tested.
6. Other malignancies: the success seen with TIL therapy in patients with metastatic melanoma has not yet translated into effective treatments for patients with other solid malignancies. A phase III trial was performed in which subjects with metastatic RCC received low-dose IL-2 alone or in combination with TIL. Of 81 patients randomized, 33 patients did not receive TIL due to insufficient TIL product. ORR and overall survival was not significantly different between the arms [38]. Regarding efficacy, however, this study did not treat subjects with lymphodepleting chemotherapy prior to cell infusion, which could account for decreased activity. In addition, with modern TIL expansion methods, 90% of RCC samples can produce TIL [39]. Currently, ACT with TIL is being tested in lung cancer and other malignancies.

8 Conclusions

ACT is an old therapy that has changed dramatically over the last 30 years but now has the potential for much greater innovation. Between newer protocols for cell growth, genetic engineering of immune cells, and combinations with other therapies, the potential is great for ACT to revolutionize cancer therapy.

Acknowledgements

Disclosures: M.H. is on the advisory boards of Nektar Therapeutics, Janssen Pharmaceuticals and CRISPR Therapeutics and receives research funding from Apexigen, Astellas, AstraZeneca, Bayer, Bristol Myer Squibb, Clovis, Corvus, Eli Lilly, Endocyte, Genentech, Genmab, Innocrin, Iovance, MedImmune, Merck, Nektar Therapeutics, Novartis, Pfizer, Progenics, Roche Laboratories, Sanofi Aventis, and Seattle Genetics.

References

1. Wolchok JD, Chiarion-Sileni V, Gonzalez R, Rutkowski P, Grob JJ, Cowey CL, Lao CD, Wagstaff J, Schadendorf D, Ferrucci PF, Smylie M, Dummer R, Hill A, Hogg D, Haanen J, Carlino MS, Bechter O, Maio M, Marquez-Rodas I, Guidoboni M, McArthur G, Lebbé C, Ascierto PA, Long GV, Cebon J, Sosman J, Postow MA, Callahan MK, Walker D, Rollin L, Bhore R, Hodi FS, Larkin J (2017) Overall survival with combined nivolumab and ipilimumab in advanced melanoma. N Engl J Med 377:1345–1356. https://doi.org/10.1056/NEJMoa1709684
2. Rosenberg S (1985) Lymphokine-activated killer cells: a new approach to immunotherapy of cancer. J Natl Cancer Inst 75:595–603
3. Topalian SL, Muul LM, Solomon D, Rosenberg SA (1987) Expansion of human tumor infiltrating lymphocytes for use in immunotherapy trials. J Immunol Methods 102:127–141
4. Topalian SL, Solomon D, Avis FP, Chang AE, Freerksen DL, Linehan WM, Lotze MT, Robertson CN, Seipp CA, Simon P (1988) Immunotherapy of patients with advanced cancer using tumor-infiltrating lymphocytes and recombinant interleukin-2: a pilot study. J Clin Oncol 6:839–853
5. Kradin RL, Kurnick JT, Lazarus DS, Preffer FI, Dubinett SM, Pinto CE, Gifford J, Davidson E, Grove B, Callahan RJ (1989) Tumour-infiltrating lymphocytes and interleukin-2 in treatment of advanced cancer. Lancet 1:577–580
6. Kradin RL, Boyle LA, Preffer FI, Callahan RJ, Barlai-Kovach M, Strauss HW, Dubinett S, Kurnick JT (1987) Tumor-derived interleukin-2-dependent lymphocytes in adoptive immunotherapy of lung cancer. Cancer Immunol Immunother 24:76–85
7. Rosenberg SA, Packard BS, Aebersold PM, Solomon D, Topalian SL, Toy ST, Simon P, Lotze MT, Yang JC, Seipp CA (1988) Use of tumor-infiltrating lymphocytes and interleukin-2 in the immunotherapy of patients with metastatic melanoma. A preliminary report. N Engl J Med 319:1676–1680. https://doi.org/10.1056/NEJM198812223192527
8. Antony PA, Piccirillo CA, Akpinarli A, Finkelstein SE, Speiss PJ, Surman DR, Palmer DC, Chan C-C, Klebanoff CA, Overwijk WW, Rosenberg SA, Restifo NP (2005) CD8+ T cell immunity against a tumor/self-antigen is augmented by CD4+ T helper cells and hindered by naturally occurring T regulatory cells. J Immunol 174:2591–2601
9. Gattinoni L, Klebanoff CA, Palmer DC, Wrzesinski C, Kerstann K, Yu Z, Finkelstein SE, Theoret MR, Rosenberg SA, Restifo NP (2005) Acquisition of full effector function in vitro paradoxically impairs the in vivo antitumor efficacy of adoptively transferred CD8+ T cells. J Clin Invest 115:1616–1626. https://doi.org/10.1172/JCI24480
10. Wrzesinski C, Paulos CM, Gattinoni L, Palmer DC, Kaiser A, Yu Z, Rosenberg SA, Restifo NP (2007) Hematopoietic stem cells promote the expansion and function of adoptively transferred antitumor CD8 T cells. J Clin Invest 117:492–501. https://doi.org/10.1172/JCI30414
11. Wrzesinski C, Paulos CM, Kaiser A, Muranski P, Palmer DC, Gattinoni L, Yu Z, Rosenberg SA, Restifo NP (2010) Increased intensity lymphodepletion enhances tumor treatment efficacy of adoptively transferred tumor-specific T cells. J Immunother 33:1–7. https://doi.org/10.1097/CJI.0b013e3181b88ffc
12. Viaud S, Saccheri F, Mignot G, Yamazaki T, Daillère R, Hannani D, Enot DP, Pfirschke C, Engblom C, Pittet MJ, Schlitzer A, Ginhoux F, Apetoh L, Chachaty E, Woerther P-L, Eberl G, Bérard M, Ecobichon C, Clermont D, Bizet C, Gaboriau-Routhiau V, Cerf-Bensussan N, Opolon P, Yessaad N, Vivier E, Ryffel B, Elson CO, Doré J, Kroemer G, Lepage P, Boneca IG, Ghiringhelli F, Zitvogel L (2013) The intestinal microbiota modulates the anticancer immune effects of cyclophosphamide. Science 342:971–976. https://doi.org/10.1126/science.1240537
13. Dudley ME, Wunderlich JR, Yang JC, Hwu P, Schwartzentruber DJ, Topalian SL, Sherry RM, Marincola FM, Leitman SF, Seipp CA, Rogers-Freezer L, Morton KE, Nahvi A, Mavroukakis SA, White DE, Rosenberg SA (2002) A phase I study of nonmyeloablative chemotherapy and adoptive transfer of autologous tumor antigen-specific T lymphocytes in patients with metastatic melanoma. J Immunother 25:243–251
14. Dudley ME, Yang JC, Sherry R, Hughes MS, Royal R, Kammula U, Robbins PF, Huang J, Citrin DE, Leitman SF, Wunderlich J, Restifo NP, Thomasian A, Downey SG, Smith FO, Klapper J, Morton K, Laurencot C, White DE, Rosenberg SA (2008) Adoptive cell therapy for patients with metastatic melanoma: evaluation of intensive myeloablative chemoradiation preparative regimens. J Clin Oncol 26:5233–5239. https://doi.org/10.1200/JCO.2008.16.5449

15. Goff SL, Dudley ME, Citrin DE, Somerville RP, Wunderlich JR, Danforth DN, Zlott DA, Yang JC, Sherry RM, Kammula US, Klebanoff CA, Hughes MS, Restifo NP, Langhan MM, Shelton TE, Lu L, Kwong MLM, Ilyas S, Klemen ND, Payabyab EC, Morton KE, Toomey MA, Steinberg SM, White DE, Rosenberg SA (2016) Randomized, prospective evaluation comparing intensity of lymphodepletion before adoptive transfer of tumor-infiltrating lymphocytes for patients with metastatic melanoma. J Clin Oncol 34:2389–2397. https://doi.org/10.1200/JCO.2016.66.7220
16. Bukowski RM (1997) Natural history and therapy of metastatic renal cell carcinoma: the role of interleukin-2. Cancer 80:1198–1220
17. Arenas-Ramirez N, Woytschak J, Boyman O (2015) Interleukin-2: biology, design and application. Trends Immunol 36:763–777. https://doi.org/10.1016/j.it.2015.10.003
18. Yang JC, Sherry RM, Steinberg SM, Topalian SL, Schwartzentruber DJ, Hwu P, Seipp CA, Rogers-Freezer L, Morton KE, White DE, Liewehr DJ, Merino MJ, Rosenberg SA (2003) Randomized study of high-dose and low-dose interleukin-2 in patients with metastatic renal cancer. J Clin Oncol 21:3127–3132. https://doi.org/10.1200/JCO.2003.02.122
19. McDermott DF, Regan MM, Clark JI, Flaherty LE, Weiss GR, Logan TF, Kirkwood JM, Gordon MS, Sosman JA, Ernstoff MS, Tretter CPG, Urba WJ, Smith JW, Margolin KA, Mier JW, Gollob JA, Dutcher JP, Atkins MB (2005) Randomized phase III trial of high-dose interleukin-2 versus subcutaneous interleukin-2 and interferon in patients with metastatic renal cell carcinoma. J Clin Oncol 23:133–141. https://doi.org/10.1200/JCO.2005.03.206
20. Dutcher JP, Schwartzentruber DJ, Kaufman HL, Agarwala SS, Tarhini AA, Lowder JN, Atkins MB (2014) High dose interleukin-2 (Aldesleukin) – expert consensus on best management practices-2014. J Immunother Cancer 2:3127–3123. https://doi.org/10.1186/s40425-014-0026-0
21. Fyfe GA, Fisher RI, Rosenberg SA, Sznol M, Parkinson DR, Louie AC (1996) Long-term response data for 255 patients with metastatic renal cell carcinoma treated with high-dose recombinant interleukin-2 therapy. J Clin Oncol 14:2410–2411
22. Ellebaek E, Iversen TZ, Junker N, Donia M, Engell-Noerregaard L, Met O, Hölmich LR, Andersen RS, Hadrup SR, Andersen MH, Thor Straten P, Svane IM (2012) Adoptive cell therapy with autologous tumor infiltrating lymphocytes and low-dose Interleukin-2 in metastatic melanoma patients. J Transl Med 10:169. https://doi.org/10.1186/1479-5876-10-169
23. Andersen R, Donia M, Ellebaek E, Borch TH, Kongsted P, Iversen TZ, Hölmich LR, Hendel HW, Met O, Andersen MH, Thor Straten P, Svane IM (2016) Long-lasting complete responses in patients with metastatic melanoma after adoptive cell therapy with tumor-infiltrating lymphocytes and an attenuated IL2 regimen. Clin Cancer Res 22:3734–3745. https://doi.org/10.1158/1078-0432.CCR-15-1879
24. Itzhaki O, Hovav E, Ziporen Y, Levy D, Kubi A, Zikich D, Hershkovitz L, Treves AJ, Shalmon B, Zippel D, Markel G, Shapira-Frommer R, Schachter J, Besser MJ (2011) Establishment and large-scale expansion of minimally cultured "young" tumor infiltrating lymphocytes for adoptive transfer therapy. J Immunother 34:212–220. https://doi.org/10.1097/CJI.0b013e318209c94c
25. Besser MJ, Shapira-Frommer R, Treves AJ, Zippel D, Itzhaki O, Schallmach E, Kubi A, Shalmon B, Hardan I, Catane R, Segal E, Markel G, Apter S, Nun AB, Kuchuk I, Shimoni A, Nagler A, Schachter J (2009) Minimally cultured or selected autologous tumor-infiltrating lymphocytes after a lympho-depleting chemotherapy regimen in metastatic melanoma patients. J Immunother 32:415–423. https://doi.org/10.1097/CJI.0b013e31819c8bda
26. Dudley ME, Gross CA, Langhan MM, Garcia MR, Sherry RM, Yang JC, Phan GQ, Kammula US, Hughes MS, Citrin DE, Restifo NP, Wunderlich JR, Prieto PA, Hong JJ, Langan RC, Zlott DA, Morton KE, White DE, Laurencot CM, Rosenberg SA (2010) CD8+ enriched "young" tumor infiltrating lymphocytes can mediate regression of metastatic melanoma. Clin Cancer Res 16:6122–6131. https://doi.org/10.1158/1078-0432.CCR-10-1297
27. Donia M, Junker N, Ellebaek E, Andersen MH, Straten PT, Svane IM (2012) Characterization and comparison of 'standard' and 'young' tumour-infiltrating lymphocytes for adoptive cell therapy at a Danish translational research institution. Scand J Immunol 75:157–167. https://doi.org/10.1111/j.1365-3083.2011.02640.x
28. Besser MJ, Shapira-Frommer R, Itzhaki O, Treves AJ, Zippel DB, Levy D, Kubi A, Shoshani N, Zikich D, Ohayon Y, Ohayon D, Shalmon B, Markel G, Yerushalmi R, Apter S, Ben-Nun A, Ben-Ami E, Shimoni A, Nagler A, Schachter J (2013) Adoptive transfer of tumor-

infiltrating lymphocytes in patients with metastatic melanoma: intent-to-treat analysis and efficacy after failure to prior immunotherapies. Clin Cancer Res 19:4792–4800. https://doi.org/10.1158/1078-0432.CCR-13-0380

29. Dudley ME, Gross CA, Somerville RPT, Hong Y, Schaub NP, Rosati SF, White DE, Nathan D, Restifo NP, Steinberg SM, Wunderlich JR, Kammula US, Sherry RM, Yang JC, Phan GQ, Hughes MS, Laurencot CM, Rosenberg SA (2013) Randomized selection design trial evaluating CD8+-enriched versus unselected tumor-infiltrating lymphocytes for adoptive cell therapy for patients with melanoma. J Clin Oncol 13:2152–2159

30. Radvanyi LG, Bernatchez C, Zhang M, Fox PS, Miller P, Chacon J, Wu R, Lizée G, Mahoney S, Alvarado G, Glass M, Johnson VE, McMannis JD, Shpall E, Prieto V, Papadopoulos N, Kim K, Homsi J, Bedikian A, Hwu W-J, Patel S, Ross MI, Lee JE, Gershenwald JE, Lucci A, Royal R, Cormier JN, Davies MA, Mansaray R, Fulbright OJ, Toth C, Ramachandran R, Wardell S, Gonzalez A, Hwu P (2012) Specific lymphocyte subsets predict response to adoptive cell therapy using expanded autologous tumor-infiltrating lymphocytes in metastatic melanoma patients. Clin Cancer Res 18:6758–6770. https://doi.org/10.1158/1078-0432.CCR-12-1177

31. Yao X, Ahmadzadeh M, Lu Y-C, Liewehr DJ, Dudley ME, Liu F, Schrump DS, Steinberg SM, Rosenberg SA, Robbins PF (2012) Levels of peripheral CD4(+)FoxP3(+) regulatory T cells are negatively associated with clinical response to adoptive immunotherapy of human cancer. Blood 119:5688–5696. https://doi.org/10.1182/blood-2011-10-386482

32. Forget M-A, Haymaker C, Dennison JB, Toth C, Maiti S, Fulbright OJ, Cooper LJN, Hwu P, Radvanyi LG, Bernatchez C (2016) The beneficial effects of a gas-permeable flask for expansion of tumor-infiltrating lymphocytes as reflected in their mitochondrial function and respiration capacity. Oncoimmunology 5:e1057386. https://doi.org/10.1080/2162402X.2015.1057386

33. Tavera RJ, Forget M-A, Kim YU, Sakellariou-Thompson D, Creasy CA, Bhatta A, Fulbright OJ, Ramachandran R, Thorsen ST, Flores E, Wahl A, Gonzalez AM, Toth C, Wardell S, Mansaray R, Radvanyi LG, Gombos DS, Patel SP, Hwu P, Amaria RN, Bernatchez C, Haymaker C (2018) Utilizing T-cell activation signals 1, 2, and 3 for tumor-infiltrating lymphocytes (TIL) expansion: the advantage over the sole use of interleukin-2 in cutaneous and uveal melanoma. J Immunother 41:399–405. https://doi.org/10.1097/CJI.0000000000000230

34. Inozume T, Hanada K-I, Wang QJ, Ahmadzadeh M, Wunderlich JR, Rosenberg SA, Yang JC (2010) Selection of CD8+PD-1+ lymphocytes in fresh human melanomas enriches for tumor-reactive T cells. J Immunother 33:956–964. https://doi.org/10.1097/CJI.0b013e3181fad2b0

35. Gros A, Robbins PF, Yao X, Li YF, Turcotte S, Tran E, Wunderlich JR, Mixon A, Farid S, Dudley ME, Hanada K-I, Almeida JR, Darko S, Douek DC, Yang JC, Rosenberg SA (2014) PD-1 identifies the patient-specific CD8$^+$ tumor-reactive repertoire infiltrating human tumors. J Clin Invest 124:2246–2259. https://doi.org/10.1172/JCI73639

36. Tran E, Turcotte S, Gros A, Robbins PF, Lu Y-C, Dudley ME, Wunderlich JR, Somerville RP, Hogan K, Hinrichs CS, Parkhurst MR, Yang JC, Rosenberg SA (2014) Cancer immunotherapy based on mutation-specific CD4+ T cells in a patient with epithelial cancer. Science 344:641–645. https://doi.org/10.1126/science.1251102

37. Besser MJ, Shoham T, Harari-Steinberg O, Zabari N, Ortenberg R, Yakirevitch A, Nagler A, Loewenthal R, Schachter J, Markel G (2013) Development of allogeneic NK cell adoptive transfer therapy in metastatic melanoma patients: in vitro preclinical optimization studies. PLoS One 8:e57922. https://doi.org/10.1371/journal.pone.0057922

38. Figlin RA, Thompson JA, Bukowski RM, Vogelzang NJ, Novick AC, Lange P, Steinberg GD, Belldegrun AS (1999) Multicenter, randomized, phase III trial of CD8(+) tumor-infiltrating lymphocytes in combination with recombinant interleukin-2 in metastatic renal cell carcinoma. J Clin Oncol 17:2521–2529

39. Baldan V, Griffiths R, Hawkins RE, Gilham DE (2015) Efficient and reproducible generation of tumour-infiltrating lymphocytes for renal cell carcinoma. Br J Cancer 112:1510–1518. https://doi.org/10.1038/bjc.2015.96

40. Rosenberg SA, Yang JC, Topalian SL, Schwartzentruber DJ, Weber JS, Parkinson DR, Seipp CA, Einhorn JH, White DE (1994) Treatment of 283 consecutive patients with metastatic melanoma or renal cell cancer using high-dose bolus interleukin 2. JAMA 271:907–913

41. Dudley ME, Wunderlich J, Nishimura MI, Yu D, Yang JC, Topalian SL, Schwartzentruber DJ, Hwu P, Marincola FM, Sherry R, Leitman SF, Rosenberg SA (2001) Adoptive transfer of

cloned melanoma-reactive T lymphocytes for the treatment of patients with metastatic melanoma. J Immunother 24:363–373

42. Dudley ME, Wunderlich JR, Robbins PF, Yang JC, Hwu P, Schwartzentruber DJ, Topalian SL, Sherry R, Restifo NP, Hubicki AM, Robinson MR, Raffeld M, Duray P, Seipp CA, Rogers-Freezer L, Morton KE, Mavroukakis SA, White DE, Rosenberg SA (2002) Cancer regression and autoimmunity in patients after clonal repopulation with antitumor lymphocytes. Science 298:850–854. https://doi.org/10.1126/science.1076514
43. Dudley ME, Wunderlich JR, Yang JC, Sherry RM, Topalian SL, Restifo NP, Royal RE, Kammula U, White DE, Mavroukakis SA, Rogers LJ, Gracia GJ, Jones SA, Mangiameli DP, Pelletier MM, Gea-Banacloche J, Robinson MR, Berman DM, Filie AC, Abati A, Rosenberg SA (2005) Adoptive cell transfer therapy following non-myeloablative but lymphodepleting chemotherapy for the treatment of patients with refractory metastatic melanoma. J Clin Oncol 23:2346–2357. https://doi.org/10.1200/JCO.2005.00.240
44. Kershaw MH, Westwood JA, Parker LL, Wang G, Eshhar Z, Mavroukakis SA, White DE, Wunderlich JR, Canevari S, Rogers-Freezer L, Chen CC, Yang JC, Rosenberg SA, Hwu P (2006) A phase I study on adoptive immunotherapy using gene-modified T cells for ovarian cancer. Clin Cancer Res 12:6106–6115. https://doi.org/10.1158/1078-0432.CCR-06-1183
45. Johnson LA, Morgan RA, Dudley ME, Cassard L, Yang JC, Hughes MS, Kammula US, Royal RE, Sherry RM, Wunderlich JR, Lee C-CR, Restifo NP, Schwarz SL, Cogdill AP, Bishop RJ, Kim H, Brewer CC, Rudy SF, VanWaes C, Davis JL, Mathur A, Ripley RT, Nathan DA, Laurencot CM, Rosenberg SA (2009) Gene therapy with human and mouse T-cell receptors mediates cancer regression and targets normal tissues expressing cognate antigen. Blood 114:535–546. https://doi.org/10.1182/blood-2009-03-211714
46. Besser MJ, Shapira-Frommer R, Treves AJ, Zippel D, Itzhaki O, Hershkovitz L, Levy D, Kubi A, Hovav E, Chermoshniuk N, Shalmon B, Hardan I, Catane R, Markel G, Apter S, Ben-Nun A, Kuchuk I, Shimoni A, Nagler A, Schachter J (2010) Clinical responses in a phase II study using adoptive transfer of short-term cultured tumor infiltration lymphocytes in metastatic melanoma patients. Clin Cancer Res 16:2646–2655. https://doi.org/10.1158/1078-0432.CCR-10-0041
47. Robbins PF, Morgan RA, Feldman SA, Yang JC, Sherry RM, Dudley ME, Wunderlich JR, Nahvi AV, Helman LJ, Mackall CL, Kammula US, Hughes MS, Restifo NP, Raffeld M, Lee CCR, Levy CL, Li YF, El-Gamil M, Schwarz SL, Laurencot C, Rosenberg SA (2011) Tumor regression in patients with metastatic synovial cell sarcoma and melanoma using genetically engineered lymphocytes reactive with NY-ESO-1. J Clin Oncol 29:917–924. https://doi.org/10.1200/JCO.2010.32.2537
48. Chandran SS, Somerville RPT, Yang JC, Sherry RM, Klebanoff CA, Goff SL, Wunderlich JR, Danforth DN, Zlott D, Paria BC, Sabesan AC, Srivastava AK, Xi L, Pham TH, Raffeld M, White DE, Toomey MA, Rosenberg SA, Kammula US (2017) Treatment of metastatic uveal melanoma with adoptive transfer of tumour-infiltrating lymphocytes: a single-centre, two-stage, single-arm, phase 2 study. Lancet Oncol 18:792–802. https://doi.org/10.1016/S1470-2045(17)30251-6
49. Mullinax JE, Hall M, Prabhakaran S, Weber J, Khushalani N, Eroglu Z, Brohl AS, Markowitz J, Royster E, Richards A, Stark V, Zager JS, Kelley L, Cox C, Sondak VK, Mulé JJ, Pilon-Thomas S, Sarnaik AA (2018) Combination of ipilimumab and adoptive cell therapy with tumor-infiltrating lymphocytes for patients with metastatic melanoma. Front Oncol 8:44. https://doi.org/10.3389/fonc.2018.00044

Chapter 21

Place of Academic GMP Facilities in Modern Cell Therapy

Alexey Bersenev and Andrew Fesnak

Abstract

Academic medicine, in general, serves a dual role in advancing the scientific field as well as providing the highest quality clinical care. In the last decade we have observed significant development of commercial cell and gene therapy products with rapid growth of the industry. Currently, hospital-based Good Manufacturing Practice (GMP) facilities, which are used to support primarily academic investigator-initiated clinical trials, are increasingly involved in interactions with industry.

The purpose of this piece is to review the role of academic GMP facilities in development of cellular therapies. For the purpose of this review we will discuss manufacturing facilities located in hospitals or academic medical centers and compare them with commercial GMP facilities. Although the missions of academic and commercial GMP facilities are different, both are bound by industry standards and often engage in technology transfer with industry partners. Therefore, successful operation of an academic GMP facility requires striking a unique balance between commercial and academic priorities. Finally, we highlight some of the most challenging aspects of academic GMP facility operation and point to potential solutions.

Key words Cell therapy, Cell manufacturing, GMP facility, Academic GMP, Academic facility

1 Regulation

In the United States (US), European Union (EU), and other developed countries cellular and tissue-based products are regulated as industrially produced medicinal products or drugs if (1) more than minimal (substantial) manipulation was applied and (2) products are not intended for homologous use (cells perform the same critical function in the recipient as in the donor). Some regulatory exceptions may be applied, mostly for autologous products (i.e. the "same surgical procedure" in US and "hospital exemption" in EU). Manufacturing of cell/gene-based products, classified as "Advanced Therapy Medicinal Product" (ATMP) in EU and as 351 Human Cell Tissue Products (HCT/P) in US, requires compliance with Good Manufacturing Practice (GMP) rules,

Samuel G. Katz and Peter M. Rabinovich (eds.), *Cell Reprogramming for Immunotherapy: Methods and Protocols*, Methods in Molecular Biology, vol. 2097, https://doi.org/10.1007/978-1-0716-0203-4_21,

established for conventional drugs and biologics. By definition, "current GMPs provide for systems that assure proper design, monitoring, and control of manufacturing processes and facilities" (https://www.fda.gov/drugs/developmentapprovalprocess/manufacturing/ucm169105.htm). GMP compliance assures consistent identity, potency, quality, and purity of drug products.

In Europe, ATMPs must be produced in authorized (licensed) manufacturing facilities. Successful inspections are necessary for authorization. Unlike the EU, the FDA does not require to register facilities that manufacture cell products (drugs) under an investigational new drug application (IND) (21 CFR Part 312) until the HCT/P is approved through a biologics license application (BLA).

2 Comparison of Academic and Industry-Based Cell Therapy Manufacturing Facilities

The differences between academic and industrial facilities are summarized in Table 1. Definitive advantages of academic cell therapy GMP facilities over industry include:

1. Close proximity to starting material procurement facility and treatment site (usually within the same institution/campus).
2. Lower cost of manufacturing due to savings on workforce, shipments and cold chain logistics, utilization of hospital resources and services.
3. Utilization of hospital internal laboratories for in-process and product release testing (i.e. microbiology, hematology, flow cytometry).

3 The Major Functions and Operations Provided by an Academic GMP Facility

The major activities performed by academic facilities are presented in Table 2. We divide them into "internal" (performed for institution) and "external" (sponsored by industry) activities. Many of these functions depend on business decisions, clinical needs, and sometimes historical accidents. Therefore, the array of academic GMP capabilities might vary considerably from institution to institution. Given such differences, we aim to highlight some of the most common issues and challenges, rather than exhaustively describe the laboratory landscapes.

3.1 Process Development

Process development is one of the major activities in an academic GMP manufacturing facility. Usually, an academic investigator has a research prototype of the genetic and/or cellular product, which requires a clinical-grade manufacturing process before starting a

Table 1
Major differences between academic/hospital-based and industry GMP facilities

A difference	Academic/hospital facility	Industry facility
End goal of development	Translation of "cutting edge" science into the clinical product. Generation of pioneering, "first-in-human" data. Support and training of new principal investigators. Licensing out and technology transfer to industry	Commercialization and wide clinical adoption, profit from market sales
Phase of development	First-in-human, Phase 1 (feasibility, safety)	Phase 2–3 (efficacy, long-term safety), commercialization
Product development	Multiple products, multiple trials in multiple therapeutic areas (oncology, hematology, regenerative medicine)	Focused on 1–2 technological platform, 1–3 products in 1–2 therapeutic areas
Scale	5–20 patients per trial, up to 200 products per year	30–100 patients per trial, thousands of products per year
Manufacturing focus	Product development, process development, validation of new technologies, process optimization, comparability	Industrialization: Scalability, automation, cost, secure supply chain, logistics and delivery, commercialization
Proximity to treatment center	A part of treatment center (hospital), no logistical/delivery issues. Very rarely serve other hospitals/institutes. Contact with patients at bed side, quick follow-up on trial results	Far from treatment centers. Possible logistical issues. Distribution and delivery to the hospital is a significant cost driver. No contact with patients. No clinical feedback
Facility design	Embedded into the hospital or/and academic institution, small: 2–20 clean rooms or isolators	Stand-alone central plant, very well departmentalized, modular, large: up to 100 clean rooms or suites
Regulation	Not required for registration, inspections and accreditations (in US). The same cGMP rules and other federal regulations apply as for industry	Must be registered and inspected by governmental agency before product commercialization (marketing authorization). Frequent interactions with regulatory agency
Training and staffing	Frequently poor segregation for departments allows cross-training. Significant part of workforce may be acquired from academic research labs	No cross-training due to strict segregation for departments. Workforce acquired predominantly from industry, less from academia
Quality	The role of quality assurance is frequently underappreciated and underestimated	Very strong quality system in place

clinical trial. As with many other aspects of cell therapy (Table 2), the process development needs cell therapy agents that differ from those of traditional pharmaceutical agents. At all stages of development, cell therapy agents and in particular autologous CT (cell and/or tissue) raw materials must contend with high levels of

Table 2
Major types of activities, performed in academic cell therapy GMP facilities

Activity	Explanation and examples
Internal	
Product manufacturing (including quality control (QC) testing, and quality assurance (QA))	To support internal PI-initiated clinical trials, sponsored by government or non-profit organizations
Raw materials preparation	Genetic material manufacturing, bulk media preparation, cell lines, and feeder cells preparation for multiple trials
Process development	Pre-IND studies, validations, process optimization, engineering runs
Regulatory	Chemistry, Manufacturing, and Controls (CMC) section of IND preparation, regulatory communications
Cell banks inventory	Creation an inventory of clinical-grade primary cell banks to support early product development or any research with human primary cells
Academic research	Collaboration with other academic labs to develop new cell or gene therapy products
External (sponsored by industry)	
Clinical trials supported by industry	Commercial product-candidates manufactured by industry. Academic facility handles (receives, samples and ships) starting material, and final product (receives/ships, samples, stores, prepares, dispenses, and administers)
Contract development and manufacturing	Academic facility hired by a company through contract for process development and manufacturing of a commercial investigational product for early phases of clinical trials
Technology transfer	Transfer of manufacturing know-how (via training of company personnel, comparability runs, document sharing, data mining) to the company as part of licensing agreement
Testing and validation of new tools, devices, and test systems	Pre-market beta testing and validation of new tools (i.e. cell sorter, cell washer, bioreactor, rapid microbiology testing system, antibody, reagents, ...), developed by industry
Sponsored research	Contracting academic facilities early product development. Conducted under sponsored research agreement

run-to-run variability and limited availability of optimal reagents. These challenges obviously limit academic process development compared to commercial manufacturing in achieving standardized and highly controlled pharmaceutical manufacturing cycles. Moreover, unlike industrial development, academics usually don't have a commercial product in mind, but rather are focused on feasibility and optimization of manufacturing, based on research data. With the recent wave of commercialization of cellular therapies and the big demand for smooth technology transfer from academia to industry, some academic GMP facilities are trying to develop industry-ready manufacturing process. However, it usually requires significant investment in automation, which may not be affordable for an academic center.

Pre-IND (Investigational New Drug) submission to FDA studies usually includes assays, qualifications, and validations for testing of a product's critical quality attributes, and animal studies (with a research laboratory) to test in vivo toxicity, dose, and biodistribution. The goals of pre-IND studies are to provide, among other results, the evidence of safety and potency. Only limited animal models are available for testing cell therapy agents. Even those models that more accurately reflect human biology (e.g. humanized mouse models) are still highly artificial and may miss clinically significant interactions. For example, the array of tumor targets against which CT agents are designed frequently differ between mice and humans. In vitro tissue arrays allow for high-throughput screening of potential off-target effects. However, these assays are unable to screen in vivo for reactivity against novel epitopes exposed during tissue damage and remodeling. Extrapolation from data with similar agents, such as monoclonal antibodies against the same target, do not accurately reflect the range of potential adverse effects. Finally, to date, potency assays and conclusions drawn from most pre-IND studies have shown limited correlation with clinical response.

Validation of manufacturing processes aims to ensure reproducible manufacturing outcomes with clinical runs. Therefore, ideally, validations should be performed with starting material similar or identical to that which will be used in clinical manufacturing. However, this is not always feasible with cell therapy agents. Patient material is often difficult to obtain. In addition, patient-to-patient variability is such that validations with patients with the same indication may not reflect those enrolled in clinical manufacturing. Ultimately, healthy donor material is often used as a starting material for validation studies. There are several notable differences in healthy donor ex vivo cell culture behavior that limit generalization of validation results to full-scale clinical manufacturing. These differences could be especially profound in the most frequent starting manufacturing material for cellular therapies—whole peripheral blood or apheresis product. Thus, abnormal morphology, function, composition, and characteristics of blood cells in patients with hematological malignancies could be distorted to degree, which makes it hard to compare results of validation runs with a normal blood from healthy donors. Nonetheless, validation with healthy donor material may be the only data available to support proceeding with a given manufacturing approach.

Pre-IND studies provide significant scientific rationale, and validation runs frequently support reproducible clinical manufacturing; however the translation of these data to actual clinical manufacturing cannot be guaranteed. It is important, therefore, to establish a continuous process improvement program during clinical manufacturing. Such a program includes rigorous data collection, analysis, and intervention based on findings. These programs should reflect the evolution of our understanding regarding clinical cell culture as well as an acknowledgement and effort to mitigate significant patient-to-patient variability.

3.2 Academic GMP Facilities as Drivers of Research

Academic cell manufacturing facilities should be able to embrace their main goals in that they aim to improve cell manufacturing science while making clinically beneficial therapies available to their patients. To date, all major advances in the field originated in academic institutions, while commercial enterprises are more capable of scaled up and scaled out manufacturing. Working together is critical to assist the widest possible number of patients. Thus, the role of academic facilities is not only for research, but is important for its translation as well.

Academic-industry partnerships involves a pre-existing company, purchasing or licensing technology and development from an academic group. Examples of such partnerships include the Penn-Novartis alliance and the MD Anderson Cancer Center (University of Texas)—Cellectis alliance. In these partnerships, terms are negotiated upfront, and technology is transferred from the academic facility to the commercial manufacturer. Alternatively, single-purpose industry spin-offs can be created to capitalize on academic work. Examples of this include Tmunity (University of Pennsylvania), Leucid Bio (King's College London), WindMIL Therapeutics (Johns Hopkins University), Autolus (University College London), and many others. In distinction to partnerships with existing entities, these companies are created with the sole purpose of commercializing an academically developed technology. While academic staff may simultaneously serve both commercial and academic roles, these are distinct business entities. Independent of the arrangement of academic and industry relationships, academic institutions may choose to invest in these companies regardless of how they are arranged, as far as supporting these companies benefits the academic community as well.

Conflicts of interest particularly with respect to intellectual property are a significant challenge in many of these arrangements. As mentioned, when individuals serve dual roles in both academic and commercial entities, conflicts are nearly inevitable. On the other hand, to completely eliminate conflicts, academicians would need to be barred from serving any role in commercial enterprises. This would deny critical expertise needed to advance commercial development and ultimately treat more patients. Therefore, conflict management is critical to ensure both appropriate academic and commercial development. Conflict management involves transparency where possible and may include an objective third party assessment to resolve conflicts and legal problems. Conflict management committees can originate from academic institutions, commercial entities, or health systems. The key feature is that these committees work to reach a mutually acceptable consensus with all parties involved.

3.3 Academic GMP Facilities as Contract Manufacturers

The partnership with external entities can be especially productive in certain settings. Commercial manufacturers may find it desirable to try a novel product in an academic lab with GMP capability prior to making significant investment in scaled out or scaled up manufacturing. These arrangements are most productive when forethought is given to potential pitfalls prior to formally entering into agreements. For example, the technology transfer process and intellectual property ownership should be established well in advance of any hands-on work being performed. In addition, while the academic facility serves the commercial interests of the sponsor, they also are bound to their academic mission, which may include publication or relevant data. The mission of academic cell manufacturing facilities in general serves a dual role in advancing the scientific field as well as providing the highest quality clinical care possible. While there is no doubt that contract manufacturers also contribute greatly to patient care, as an entity the contract manufacturer's mission is to serve client needs first and foremost. Careful, deliberate, and proactive planning can ensure that these relationships are beneficial to both academic and commercial partners.

4 Challenges Specific to Academic GMP Facilities

The need for academic GMP facilities has grown extraordinarily over the past decade, and, therefore, many of their challenges are only now being discovered. Many of these challenges originate from the nature of science and the interplay of regulatory, academic, and commercial interests. One definite obstacle stems from the fact that the GMP rules were created for manufacturing of conventional drugs and biologics, and their adaptation for cell-based products could be challenging. Some of the main challenges and solutions are summarized in Table 3. Solutions are largely found by trial and error and may change over the coming years.

4.1 Personnel

Finding adequate and well-qualified personnel to staff academic manufacturing facilities is a common issue [1]. There are very few cell therapy specific training programs that build this segment of the workforce. Recruiting from industry is difficult in that academic facilities can rarely offer competitive compensation com-

Table 3
Serious GMP challenges facing academic facilities and potential solutions

GMP subparts	Challenges specific for academic/hospital facilities	Potential solutions
Quality System (QS)	Hospital-based QS is not intended for manufacturing of therapeutic products. Lack of specific knowledge of GMP regulation by hospital QS. Misunderstanding of the role of QS in GMP facility. General lack of understanding of necessity of QA Changes in hospital QS may be dictated by GMP compliance of single facility (cell therapy)	Highlight importance of quality compliance and role of QA Hire QA and QS specialists with industry experience Provide training opportunities for QS specialists
Personnel and training	Lack of qualified industry-trained workforce Absence of clear career ladder within facility Abundance of over-qualified job candidates—scientists with PhD degree Industry competition poaching staff Long lead-time during training to proficiency	Involve recruiting agencies for talent acquisition Create incentives to attract industry-trained candidates (better benefits, multiple small-scale manufacturing projects, cross-training, …) Broad workforce development programs and pipelines
Process controls—product testing	Lack of resources for multiple qualifications and validations Lack of qualification of internal hospital laboratories, performing testing Lack of samples stability data and labeling	Applying industry standards for assay development, in-process, and final product testing Qualification of internal laboratories and agreements
Facility	It is difficult to embed clean GMP rooms in existent hospital building structure	Build stand-alone facility in the medical campus Assess multiple construction options Hire consultants for better facility design and architect. Have a business plan
Records and reports	Lack of good documentation practice Lack of electronic record systems Interface cell therapy software with hospital electronic system	Acquisition and implementation of electronic record systems (SOPs management, batch records, inventory, labeling, …)

pared to industrial competitors. Long training times complicate hiring further. It is difficult to find funds to train a new staff member who does not contribute to revenue for months. Employee turnover can then be costly after extended training. Altogether, too few qualified staff are available, and a growing industrial market is a strong draw for those who are qualified.

It is also unclear what are the minimum necessary qualifications for cell manufacturing staff. Standards require training be documented, relevant, and to the appropriate level. As mentioned above, cell therapy specific training programs are rare, and therefore the expectations for how to fulfill these standards are unclear.

Workforce development efforts, such as that supported through NIIMBL/NIST are sorely needed to grow the pool of labor who can manufacture these products. Without a significant increase in the number of graduates who enter the field, this will continue to be a major challenge and capacity limiter.

4.2 Adherence to Appropriate Compliance Standards

As mentioned above, the adaptation of GMP rules for cell-based products in the academic/hospital setting is not a trivial task. For example, the creation of a fully compliant GMP structure (aka small industrial plant) inside of a standard hospital environment is oftentimes prohibitively expensive or logistically impossible. Ensuring a balance of appropriate compliance while facilitating research and therapy products is an evolving process for manufacturers, quality assurance, and regulators.

Phase appropriate GMP standards, as referred to in FDA guidance documents, can be somewhat amorphous. Most academic manufacturing facilities are not well suited to adhere to requirements for biologic license application (BLA) or marketing authorization; however, it is not always clear what is acceptable in the setting of Phase I/II studies. This can be further complicated by the intended future plans for the manufactured agent. Some compliance standards must only be met prior to or during pivotal trials as a commercial entity progresses toward a BLA. If early assays were not performed to meet these standards during the Phase I study, it may be difficult if not impossible to retrospectively meet compliance standards required for later commercialization. Therefore, academic facilities should carefully plan the degree of quality compliance based on feasibility, safety, and future plans for commercialization.

5 Financial Sustainability

One of the biggest challenges for academic GMP facilities is financial sustainability.

The cost of GMP compliance generally is unappreciated and underestimated in academic institutions or hospitals. So, relying on a single funding source may put the facility into significant financial risk. No matter whether the academic GMP facility is a university's core or hospital clinical laboratory, the base funding is included in the organizational budget. However, a key to financial sustainability of an academic GMP facility is diversification of services and funding sources. Institutional funding could be received from different sources, such as the hospital or cancer center, the university's medical school or dean's office, and particular departments. Once established, some facilities may receive philanthropic donations and apply for manufacture-related research grants. Recent influx of industry clients allows to diversify further a structure of paid services and create a portfolio, which may attract more potential clients.

According to a recent survey [1] of academic GMP facilities, institutional funding represents less than 30% of the whole funding share, whereas industry supports about 30%, and research grants support the remaining 30% [1].

Establishing a fee for service policy is not a trivial task. Ideally, it should be structured, based on type of the client (internal investigators, other academic institutions, industry) and type of services. The facility manager should create a budget for each clinical trial or type of activity. With industry clients, a negotiated fee should be included in the agreement. Possible cancelations, risks, and liabilities should be considered as well.

Despite the high cost of facility maintenance (environmental monitoring, cleaning, equipment calibrations, and QC), it could be self-sustainable with fee for services, provided to industry. The source of long-term stable revenue is the generation of intellectual property that is licensed to industry. The caveat here is that possible revenue, generated by the GMP facility, could be directed to the department and get diluted in the hospital's general revenue stream.

6 Capacity Planning and Facility Design

Due to the current explosive growth of the cell therapy industry, the number and size of academic GMP facilities continue to increase. However, this trend may change over time with (1) shifting early cell therapy development to industry if large amount of capital will be available, (2) prevailing centralized industrial manufacturing model, and (3) technological advances, which will allow cell/gene therapy to be done by medical staff at point of care outside of a classified environment. Given the focus of industrial manufacturers on high-capacity, closed system technology, the future of academic GMP facilities is uncertain. Efficiency of centralized manufacturing far exceeds decentralized efforts. On the other hand, the personalized and unpredictable nature of engineered cell therapy manufacturing seems better suited to local control of decentralized manufacturing. Regardless, we are currently witnessing a massive investment in new academic GMP facility construction. Such investment and rapid development holds a number of risks over time. For example:

1. *Overestimating demand*: if facility capacity far exceeds clinical demand and/or funding for manufacturing, facility occupancy will be low and costs will accrue with little revenue making this financially unsustainable in the long-run.

2. *Underestimating demand*: if facility capacity is far surpassed by clinical demand, then investigators will be incentivized to look outside of the academic institution for alternative manufacturing sources and once workflows are establish, academic facilities may not be able to compete with these alternative contract manufacturers.
3. *Consistent funds flow*: while many facilities have a large, up-front, institutional investment to establish the lab, consistent funding to maintain continuous operations is critical to success and with the ebb and flow of grant-funded research dollars, a non-diversified facility runs the risk of having to shut down or limit operations at great expense.
4. *Qualified personnel*: maintaining a large enough pool of trained and qualified personnel to meet manufacturing demand is challenging with lengthy time to proficiency for complex processes and constant competition with industry poaching of staff.

Acceptable solutions to these challenges must be implemented for the academic GMP model to continue to be effective at facilitating research and patient care.

One of the preventative mechanisms for over-investment in an academic GMP facility is an adaptive approach to its design. A facility with distinct, modular GMP suites allows for re-purposing of unused clean rooms, based on capacity. Another approach could be utilization of fully closed automated devices or isolators outside of ISO 5-7 (EU grade A/B) environment. However, moving manufacturing outside of the cleanest environment within a GMP facility will be only possible with implementation of full automation and a permissive regulatory framework.

Small to mid-size facilities with three to ten cleanrooms may be a better strategic investment compared to large facilities. In order to have optimal facility occupancy (constant 80–90%), effective mechanisms should be created in the institution, where a committee of experts will discuss all potential cell/gene therapy and choose the most promising ones.

Reference

1. Digiusto DL, Melsop K, Srivastava R, Tran CT (2018) Proceedings of the first academic symposium on developing, qualifying and operating a cell and gene therapy manufacturing facility. Cytotherapy 20(12):1486–1494

INDEX

Samuel G. Katz and Peter M. Rabinovich (eds.), *Cell Reprogramming for Immunotherapy: Methods and Protocols*, Methods in Molecular Biology, vol. 2097, https://doi.org/10.1007/978-1-0716-0203-4,